# Lecture Notes in Computer Science    9159

Commenced Publication in 1973
Founding and Former Series Editors:
Gerhard Goos, Juris Hartmanis, and Jan van Leeuwen

More information about this series at http://www.springer.com/series/7407

Osvaldo Gervasi · Beniamino Murgante
Sanjay Misra · Marina L. Gavrilova
Ana Maria Alves Coutinho Rocha · Carmelo Torre
David Taniar · Bernady O. Apduhan (Eds.)

# Computational Science and Its Applications – ICCSA 2015

15th International Conference
Banff, AB, Canada, June 22–25, 2015
Proceedings, Part V

 Springer

*Editors*

Osvaldo Gervasi
University of Perugia
Perugia
Italy

Beniamino Murgante
University of Basilicata
Potenza
Italy

Sanjay Misra
Covenant University
Canaanland
Nigeria

Marina L. Gavrilova
University of Calgary
Calgary, AB
Canada

Ana Maria Alves Coutinho Rocha
University of Minho
Braga
Portugal

Carmelo Torre
Polytechnic University
Bari
Italy

David Taniar
Monash University
Clayton, VIC
Australia

Bernady O. Apduhan
Kyushu Sangyo University
Fukuoka
Japan

ISSN 0302-9743                     ISSN 1611-3349  (electronic)
Lecture Notes in Computer Science
ISBN 978-3-319-21412-2          ISBN 978-3-319-21413-9  (eBook)
DOI 10.1007/978-3-319-21413-9

Library of Congress Control Number: 2015943360

LNCS Sublibrary: SL1 – Theoretical Computer Science and General Issues

Printed on acid-free paper

Springer International Publishing AG Switzerland is part of Springer Science+Business Media
(www.springer.com)

# Preface

The year 2015 is a memorable year for the International Conference on Computational Science and Its Applications. In 2003, the First International Conference on Computational Science and Its Applications (chaired by C.J.K. Tan and M. Gavrilova) took place in Montreal, Canada (2003), and the following year it was hosted by A. Laganà and O. Gervasi in Assisi, Italy (2004). It then moved to Singapore (2005), Glasgow, UK (2006), Kuala-Lumpur, Malaysia (2007), Perugia, Italy (2008), Seoul, Korea (2009), Fukuoka, Japan (2010), Santander, Spain (2011), Salvador de Bahia, Brazil (2012), Ho Chi Minh City, Vietnam (2013), and Guimarães, Portugal (2014). The current installment of ICCSA 2015 took place in majestic Banff National Park, Banff, Alberta, Canada, during June 22–25, 2015.

The event received approximately 780 submissions from over 45 countries, evaluated by over 600 reviewers worldwide.

Its main track acceptance rate was approximately 29.7 % for full papers. In addition to full papers, published by Springer, the event accepted short papers, poster papers, and PhD student showcase works that are published in the IEEE CPS proceedings.

It also runs a number of parallel workshops, some for over 10 years, with new ones appearing for the first time this year. The success of ICCSA is largely contributed to the continuous support of the computational sciences community as well as researchers working in the applied relevant fields, such as graphics, image processing, biometrics, optimization, computer modeling, information systems, geographical sciences, physics, biology, astronomy, biometrics, virtual reality, and robotics, to name a few.

Over the past decade, the vibrant and promising area focusing on performance-driven computing and big data has became one of the key points of research enhancing the performance of information systems and supported processes. In addition to high-quality research at the frontier of these fields, consistently presented at ICCSA, a number of special journal issues are being planned following ICCSA 2015, including TCS Springer (*Transactions on Computational Sciences*, LNCS).

The contribution of the International Steering Committee and the International Program Committee are invaluable in the conference success. The dedication of members of these committees, the majority of whom have fulfilled this difficult role for the last 10 years, is astounding. Our warm appreciation also goes to the invited speakers, all event sponsors, supporting organizations, and volunteers. Finally, we thank all the authors for their submissions making the ICCSA conference series a well recognized and a highly successful event year after year.

June 2015

Marina L. Gavrilova
Osvaldo Gervasi
Bernady O. Apduhan

# Organization

ICCSA 2015 was organized by the University of Calgary (Canada), the University of Perugia (Italy), the University of Basilicata (Italy), Monash University (Australia), Kyushu Sangyo University (Japan), and the University of Minho, (Portugal)

## Honorary General Chairs

| | |
|---|---|
| Antonio Laganà | University of Perugia, Italy |
| Norio Shiratori | Tohoku University, Japan |
| Kenneth C.J. Tan | Sardina Systems, Estonia |

## General Chairs

| | |
|---|---|
| Marina L. Gavrilova | University of Calgary, Canada |
| Osvaldo Gervasi | University of Perugia, Italy |
| Bernady O. Apduhan | Kyushu Sangyo University, Japan |

## Program Committee Chairs

| | |
|---|---|
| Beniamino Murgante | University of Basilicata, Italy |
| Ana Maria A.C. Rocha | University of Minho, Portugal |
| David Taniar | Monash University, Australia |

## International Advisory Committee

| | |
|---|---|
| Jemal Abawajy | Deakin University, Australia |
| Dharma P. Agrawal | University of Cincinnati, USA |
| Claudia Bauzer Medeiros | University of Campinas, Brazil |
| Manfred M. Fisher | Vienna University of Economics and Business, Austria |
| Yee Leung | Chinese University of Hong Kong, SAR China |

## International Liaison Chairs

| | |
|---|---|
| Ana Carla P. Bitencourt | Universidade Federal do Reconcavo da Bahia, Brazil |
| Alfredo Cuzzocrea | ICAR-CNR and University of Calabria, Italy |
| Maria Irene Falcão | University of Minho, Portugal |
| Marina L. Gavrilova | University of Calgary, Canada |
| Robert C.H. Hsu | Chung Hua University, Taiwan |
| Andrés Iglesias | University of Cantabria, Spain |
| Tai-Hoon Kim | Hannam University, Korea |
| Sanjay Misra | University of Minna, Nigeria |
| Takashi Naka | Kyushu Sangyo University, Japan |

Rafael D.C. Santos          Brazilian National Institute for Space Research, Brazil
Maribel Yasmina Santos     University of Minho, Portugal

## Workshop and Session Organizing Chairs

Beniamino Murgante         University of Basilicata, Italy
Jorge Gustavo Rocha        University of Minho, Portugal

## Local Arrangement Chairs

Marina Gavrilova           University of Calgary, Canada (Chair)
Madeena Sultana            University of Calgary, Canada
Padma Polash Paul          University of Calgary, Canada
Faisal Ahmed               University of Calgary, Canada
Hossein Talebi             University of Calgary, Canada
Camille Sinanan            University of Calgary, Canada

## Venue

ICCSA 2015 took place in the Banff Park Lodge Conference Center, Alberta (Canada).

## Workshop Organizers

### Agricultural and Environment Information and Decision Support Systems (AEIDSS 2015)

Sandro Bimonte             IRSTEA, France
André Miralles             IRSTEA, France
Frederic Hubert            University of Laval, Canada
François Pinet             IRSTEA, France

### Approaches or Methods of Security Engineering (AMSE 2015)

TaiHoon Kim                Sungshin W. University, Korea

### Advances in Information Systems and Technologies for Emergency Preparedness and Risk Assessment (ASTER 2015)

Maurizio Pollino           ENEA, Italy
Marco Vona                 University of Basilicata, Italy
Beniamino Murgante         University of Basilicata, Italy

### Advances in Web-Based Learning (AWBL 2015)

Mustafa Murat Inceoglu     Ege University, Turkey

## Bio-inspired Computing and Applications (BIOCA 2015)

| | |
|---|---|
| Nadia Nedjah | State University of Rio de Janeiro, Brazil |
| Luiza de Macedo Mourell | State University of Rio de Janeiro, Brazil |

## Computer-Aided Modeling, Simulation, and Analysis (CAMSA 2015)

| | |
|---|---|
| Jie Shen | University of Michigan, USA, and Jilin University, China |
| Hao Chen | Shanghai University of Engineering Science, China |
| Xiaoqiang Liun | Donghua University, China |
| Weichun Shi | Shanghai Maritime University, China |

## Computational and Applied Statistics (CAS 2015)

| | |
|---|---|
| Ana Cristina Braga | University of Minho, Portugal |
| Ana Paula Costa Conceicao Amorim | University of Minho, Portugal |

## Computational Geometry and Security Applications (CGSA 2015)

| | |
|---|---|
| Marina L. Gavrilova | University of Calgary, Canada |

## Computational Algorithms and Sustainable Assessment (CLASS 2015)

| | |
|---|---|
| Antonino Marvuglia | Public Research Centre Henri Tudor, Luxembourg |
| Beniamino Murgante | University of Basilicata, Italy |

## Chemistry and Materials Sciences and Technologies (CMST 2015)

| | |
|---|---|
| Antonio Laganà | University of Perugia, Italy |
| Alessandro Costantini | INFN, Italy |
| Noelia Faginas Lago | University of Perugia, Italy |
| Leonardo Pacifici | University of Perugia, Italy |

## Computational Optimization and Applications (COA 2015)

| | |
|---|---|
| Ana Maria Rocha | University of Minho, Portugal |
| Humberto Rocha | University of Coimbra, Portugal |

## Cities, Technologies and Planning (CTP 2015)

| | |
|---|---|
| Giuseppe Borruso | University of Trieste, Italy |
| Beniamino Murgante | University of Basilicata, Italy |

## Econometrics and Multidimensional Evaluation in the Urban Environment (EMEUE 2015)

| | |
|---|---|
| Carmelo M. Torre | Polytechnic of Bari, Italy |
| Maria Cerreta | University of Naples Federico II, Italy |
| Paola Perchinunno | University of Bari, Italy |

Simona Panaro            University of Naples Federico II, Italy
Raffaele Attardi         University of Naples Federico II, Italy
Claudia Ceppi            Polytechnic of Bari, Italy

## Future Computing Systems, Technologies, and Applications (FISTA 2015)

Bernady O. Apduhan       Kyushu Sangyo University, Japan
Rafael Santos            Brazilian National Institute for Space Research, Brazil
Jianhua Ma               Hosei University, Japan
Qun Jin                  Waseda University, Japan

## Geographical Analysis, Urban Modeling, Spatial Statistics (GEOGAN-MOD 2015)

Giuseppe Borruso         University of Trieste, Italy
Beniamino Murgante       University of Basilicata, Italy
Hartmut Asche            University of Potsdam, Germany

## Land Use Monitoring for Soil Consumption Reduction (LUMS 2015)

Carmelo M. Torre         Polytechnic of Bari, Italy
Alessandro Bonifazi      Polytechnic of Bari, Italy
Valentina Sannicandro    University Federico II of Naples, Italy
Massimiliano             University of Salerno, Italy
  Bencardino
Gianluca di Cugno        Polytechnic of Bari, Italy
Beniamino Murgante       University of Basilicata, Italy

## Mobile Communications (MC 2015)

Hyunseung Choo           Sungkyunkwan University, Korea

## Mobile Computing, Sensing, and Actuation for Cyber Physical Systems (MSA4CPS 2015)

Saad Qaisar              NUST School of Electrical Engineering and Computer
                           Science, Pakistan
Moonseong Kim            Korean Intellectual Property Office, Korea

## Quantum Mechanics: Computational Strategies and Applications (QMCSA 2015)

Mirco Ragni              Universidad Federal de Bahia, Brazil
Ana Carla Peixoto        Universidade Estadual de Feira de Santana, Brazil
  Bitencourt
Roger Anderson           University of California, USA
Vincenzo Aquilanti       University of Perugia, Italy
Frederico Vasconcellos   Universidad Federal de Bahia, Brazil
  Prudente

**Remote Sensing Data Analysis, Modeling, Interpretation and Applications: From a Global View to a Local Analysis (RS2015)**

Rosa Lasaponara      Institute of Methodologies for Environmental Analysis, National Research Council, Italy

**Scientific Computing Infrastructure (SCI 2015)**

Alexander Bodganov      St. Petersburg State University, Russia
Elena Stankova      St. Petersburg State University, Russia

**Software Engineering Processes and Applications (SEPA 2015)**

Sanjay Misra      Covenant University, Nigeria

**Software Quality (SQ 2015)**

Sanjay Misra      Covenant University, Nigeria

**Advances in Spatio-Temporal Analytics (ST-Analytics 2015)**

Joao Moura Pires      New University of Lisbon, Portugal
Maribel Yasmina Santos      New University of Lisbon, Portugal

**Tools and Techniques in Software Development Processes (TTSDP 2015)**

Sanjay Misra      Covenant University, Nigeria

**Virtual Reality and Its Applications (VRA 2015)**

Osvaldo Gervasi      University of Perugia, Italy
Lucio Depaolis      University of Salento, Italy

## Program Committee

| | |
|---|---|
| Jemal Abawajy | Deakin University, Australia |
| Kenny Adamson | University of Ulster, UK |
| Filipe Alvelos | University of Minho, Portugal |
| Paula Amaral | Universidade Nova de Lisboa, Portugal |
| Hartmut Asche | University of Potsdam, Germany |
| Md. Abul Kalam Azad | University of Minho, Portugal |
| Michela Bertolotto | University College Dublin, Ireland |
| Sandro Bimonte | CEMAGREF, TSCF, France |
| Rod Blais | University of Calgary, Canada |
| Ivan Blecic | University of Sassari, Italy |
| Giuseppe Borruso | University of Trieste, Italy |
| Yves Caniou | Lyon University, France |
| José A. Cardoso e Cunha | Universidade Nova de Lisboa, Portugal |
| Leocadio G. Casado | University of Almeria, Spain |

| | |
|---|---|
| Carlo Cattani | University of Salerno, Italy |
| Mete Celik | Erciyes University, Turkey |
| Alexander Chemeris | National Technical University of Ukraine KPI, Ukraine |
| Min Young Chung | Sungkyunkwan University, Korea |
| Gilberto Corso Pereira | Federal University of Bahia, Brazil |
| M. Fernanda Costa | University of Minho, Portugal |
| Gaspar Cunha | University of Minho, Portugal |
| Alfredo Cuzzocrea | ICAR-CNR and University of Calabria, Italy |
| Carla Dal Sasso Freitas | Universidade Federal do Rio Grande do Sul, Brazil |
| Pradesh Debba | The Council for Scientific and Industrial Research (CSIR), South Africa |
| Hendrik Decker | Instituto Tecnológico de Informática, Spain |
| Frank Devai | London South Bank University, UK |
| Rodolphe Devillers | Memorial University of Newfoundland, Canada |
| Prabu Dorairaj | NetApp, India/USA |
| M. Irene Falcao | University of Minho, Portugal |
| Cherry Liu Fang | U.S. DOE Ames Laboratory, USA |
| Edite M.G.P. Fernandes | University of Minho, Portugal |
| Jose-Jesus Fernandez | National Centre for Biotechnology, CSIS, Spain |
| Maria Antonia Forjaz | University of Minho, Portugal |
| Maria Celia Furtado Rocha | PRODEB/UFBA, Brazil |
| Akemi Galvez | University of Cantabria, Spain |
| Paulino Jose Garcia Nieto | University of Oviedo, Spain |
| Marina Gavrilova | University of Calgary, Canada |
| Jerome Gensel | LSR-IMAG, France |
| Maria Giaoutzi | National Technical University, Athens, Greece |
| Andrzej M. Goscinski | Deakin University, Australia |
| Alex Hagen-Zanker | University of Cambridge, UK |
| Malgorzata Hanzl | Technical University of Lodz, Poland |
| Shanmugasundaram Hariharan | B.S. Abdur Rahman University, India |
| Eligius M.T. Hendrix | University of Malaga/Wageningen University, Spain/The Netherlands |
| Tutut Herawan | Universitas Teknologi Yogyakarta, Indonesia |
| Hisamoto Hiyoshi | Gunma University, Japan |
| Fermin Huarte | University of Barcelona, Spain |
| Andres Iglesias | University of Cantabria, Spain |
| Mustafa Inceoglu | EGE University, Turkey |
| Peter Jimack | University of Leeds, UK |
| Qun Jin | Waseda University, Japan |
| Farid Karimipour | Vienna University of Technology, Austria |
| Baris Kazar | Oracle Corp., USA |
| DongSeong Kim | University of Canterbury, New Zealand |
| Taihoon Kim | Hannam University, Korea |

| | |
|---|---|
| Ivana Kolingerova | University of West Bohemia, Czech Republic |
| Dieter Kranzlmueller | LMU and LRZ Munich, Germany |
| Antonio Laganà | University of Perugia, Italy |
| Rosa Lasaponara | National Research Council, Italy |
| Maurizio Lazzari | National Research Council, Italy |
| Cheng Siong Lee | Monash University, Australia |
| Sangyoun Lee | Yonsei University, Korea |
| Jongchan Lee | Kunsan National University, Korea |
| Clement Leung | Hong Kong Baptist University, Hong Kong, SAR China |
| Chendong Li | University of Connecticut, USA |
| Gang Li | Deakin University, Australia |
| Ming Li | East China Normal University, China |
| Fang Liu | AMES Laboratories, USA |
| Xin Liu | University of Calgary, Canada |
| Savino Longo | University of Bari, Italy |
| Tinghuai Ma | NanJing University of Information Science and Technology, China |
| Sergio Maffioletti | University of Zurich, Switzerland |
| Ernesto Marcheggiani | Katholieke Universiteit Leuven, Belgium |
| Antonino Marvuglia | Research Centre Henri Tudor, Luxembourg |
| Nicola Masini | National Research Council, Italy |
| Nirvana Meratnia | University of Twente, The Netherlands |
| Alfredo Milani | University of Perugia, Italy |
| Sanjay Misra | Federal University of Technology Minna, Nigeria |
| Giuseppe Modica | University of Reggio Calabria, Italy |
| José Luis Montaña | University of Cantabria, Spain |
| Beniamino Murgante | University of Basilicata, Italy |
| Jiri Nedoma | Academy of Sciences of the Czech Republic, Czech Republic |
| Laszlo Neumann | University of Girona, Spain |
| Kok-Leong Ong | Deakin University, Australia |
| Belen Palop | Universidad de Valladolid, Spain |
| Marcin Paprzycki | Polish Academy of Sciences, Poland |
| Eric Pardede | La Trobe University, Australia |
| Kwangjin Park | Wonkwang University, Korea |
| Ana Isabel Pereira | Polytechnic Institute of Braganca, Portugal |
| Maurizio Pollino | Italian National Agency for New Technologies, Energy and Sustainable Economic Development, Italy |
| Alenka Poplin | University of Hamburg, Germany |
| Vidyasagar Potdar | Curtin University of Technology, Australia |
| David C. Prosperi | Florida Atlantic University, USA |
| Wenny Rahayu | La Trobe University, Australia |
| Jerzy Respondek | Silesian University of Technology Poland |
| Ana Maria A.C. Rocha | University of Minho, Portugal |

| | |
|---|---|
| Humberto Rocha | INESC-Coimbra, Portugal |
| Alexey Rodionov | Institute of Computational Mathematics and Mathematical Geophysics, Russia |
| Cristina S. Rodrigues | University of Minho, Portugal |
| Octavio Roncero | CSIC, Spain |
| Maytham Safar | Kuwait University, Kuwait |
| Chiara Saracino | A.O. Ospedale Niguarda Ca' Granda - Milano, Italy |
| Haiduke Sarafian | The Pennsylvania State University, USA |
| Jie Shen | University of Michigan, USA |
| Qi Shi | Liverpool John Moores University, UK |
| Dale Shires | U.S. Army Research Laboratory, USA |
| Takuo Suganuma | Tohoku University, Japan |
| Sergio Tasso | University of Perugia, Italy |
| Ana Paula Teixeira | University of Tras-os-Montes and Alto Douro, Portugal |
| Senhorinha Teixeira | University of Minho, Portugal |
| Parimala Thulasiraman | University of Manitoba, Canada |
| Carmelo Torre | Polytechnic of Bari, Italy |
| Javier Martinez Torres | Centro Universitario de la Defensa Zaragoza, Spain |
| Giuseppe A. Trunfio | University of Sassari, Italy |
| Unal Ufuktepe | Izmir University of Economics, Turkey |
| Toshihiro Uchibayashi | Kyushu Sangyo University, Japan |
| Mario Valle | Swiss National Supercomputing Centre, Switzerland |
| Pablo Vanegas | University of Cuenca, Equador |
| Piero Giorgio Verdini | INFN Pisa and CERN, Italy |
| Marco Vizzari | University of Perugia, Italy |
| Koichi Wada | University of Tsukuba, Japan |
| Krzysztof Walkowiak | Wroclaw University of Technology, Poland |
| Robert Weibel | University of Zurich, Switzerland |
| Roland Wismüller | Universität Siegen, Germany |
| Mudasser Wyne | SOET National University, USA |
| Chung-Huang Yang | National Kaohsiung Normal University, Taiwan |
| Xin-She Yang | National Physical Laboratory, UK |
| Salim Zabir | France Telecom Japan Co., Japan |
| Haifeng Zhao | University of California, Davis, USA |
| Kewen Zhao | University of Qiongzhou, China |
| Albert Y. Zomaya | University of Sydney, Australia |

## Reviewers

| | |
|---|---|
| Abawajy Jemal | Deakin University, Australia |
| Abdi Samane | University College Cork, Ireland |
| Aceto Lidia | University of Pisa, Italy |
| Acharjee Shukla | Dibrugarh University, India |
| Adriano Elias | Universidade Nova de Lisboa, Portugal |
| Afreixo Vera | University of Aveiro, Portugal |
| Aguiar Ademar | Universidade do Porto, Portugal |

| | |
|---|---|
| Aguilar Antonio | University of Barcelona, Spain |
| Aguilar José Alfonso | Universidad Autónoma de Sinaloa, Mexico |
| Ahmed Faisal | University of Calgary, Canada |
| Aktas Mehmet | Yildiz Technical University, Turkey |
| Al-Juboori AliAlwan | International Islamic University Malaysia, Malaysia |
| Alarcon Vladimir | Universidad Diego Portales, Chile |
| Alberti Margarita | University of Barcelona, Spain |
| Ali Salman | NUST, Pakistan |
| Alkazemi Basem Qassim | University, Saudi Arabia |
| Alvanides Seraphim | Northumbria University, UK |
| Alvelos Filipe | University of Minho, Portugal |
| Alves Cláudio | University of Minho, Portugal |
| Alves José Luis | University of Minho, Portugal |
| Alves Maria Joo | Universidade de Coimbra, Portugal |
| Amin Benatia Mohamed | Groupe Cesi, France |
| Amorim Ana Paula | University of Minho, Portugal |
| Amorim Paulo | Federal University of Rio de Janeiro, Brazil |
| Andrade Wilkerson | Federal University of Campina Grande, Brazil |
| Andrianov Serge | Yandex, Russia |
| Aniche Mauricio | University of São Paulo, Brazil |
| Andrienko Gennady | Fraunhofer Institute for Intelligent Analysis and Informations Systems, Germany |
| Apduhan Bernady | Kyushu Sangyo University, Japan |
| Aquilanti Vincenzo | University of Perugia, Italy |
| Aquino Gibeon | UFRN, Brazil |
| Argiolas Michele | University of Cagliari, Italy |
| Asche Hartmut | Potsdam University, Germany |
| Athayde Maria Emilia Feijão Queiroz | University of Minho, Portugal |
| Attardi Raffaele | University of Napoli Federico II, Italy |
| Azad Md. Abdul | Indian Institute of Technology Kanpur, India |
| Azad Md. Abul Kalam | University of Minho, Portugal |
| Bao Fernando | Universidade Nova de Lisboa, Portugal |
| Badard Thierry | Laval University, Canada |
| Bae Ihn-Han | Catholic University of Daegu, South Korea |
| Baioletti Marco | University of Perugia, Italy |
| Balena Pasquale | Polytechnic of Bari, Italy |
| Banerjee Mahua | Xavier Institute of Social Sciences, India |
| Barroca Filho Itamir | UFRN, Brazil |
| Bartoli Daniele | University of Perugia, Italy |
| Bastanfard Azam | Islamic Azad University, Iran |
| Belanzoni Paola | University of Perugia, Italy |
| Bencardino Massimiliano | University of Salerno, Italy |
| Benigni Gladys | University of Oriente, Venezuela |

| | |
|---|---|
| Bertolotto Michela | University College Dublin, Ireland |
| Bilancia Massimo | Università di Bari, Italy |
| Blanquer Ignacio | Universitat Politècnica de València, Spain |
| Bodini Olivier | Université Pierre et Marie Curie Paris and CNRS, France |
| Bogdanov Alexander | Saint-Petersburg State University, Russia |
| Bollini Letizia | University of Milano, Italy |
| Bonifazi Alessandro | Polytechnic of Bari, Italy |
| Borruso Giuseppe | University of Trieste, Italy |
| Bostenaru Maria | "Ion Mincu" University of Architecture and Urbanism, Romania |
| Boucelma Omar | University of Marseille, France |
| Braga Ana Cristina | University of Minho, Portugal |
| Branquinho Amilcar | University of Coimbra, Portugal |
| Brás Carmo | Universidade Nova de Lisboa, Portugal |
| Cacao Isabel | University of Aveiro, Portugal |
| Cadarso-Suárez Carmen | University of Santiago de Compostela, Spain |
| Caiaffa Emanuela | ENEA, Italy |
| Calamita Giuseppe | National Research Council, Italy |
| Campagna Michele | University of Cagliari, Italy |
| Campobasso Francesco | University of Bari, Italy |
| Campos José | University of Minho, Portugal |
| Caniato Renhe Marcelo | Universidade Federal de Juiz de Fora, Brazil |
| Cannatella Daniele | University of Napoli Federico II, Italy |
| Canora Filomena | University of Basilicata, Italy |
| Cannatella Daniele | University of Napoli Federico II, Italy |
| Canora Filomena | University of Basilicata, Italy |
| Carbonara Sebastiano | University of Chieti, Italy |
| Carlini Maurizio | University of Tuscia, Italy |
| Carneiro Claudio | École Polytechnique Fédérale de Lausanne, Switzerland |
| Ceppi Claudia | Polytechnic of Bari, Italy |
| Cerreta Maria | University Federico II of Naples, Italy |
| Chen Hao | Shanghai University of Engineering Science, China |
| Choi Joonsoo | Kookmin University, South Korea |
| Choo Hyunseung | Sungkyunkwan University, South Korea |
| Chung Min Young | Sungkyunkwan University, South Korea |
| Chung Myoungbeom | Sungkyunkwan University, South Korea |
| Chung Tai-Myoung | Sungkyunkwan University, South Korea |
| Cirrincione Maurizio | Université de Technologie Belfort-Montbeliard, France |
| Clementini Eliseo | University of L'Aquila, Italy |
| Coelho Leandro dos Santos | PUC-PR, Brazil |
| Coletti Cecilia | University of Chieti, Italy |
| Conceicao Ana | Universidade do Algarve, Portugal |
| Correia Elisete | University of Trás-Os-Montes e Alto Douro, Portugal |
| Correia Filipe | FEUP, Portugal |

| | |
|---|---|
| Correia Florbela Maria da Cruz Domingues | Instituto Politécnico de Viana do Castelo, Portugal |
| Corso Pereira Gilberto | UFPA, Brazil |
| Cortés Ana | Universitat Autònoma de Barcelona, Spain |
| Cosido Oscar | Ayuntamiento de Santander, Spain |
| Costa Carlos | Faculdade Engenharia U. Porto, Portugal |
| Costa Fernanda | University of Minho, Portugal |
| Costantini Alessandro | INFN, Italy |
| Crasso Marco | National Scientific and Technical Research Council, Argentina |
| Crawford Broderick | Universidad Catolica de Valparaiso, Chile |
| Crestaz Ezio | GiScience, Italia |
| Cristia Maximiliano | CIFASIS and UNR, Argentina |
| Cunha Gaspar | University of Minho, Portugal |
| Cutini Valerio | University of Pisa, Italy |
| Danese Maria | IBAM, CNR, Italy |
| Daneshpajouh Shervin | University of Western Ontario, Canada |
| De Almeida Regina | University of Trás-os-Montes e Alto Douro, Portugal |
| de Doncker Elise | University of Michgan, USA |
| De Fino Mariella | Polytechnic of Bari, Italy |
| De Paolis Lucio Tommaso | University of Salento, Italy |
| de Rezende Pedro J. | Universidade Estadual de Campinas, Brazil |
| De Rosa Fortuna | University of Napoli Federico II, Italy |
| De Toro Pasquale | University of Napoli Federico II, Italy |
| Decker Hendrik | Instituto Tecnológico de Informática, Spain |
| Degtyarev Alexander | Saint-Petersburg State University, Russia |
| Deiana Andrea | Geoinfolab, Italia |
| Deniz Berkhan | Aselsan Electronics Inc., Turkey |
| Desjardin Eric | University of Reims, France |
| Devai Frank | London South Bank University, UK |
| Dwivedi Sanjay Kumar | Babasaheb Bhimrao Ambedkar University, India |
| Dhawale Chitra | PR Pote College, Amravati, India |
| Di Cugno Gianluca | Polytechnic of Bari, Italy |
| Di Gangi Massimo | University of Messina, Italy |
| Di Leo Margherita | JRC, European Commission, Belgium |
| Dias Joana | University of Coimbra, Portugal |
| Dias d'Almeida Filomena | University of Porto, Portugal |
| Diez Teresa | Universidad de Alcalá, Spain |
| Dilo Arta | University of Twente, The Netherlands |
| Dixit Veersain | Delhi University, India |
| Doan Anh Vu | Université Libre de Bruxelles, Belgium |
| Durrieu Sylvie | Maison de la Teledetection Montpellier, France |
| Dutra Inês | University of Porto, Portugal |
| Dyskin Arcady | The University of Western Australia, Australia |

| | |
|---|---|
| Eichelberger Hanno | University of Tübingen, Germany |
| El-Zawawy Mohamed A. | Cairo University, Egypt |
| Escalona Maria-Jose | University of Seville, Spain |
| Falcão M. Irene | University of Minho, Portugal |
| Farantos Stavros | University of Crete and FORTH, Greece |
| Faria Susana | University of Minho, Portugal |
| Fernandes Edite | University of Minho, Portugal |
| Fernandes Rosário | University of Minho, Portugal |
| Fernandez Joao P. | Universidade da Beira Interior, Portugal |
| Ferrão Maria | University of Beira Interior and CEMAPRE, Portugal |
| Ferreira Fátima | University of Trás-Os-Montes e Alto Douro, Portugal |
| Figueiredo Manuel Carlos | University of Minho, Portugal |
| Filipe Ana | University of Minho, Portugal |
| Flouvat Frederic | University New Caledonia, New Caledonia |
| Forjaz Maria Antónia | University of Minho, Portugal |
| Formosa Saviour | University of Malta, Malta |
| Fort Marta | University of Girona, Spain |
| Franciosa Alfredo | University of Napoli Federico II, Italy |
| Freitas Adelaide de Fátima Baptista Valente | University of Aveiro, Portugal |
| Frydman Claudia | Laboratoire des Sciences de l'Information et des Systèmes, France |
| Fusco Giovanni | CNRS - UMR ESPACE, France |
| Gabrani Goldie | University of Delhi, India Galleguillos Cristian, Pontificia Universidad Catlica de Valparaso, Chile |
| Gao Shang | Zhongnan University of Economics and Law, China |
| Garau Chiara | University of Cagliari, Italy |
| Garcia Ernesto | University of the Basque Country, Spain |
| Garca Omar Vicente | Universidad Autònoma de Sinaloa, Mexico |
| Garcia Tobio Javier | Centro de Supercomputación de Galicia, CESGA, Spain |
| Gavrilova Marina | University of Calgary, Canada |
| Gazzea Nicoletta | ISPRA, Italy |
| Gensel Jerome | IMAG, France |
| Geraldi Edoardo | National Research Council, Italy |
| Gervasi Osvaldo | University of Perugia, Italy |
| Giaoutzi Maria | National Technical University Athens, Greece |
| Gil Artur | University of the Azores, Portugal |
| Gizzi Fabrizio | National Research Council, Italy |
| Gomes Abel | Universidad de Beira Interior, Portugal |
| Gomes Maria Cecilia | Universidade Nova de Lisboa, Portugal |
| Gomes dos Anjos Eudisley | Federal University of Paraba, Brazil |
| Gonçalves Alexandre | Instituto Superior Tecnico Lisboa, Portugal |

| | |
|---|---|
| Gonçalves Arminda Manuela | University of Minho, Portugal |
| Gonzaga de Oliveira Sanderson Lincohn | Universidade Do Estado De Santa Catarina, Brazil |
| Gonzalez-Aguilera Diego | Universidad de Salamanca, Spain |
| Gorbachev Yuriy | Geolink Technologies, Russia |
| Govani Kishan | Darshan Institute of Engineering Technology, India |
| Grandison Tyrone | Proficiency Labs International, USA |
| Gravagnuolo Antonia | University of Napoli Federico II, Italy |
| Grilli Luca | University of Perugia, Italy |
| Guerra Eduardo | National Institute for Space Research, Brazil |
| Guo Hua | Carleton University, Canada |
| Hanazumi Simone | University of São Paulo, Brazil |
| Hanif Mohammad Abu | Chonbuk National University, South Korea |
| Hansen Henning Sten | Aalborg University, Denmark |
| Hanzl Malgorzata | University of Lodz, Poland |
| Hegedus Peter | University of Szeged, Hungary |
| Heijungs Reinout | VU University Amsterdam, The Netherlands |
| Hendrix Eligius M.T. | University of Malaga/Wageningen University, Spain/The Netherlands |
| Henriques Carla | Escola Superior de Tecnologia e Gestão, Portugal |
| Herawan Tutut | University of Malaya, Malaysia |
| Hiyoshi Hisamoto | Gunma University, Japan |
| Hodorog Madalina | Austria Academy of Science, Austria |
| Hong Choong Scon | Kyung Hee University, South Korea |
| Hsu Ching-Hsien | Chung Hua University, Taiwan |
| Hsu Hui-Huang | Tamkang University, Taiwan |
| Hu Hong | The Honk Kong Polytechnic University, China |
| Huang Jen-Fa | National Cheng Kung University, Taiwan |
| Hubert Frederic | Université Laval, Canada |
| Iglesias Andres | University of Cantabria, Spain |
| Jamal Amna | National University of Singapore, Singapore |
| Jank Gerhard | Aachen University, Germany |
| Jeong Jongpil | Sungkyunkwan University, South Korea |
| Jiang Bin | University of Gävle, Sweden |
| Johnson Franklin | Universidad de Playa Ancha, Chile |
| Kalogirou Stamatis | Harokopio University of Athens, Greece |
| Kamoun Farouk | Université de la Manouba, Tunisia |
| Kanchi Saroja | Kettering University, USA |
| Kanevski Mikhail | University of Lausanne, Switzerland |
| Kang Myoung-Ah | ISIMA Blaise Pascal University, France |
| Karandikar Varsha | Devi Ahilya University, Indore, India |
| Karimipour Farid | Vienna University of Technology, Austria |
| Kavouras Marinos | University of Lausanne, Switzerland |
| Kazar Baris | Oracle Corp., USA |

| | |
|---|---|
| Keramat Alireza | Jundi-Shapur Univ. of Technology, Iran |
| Khan Murtaza | NUST, Pakistan |
| Khattak Asad Masood | Kyung Hee University, Korea |
| Khazaei Hamzeh | Ryerson University, Canada |
| Khurshid Khawar | NUST, Pakistan |
| Kim Dongsoo | Indiana University-Purdue University Indianapolis, USA |
| Kim Mihui | Hankyong National University, South Korea |
| Koo Bonhyun | Samsung, South Korea |
| Korkhov Vladimir | St. Petersburg State University, Russia |
| Kotzinos Dimitrios | Université de Cergy-Pontoise, France |
| Kumar Dileep | SR Engineering College, India |
| Kurdia Anastasia | Buknell University, USA |
| Lachance-Bernard Nicolas | École Polytechnique Fédérale de Lausanne, Switzerland |
| Laganà Antonio | University of Perugia, Italy |
| Lai Sabrina | University of Cagliari, Italy |
| Lanorte Antonio | CNR-IMAA, Italy |
| Lanza Viviana | Lombardy Regional Institute for Research, Italy |
| Lasaponara Rosa | National Research Council, Italy |
| Lassoued Yassine | University College Cork, Ireland |
| Lazzari Maurizio | CNR IBAM, Italy |
| Le Duc Tai | Sungkyunkwan University, South Korea |
| Le Duc Thang | Sungkyunkwan University, South Korea |
| Le-Thi Kim-Tuyen | Sungkyunkwan University, South Korea |
| Ledoux Hugo | Delft University of Technology, The Netherlands |
| Lee Dong-Wook | INHA University, South Korea |
| Lee Hongseok | Sungkyunkwan University, South Korea |
| Lee Ickjai | James Cook University, Australia |
| Lee Junghoon | Jeju National University, South Korea |
| Lee KangWoo | Sungkyunkwan University, South Korea |
| Legatiuk Dmitrii | Bauhaus University Weimar, Germany |
| Lendvay Gyorgy | Hungarian Academy of Science, Hungary |
| Leonard Kathryn | California State University, USA |
| Li Ming | East China Normal University, China |
| Libourel Thrse | LIRMM, France |
| Lin Calvin | University of Texas at Austin, USA |
| Liu Xin | University of Calgary, Canada |
| Loconte Pierangela | Technical University of Bari, Italy |
| Lombardi Andrea | University of Perugia, Italy |
| Longo Savino | University of Bari, Italy |
| Lopes Cristina | University of California Irvine, USA |
| Lopez Cabido Ignacio | Centro de Supercomputación de Galicia, CESGA |
| Lourenço Vanda Marisa | University Nova de Lisboa, Portugal |
| Luaces Miguel | University of A Coruña, Spain |
| Lucertini Giulia | IUAV, Italy |
| Luna Esteban Robles | Universidad Nacional de la Plata, Argentina |

| | |
|---|---|
| M.M.H. Gregori Rodrigo | Universidade Tecnológica Federal do Paraná, Brazil |
| Machado Gaspar | University of Minho, Portugal |
| Machado Jose | University of Minho, Portugal |
| Mahinderjit Singh Manmeet | University Sains Malaysia, Malaysia |
| Malonek Helmuth | University of Aveiro, Portugal |
| Manfreda Salvatore | University of Basilicata, Italy |
| Manns Mary Lynn | University of North Carolina Asheville, USA |
| Manso Callejo Miguel Angel | Universidad Politécnica de Madrid, Spain |
| Marechal Bernard | Universidade Federal de Rio de Janeiro, Brazil |
| Marechal Franois | École Polytechnique Fédérale de Lausanne, Switzerland |
| Margalef Tomas | Universitat Autònoma de Barcelona, Spain |
| Marghany Maged | Universiti Teknologi Malaysia, Malaysia |
| Marsal-Llacuna Maria-Llusa | Universitat de Girona, Spain |
| Marsh Steven | University of Ontario, Canada |
| Martins Ana Mafalda | Universidade de Aveiro, Portugal |
| Martins Pedro | Universidade do Minho, Portugal |
| Marvuglia Antonino | Public Research Centre Henri Tudor, Luxembourg |
| Mateos Cristian | Universidad Nacional del Centro, Argentina |
| Matos Inés | Universidade de Aveiro, Portugal |
| Matos Jose | Instituto Politecnico do Porto, Portugal |
| Matos João | ISEP, Portugal |
| Mauro Giovanni | University of Trieste, Italy |
| Mauw Sjouke | University of Luxembourg, Luxembourg |
| Medeiros Pedro | Universidade Nova de Lisboa, Portugal |
| Melle Franco Manuel | University of Minho, Portugal |
| Melo Ana | Universidade de São Paulo, Brazil |
| Michikawa Takashi | University of Tokio, Japan |
| Milani Alfredo | University of Perugia, Italy |
| Millo Giovanni | Generali Assicurazioni, Italy |
| Min-Woo Park | SungKyunKwan University, South Korea |
| Miranda Fernando | University of Minho, Portugal |
| Misra Sanjay | Covenant University, Nigeria |
| Mo Otilia | Universidad Autonoma de Madrid, Spain |
| Modica Giuseppe | Università Mediterranea di Reggio Calabria, Italy |
| Mohd Nawi Nazri | Universiti Tun Hussein Onn Malaysia, Malaysia |
| Morais João | University of Aveiro, Portugal |
| Moreira Adriano | University of Minho, Portugal |
| Moerig Marc | University of Magdeburg, Germany |
| Morzy Mikolaj | University of Poznan, Poland |
| Mota Alexandre | Universidade Federal de Pernambuco, Brazil |
| Moura Pires João | Universidade Nova de Lisboa - FCT, Portugal |
| Mourão Maria | Polytechnic Institute of Viana do Castelo, Portugal |

| | |
|---|---|
| Mourelle Luiza de Macedo | UERJ, Brazil |
| Mukhopadhyay Asish | University of Windsor, Canada |
| Mulay Preeti | Bharti Vidyapeeth University, India |
| Murgante Beniamino | University of Basilicata, Italy |
| Naghizadeh Majid Reza | Qazvin Islamic Azad University, Iran |
| Nagy Csaba | University of Szeged, Hungary |
| Nandy Subhas | Indian Statistical Institute, India |
| Nash Andrew | Vienna Transport Strategies, Austria |
| Natário Isabel Cristina Maciel | University Nova de Lisboa, Portugal |
| Navarrete Gutierrez Tomas | Luxembourg Institute of Science and Technology, Luxembourg |
| Nedjah Nadia | State University of Rio de Janeiro, Brazil |
| Nguyen Hong-Quang | Ho Chi Minh City University, Vietnam |
| Nguyen Tien Dzung | Sungkyunkwan University, South Korea |
| Nickerson Bradford | University of New Brunswick, Canada |
| Nielsen Frank | Université Paris Saclay CNRS, France |
| NM Tuan | Ho Chi Minh City University of Technology, Vietnam |
| Nogueira Fernando | University of Coimbra, Portugal |
| Nole Gabriele | IRMAA National Research Council, Italy |
| Nourollah Ali | Amirkabir University of Technology, Iran |
| Olivares Rodrigo | UCV, Chile |
| Oliveira Irene | University of Trás-Os-Montes e Alto Douro, Portugal |
| Oliveira José A. | University of Minho, Portugal |
| Oliveira e Silva Luis | University of Lisboa, Portugal |
| Osaragi Toshihiro | Tokyo Institute of Technology, Japan |
| Ottomanelli Michele | Polytechnic of Bari, Italy |
| Ozturk Savas | TUBITAK, Turkey |
| Pagliara Francesca | University of Naples, Italy |
| Painho Marco | New University of Lisbon, Portugal |
| Pantazis Dimos | Technological Educational Institute of Athens, Greece |
| Paolotti Luisa | University of Perugia, Italy |
| Papa Enrica | University of Amsterdam, The Netherlands |
| Papathanasiou Jason | University of Macedonia, Greece |
| Pardede Eric | La Trobe University, Australia |
| Parissis Ioannis | Grenoble INP - LCIS, France |
| Park Gyung-Leen | Jeju National University, South Korea |
| Park Sooyeon | Korea Polytechnic University, South Korea |
| Pascale Stefania | University of Basilicata, Italy |
| Parker Gregory | University of Oklahoma, USA |
| Parvin Hamid | Iran University of Science and Technology, Iran |
| Passaro Pierluigi | University of Bari Aldo Moro, Italy |
| Pathan Al-Sakib Khan | International Islamic University Malaysia, Malaysia |
| Paul Padma Polash | University of Calgary, Canada |

| | |
|---|---|
| Peixoto Bitencourt Ana Carla | Universidade Estadual de Feira de Santana, Brazil |
| Peraza Juan Francisco | Autonomous University of Sinaloa, Mexico |
| Perchinunno Paola | University of Bari, Italy |
| Pereira Ana | Polytechnic Institute of Bragança, Portugal |
| Pereira Francisco | Instituto Superior de Engenharia, Portugal |
| Pereira Paulo | University of Minho, Portugal |
| Pereira Javier | Diego Portales University, Chile |
| Pereira Oscar | Universidade de Aveiro, Portugal |
| Pereira Ricardo | Portugal Telecom Inovacao, Portugal |
| Perez Gregorio | Universidad de Murcia, Spain |
| Pesantes Mery | CIMAT, Mexico |
| Pham Quoc Trung | HCMC University of Technology, Vietnam |
| Pietrantuono Roberto | University of Napoli "Federico II", Italy |
| Pimentel Carina | University of Aveiro, Portugal |
| Pina Antonio | University of Minho, Portugal |
| Piñar Miguel | Universidad de Granada, Spain |
| Pinciu Val | Southern Connecticut State University, USA |
| Pinet Francois | IRSTEA, France |
| Piscitelli Claudia | Polytechnic University of Bari, Italy |
| Pollino Maurizio | ENEA, Italy |
| Poplin Alenka | University of Hamburg, Germany |
| Porschen Stefan | University of Köln, Germany |
| Potena Pasqualina | University of Bergamo, Italy |
| Prata Paula | University of Beira Interior, Portugal |
| Previtali Mattia | Polytechnic of Milan, Italy |
| Prosperi David | Florida Atlantic University, USA |
| Protheroe Dave | London South Bank University, UK |
| Pusatli Tolga | Cankaya University, Turkey |
| Qaisar Saad | NURST, Pakistan |
| Qi Yu | Mesh Capital LLC, USA |
| Quan Tho | Ho Chi Minh City University of Technology, Vietnam |
| Raffaeta Alessandra | University of Venice, Italy |
| Ragni Mirco | Universidade Estadual de Feira de Santana, Brazil |
| Rahayu Wenny | La Trobe University, Australia |
| Rautenberg Carlos | University of Graz, Austria |
| Ravat Franck | IRIT, France |
| Raza Syed Muhammad | Sungkyunkwan University, South Korea |
| Rinaldi Antonio | DIETI - UNINA, Italy |
| Rinzivillo Salvatore | University of Pisa, Italy |
| Rios Gordon | University College Dublin, Ireland |
| Riva Sanseverino Eleonora | University of Palermo, Italy |
| Roanes-Lozano Eugenio | Universidad Complutense de Madrid, Spain |
| Rocca Lorena | University of Padova, Italy |
| Roccatello Eduard | 3DGIS, Italy |

| | |
|---|---|
| Rocha Ana Maria | University of Minho, Portugal |
| Rocha Humberto | University of Coimbra, Portugal |
| Rocha Jorge | University of Minho, Portugal |
| Rocha Maria Clara | ESTES Coimbra, Portugal |
| Rocha Miguel | University of Minho, Portugal |
| Rodrigues Armanda | Universidade Nova de Lisboa, Portugal |
| Rodrigues Cristina | DPS, University of Minho, Portugal |
| Rodrigues Joel | University of Minho, Portugal |
| Rodriguez Daniel | University of Alcala, Spain |
| Rodrguez Gonzlez Alejandro | Universidad Carlos III Madrid, Spain |
| Roh Yongwan | Korean IP, South Korea |
| Romano Bernardino | University of l'Aquila, Italy |
| Roncaratti Luiz | Instituto de Física, University of Brasilia, Brazil |
| Roshannejad Ali | University of Calgary, Canada |
| Rosi Marzio | University of Perugia, Italy |
| Rossi Gianfranco | University of Parma, Italy |
| Rotondo Francesco | Polytechnic of Bari, Italy |
| Roussey Catherine | IRSTEA, France |
| Ruj Sushmita | Indian Statistical Institute, India |
| S. Esteves Jorge | University of Aveiro, Portugal |
| Saeed Husnain | NUST, Pakistan |
| Sahore Mani | Lovely Professional University, India |
| Saini Jatinder Singh | Baba Banda Singh Bahadur Engineering College, India |
| Salzer Reiner | Technical University Dresden, Germany |
| Sameh Ahmed | The American University in Cairo, Egypt |
| Sampaio Alcinia Zita | Instituto Superior Tecnico Lisboa, Portugal |
| Sannicandro Valentina | Polytechnic of Bari, Italy |
| Santiago Jnior Valdivino | Instituto Nacional de Pesquisas Espaciais, Brazil |
| Santos Josué | UFABC, Brazil |
| Santos Rafael | INPE, Brazil |
| Santos Viviane | Universidade de São Paulo, Brazil |
| Santucci Valentino | University of Perugia, Italy |
| Saracino Gloria | University of Milano-Bicocca, Italy |
| Sarafian Haiduke | Pennsylvania State University, USA |
| Saraiva João | University of Minho, Portugal |
| Sarrazin Renaud | Université Libre de Bruxelles, Belgium |
| Schirone Dario Antonio | University of Bari, Italy |
| Schneider Michel | ISIMA, France |
| Schoier Gabriella | University of Trieste, Italy |
| Schuhmacher Marta | Universitat Rovira i Virgili, Spain |
| Scorza Francesco | University of Basilicata, Italy |
| Seara Carlos | Universitat Politècnica de Catalunya, Spain |
| Sellares J. Antoni | Universitat de Girona, Spain |
| Selmaoui Nazha | University of New Caledonia, New Caledonia |
| Severino Ricardo Jose | University of Minho, Portugal |

| | |
|---|---|
| Shaik Mahaboob Hussain | JNTUK Vizianagaram, A.P., India |
| Sheikho Kamel | KACST, Saudi Arabia |
| Shen Jie | University of Michigan, USA |
| Shi Xuefei | University of Science Technology Beijing, China |
| Shin Dong Hee | Sungkyunkwan University, South Korea |
| Shojaeipour Shahed | Universiti Kebangsaan Malaysia, Malaysia |
| Shon Minhan | Sungkyunkwan University, South Korea |
| Shukla Ruchi | University of Johannesburg, South Africa |
| Silva Carlos | University of Minho, Portugal |
| Silva J.C. | IPCA, Portugal |
| Silva de Souza Laudson | Federal University of Rio Grande do Norte, Brazil |
| Silva-Fortes Carina | ESTeSL-IPL, Portugal |
| Simão Adenilso | Universidade de São Paulo, Brazil |
| Singh R.K. | Delhi University, India |
| Singh V.B. | University of Delhi, India |
| Singhal Shweta | GGSIPU, India |
| Sipos Gergely | European Grid Infrastructure, The Netherlands |
| Smolik Michal | University of West Bohemia, Czech Republic |
| Soares Inês | INESC Porto, Portugal |
| Soares Michel | Federal University of Sergipe, Brazil |
| Sobral Joao | University of Minho, Portugal |
| Son Changhwan | Sungkyunkwan University, South Korea |
| Song Kexing | Henan University of Science and Technology, China |
| Sosnin Petr | Ulyanovsk State Technical University, Russia |
| Souza Eric | Universidade Nova de Lisboa, Portugal |
| Sproessig Wolfgang | Technical University Bergakademie Freiberg, Germany |
| Sreenan Cormac | University College Cork, Ireland |
| Stankova Elena | Saint-Petersburg State University, Russia |
| Starczewski Janusz | Institute of Computational Intelligence, Poland |
| Stehn Fabian | University of Bayreuth, Germany |
| Sultana Madeena | University of Calgary, Canada |
| Swarup Das | Ananda Kalinga Institute of Industrial Technology, India |
| Tahar Sofiène | Concordia University, Canada |
| Takato Setsuo | Toho University, Japan |
| Talebi Hossein | University of Calgary, Canada |
| Tanaka Kazuaki | Kyushu Institute of Technology, Japan |
| Taniar David | Monash University, Australia |
| Taramelli Andrea | Columbia University, USA |
| Tarantino Eufemia | Polytechnic of Bari, Italy |
| Tariq Haroon | Connekt Lab, Pakistan |
| Tasso Sergio | University of Perugia, Italy |
| Teixeira Ana Paula | University of Trás-Os-Montes e Alto Douro, Portugal |
| Tesseire Maguelonne | IRSTEA, France |
| Thi Thanh Huyen Phan | Japan Advanced Institute of Science and Technology, Japan |

| | |
|---|---|
| Thorat Pankaj | Sungkyunkwan University, South Korea |
| Tilio Lucia | University of Basilicata, Italy |
| Tiwari Rupa | University of Minnesota, USA |
| Toma Cristian | Polytechnic University of Bucarest, Romania |
| Tomaz Graça | Polytechnic Institute of Guarda, Portugal |
| Tortosa Leandro | University of Alicante, Spain |
| Tran Nguyen | Kyung Hee University, South Korea |
| Tripp Barba, Carolina | Universidad Autnoma de Sinaloa, Mexico |
| Trunfio Giuseppe A. | University of Sassari, Italy |
| Uchibayashi Toshihiro | Kyushu Sangyo University, Japan |
| Ugalde Jesus | Universidad del Pais Vasco, Spain |
| Urbano Joana | LIACC University of Porto, Portugal |
| Van de Weghe Nico | Ghent University, Belgium |
| Varella Evangelia | Aristotle University of Thessaloniki, Greece |
| Vasconcelos Paulo | University of Porto, Portugal |
| Vella Flavio | University of Rome La Sapienza, Italy |
| Velloso Pedro | Universidade Federal Fluminense, Brazil |
| Viana Ana | INESC Porto, Portugal |
| Vidacs Laszlo | MTA-SZTE, Hungary |
| Vieira Ramadas Gisela | Polytechnic of Porto, Portugal |
| Vijay NLankalapalli | National Institute for Space Research, Brazil |
| Vijaykumar Nandamudi | INPE, Brazil |
| Viqueira José R.R. | University of Santiago de Compostela, Spain |
| Vitellio Ilaria | University of Naples, Italy |
| Vizzari Marco | University of Perugia, Italy |
| Wachowicz Monica | University of New Brunswick, Canada |
| Walentynski Ryszard | Silesian University of Technology, Poland |
| Walkowiak Krzysztof | Wroclav University of Technology, Poland |
| Wallace Richard J. | University College Cork, Ireland |
| Waluyo Agustinus Borgy | Monash University, Australia |
| Wanderley Fernando | FCT/UNL, Portugal |
| Wang Chao | University of Science and Technology of China, China |
| Wang Yanghui | Beijing Jiaotong University, China |
| Wei Hoo Chong | Motorola, USA |
| Won Dongho | Sungkyunkwan University, South Korea |
| Wu Jian-Da | National Changhua University of Education, Taiwan |
| Xin Liu | École Polytechnique Fédérale de Lausanne, Switzerland |
| Yadav Nikita | Delhi Universty, India |
| Yamauchi Toshihiro | Okayama University, Japan |
| Yao Fenghui | Tennessee State University, USA |
| Yatskevich Mikalai | Assioma, Italy |
| Yeoum Sanggil | Sungkyunkwan University, South Korea |
| Yoder Joseph | Refactory Inc., USA |
| Zalyubovskiy Vyacheslav | Russian Academy of Sciences, Russia |

## Sponsoring Organizations

ICCSA 2015 would not have been possible without the tremendous support of many organizations and institutions, for which all organizers and participants of ICCSA 2015 express their sincere gratitude:

University of Calgary, Canada (http://www.ucalgary.ca)

University of Perugia, Italy (http://www.unipg.it)

University of Basilicata, Italy (http://www.unibas.it)

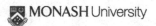

Monash University, Australia (http://monash.edu)

Kyushu Sangyo University, Japan (www.kyusan-u.ac.jp)

Universidade do Minho, Portugal (http://www.uminho.pt)

# Contents – Part V

# Workshop on Software Quality
# (SQ 2015)

# Code Ownership: Impact on Maintainability

Csaba Faragó, Péter Hegedűs[✉], and Rudolf Ferenc

Department of Software Engineering, University of Szeged, Dugonics tér 13,
H-6720 Szeged, Hungary
{farago,hpeter,ferenc}@inf.u-szeged.hu

**Abstract.** Software systems erode during development, which results in
high maintenance costs in the long term. Is it possible to narrow down
where exactly this erosion happens? Can we infer the future erosion based
on past code changes?

In this paper we investigate code ownership and show that a further
step of code quality decrease is more likely to happen due to the changes
in source files modified by several developers in the past, compared to
files with clear ownership. We estimate the level of code ownership and
maintainability changes for every commit of three open-source and one
proprietary software systems. With the help of Wilcoxon rank test we
compare the ownership values of the files in commits resulting maintainability increase with those of decreasing the maintainability. Three tests
out of the four gave strong results and the fourth one did not contradict them either. The conclusion of this study is a generalization of the
already known fact that common code is more error-prone than those of
developed by fewer developers.

This result could be utilized in identifying the "hot spots" of the
source code from maintainability point of view. A possible IDE plug-in,
which indicates the risk of decreasing the maintainability of the source
code, could help the architect and warn the developers.

**Keywords:** Code ownership · ISO/IEC 25010 · Source code maintainability · Wilcoxon test

## 1 Introduction

Software quality plays a crucial role in modern development projects. Maintainability is one of the six sub-characteristics of software quality, as defined
originally in the ISO/IEC 9126 standard [14]. Software maintenance consumes
huge efforts: based on the experiences, at least half of the total amount of software development costs are spent on this activity. As maintainability is in direct
connection with maintenance costs, our motivation is to investigate the effect
of the development process on the maintainability of the code. Our goal is to
explore typical patterns causing similar changes in software quality, which could
either help to avoid software erosion, or provide information about how to better
allocate efforts spent on improving software quality.

© Springer International Publishing Switzerland 2015
O. Gervasi et al. (Eds.): ICCSA 2015, Part V, LNCS 9159, pp. 3–19, 2015.
DOI: 10.1007/978-3-319-21413-9_1

In a recent paper [9], we presented that there is a strong connection between the version control operations and the maintainability of the source code. We also performed a study [8] that revealed the connection of the version control operations and maintainability. It turned out that file additions in a project have rather positive, file updates (i.e. propagation of the changes in the existing source code) have rather negative effect on maintainability, while a clear effect of file deletions was not identified. Furthermore, in a more recent work [7] we presented the results of a variance analysis. File additions and file deletions increase the variance of the maintainability, and the update operation decreases it.

In this work we make use of the author information coming from the version control system and check the effect of code ownership on maintainability. We performed the analysis on commit basis. We collected historical data form SVN version control system and estimated the maintainability with the help of ColumbusQM probabilistic software quality model [1]. We defined the code ownership values based on the historical changes.

Formally, we investigated the following research question:

**Research Question:** *Does the number of developers modifying the same code in the past have any affect on the maintainability change of future commits?*

**Null hypothesis:** *The past does not have any influence on the future: the future maintainability time line of the source code is totally independent of the number of developers modifying a code in the past.*

**Assumed alternative hypothesis:** *Modifying files without clear ownership (i.e. those of which have been modified by several different developers in the past) is more likely to result in further maintainability decrease than modifying files with clear ownership (i.e. those modified by only one or by a very few number of developers).*

We investigated this question by studying three open-source systems and an industrial one. According to the results presented later in this paper, we rejected the null hypothesis. The performed Wilcoxon rank test varied in significance (p-values of 0.000014, 0.034, 0.060 and 0.21). The only not significant result was caused by a software system where more than 80% of the total commits were performed by a single developer, therefore we can consider that as an outlier from this respect. This supports our assumption that modifying common code (i.e. source files which have been modified by several developers in the past, according to the source control log) is more likely to cause software degradation and decrease the maintainability compared to those modifications affecting a code with clear ownership.

The remaining of the paper is organized as follows. Section 2 provides a brief overview of works that are related to this research. In Section 3 we present the methodology – how we collected the data, what kinds of tests were performed and how we illustrated the results. In Section 4.1 we provide details about the software systems we used for our study. In Section 4 we present the results of the statistical tests. In Section 5 we list the possible threats to the validity of the results, while we conclude the paper in Section 6.

# 2   Related Work

There are several papers dealing with the topic of code ownership or developer related issues.

In their work Mockus et al. [16] present a case study of the Apache Server open source development. Among others they considered the topic of code ownership as well. They analyzed a single project which had nearly 400 contributors and concluded that in the analyzed project no real code ownership was evolved. We analyzed 4 systems, with the magnitude of 1-3 dozens of developers each and analyzed the effects of the code ownership on future maintainability.

In a study, Nordberg [17] describe four types of code ownership: product specialist, subsystem ownership, chief architect and collective ownership. They discusses the advantages and disadvantages of each models. Our findings support the base assumption of this study: in case of lack of well defined code ownership the code quality is likely to decrease. We did not consider code ownership models in such detail, but present the most obvious developer related facts for the 4 analyzed systems.

LaToza et al. present the results of two surveys and eleven interviews [15] conducted by software developers at Microsoft, regarding software development questions. Some of the questions were related to code ownership as well. An interesting statement of this article is that code ownership can also be wrong, as if a code is understood and maintained by a single developer, it makes individuals too indispensable. As an alternative of individual code ownership, the team code ownership was also investigated. Contrary to them, we examined the effect of the code ownership on the maintainability, i.e. studied why code ownership is good, but from the organizational level the aspects can be different in the longer term.

Fritz et al. investigated the frequency and recency of interactions [10] on the code by developers: questions were asked to find out if they can recall types of variables, types of parameters, method names, another method calls and methods which calls a specified method. They showed that according to the assumed hypothesis, the developers know their own code better (that was modified by him/her frequently and recently) compared to foreign code. We, on the other hand, analyze code ownership instead of code knowledge.

Weyuker et al. investigate if their already presented fault prediction model can be enhanced by including the number of developers [20]. They found that the achieved improvement is negligible, which might be surprising at a first glance. We, on the other hand, found a significant correlation by examining the number of different developers' effect on maintainability. The contradiction could be resolved by the following: an already well established model cannot be enhanced further significantly by including the number of developers predictor; but it itself is a good predictor of maintainability change and of defects as well.

In their study Bird et al. [3] investigate if there are significant differences in software quality following a distributed development model compared to a collocated development. They analyzed the development of Windows Vista and argue that the differences are hardly notable. As a complementary result they found a positive correlation between the number of developers and defects, which result is similar to ours. In our current work, we did not consider the distance among development team members, but analyze the effect of ownership on software maintainability.

The same authors present a fault prediction method [4], which combines social factors in development organizations and program dependency information. They found that this was a better model than considering only one of the factors. They proved their concept on 2 huge projects: Windows Vista and Eclipse. We also used both social and technical networks implicitly: the social one is the number of developers of a module, and the technical one is the sources committed together.

The problem of code ownership, especially finding the hidden co-authors, is analyzed by Hattori et al. [12]. They created a tool called Syde, which records every change by every commit, and with the help of this information they were able to determine the code ownership more precisely. They validated their concept using a commercial system. We also analyzed code ownership, bud did not consider developer interaction information.

Rahman et al. [19] introduced a code ownership and experience based defect prediction model, but instead of just considering the modifications performed on source file itself, they introduced a fine-grained level by analyzing the contributions to code fragments. We on the other hand performed our analysis conventionally on source file basis.

The study by Bird et al. [5] targets similar goal to ours; as its title says: the effects of ownership on software quality. The authors investigated 2 huge projects: Windows Vista and Windows 7. We, on the other hand, investigated 4 smaller projects. They considered software quality in terms of pre-release faults and post-release failures; we consider code maintainability as an aggregated value of complexity metrics. They performed the analysis on binary and release level; our study is based on source code and commits. For a binary they defined the terms minor contributor (developers who contributed at most 5% of the total commits), major contributor (above 5%) and ownership (proportion of the commits of the highest contributor). Among others, they found that software components with many minor contributors had more failures than other software components. Moreover, the high level of ownership resulted in less defects. These findings are very similar to ours: by increasing the number of developers and therefore decreasing the ownership the software quality tends to reduce.

## 3   Methodology

### 3.1   Overview

In order to be able to analyze the connection between code ownership and maintainability, we need a method to express them numerically. Neither of them are trivial concepts, and there are no exact definitions on how to compute them.

For the maintainability we used the same calculation method that we applied in our previous studies. Therefore these values were available even before preparing the current work. The maintainability estimation method is described in detail in Section 3.3. As the used quality model works on a revision basis (it analyzes a certain revision of a system), we had to work on a per commit basis.

One of the most important tasks of this research was to find a proper way to numerically express the code ownership of a commit. The details on how we did this is described in Section 3.4. It was a constraint to deal with code ownership on a commit basis because working with files would have been more natural. Nonetheless, we are convinced that this loss of precision might only weakened the results, and our main conclusions would remain valid using file level ownership values.

In Section 3.5 we argue that the two number series (maintainability and ownership values) are totally independent. Section 3.6 describes the statistical tests we used to verify our hypothesis.

## 3.2   Preliminary Steps

Before collecting the required information we did some data cleaning. The analyzed software systems were all written in Java and the quality model we used (see below) considers Java source files only. Therefore we removed the data related to the non Java source files (e.g. xml files) from the input. We also removed the commits that became empty (i.e. which contained no Java source files at all). In this way we worked on an input commit set containing exclusively Java source files, and each analyzed revision contained at least one affected Java file.

## 3.3   Estimation of the Maintainability Change

We estimated the maintainability value of every revision with help of the ColumbusQM probabilistic software quality model [1]. This model is among others based on the fact that the increase of software metrics (e.g. object-oriented metrics defined by Chidamber and Kemerer [6]) decreases the maintainability. Gyimóthy et al. [11] empirically validated that the increase of some of these metrics increase the probability of faults.

The model itself considers the following metrics: logical lines of codes, the number of ancestors, the maximum nesting level, the coupling between object classes, clone coverage, number of parameters, McCabe's cyclomatic complexity, number of incoming invocations, number of outgoing invocations, and number of coding rule violations. These metrics are compared with those of other systems in a benchmark, and then the results of the comparisons are aggregated by utilizing also weights provided by developers.

From this study's viewpoint we treat this quality model as a black box. Details were published in the works of Bakota et al. [1], and they also showed that there is a correlation between the estimated quality value and the real development costs [1,2]. For this level of abstraction it is enough to know that if all of the metrics increase, then the maintainability decreases; if they all decrease, then the maintainability increases; and if some of them increase and others decrease, then the direction of the maintainability change depends on the benchmark and the aggregation. As a result, we get a sign (positive, zero, or negative) for each commit.

## 3.4 Code Ownership Calculation

We used the following method to express the code ownership numerically. In a particular commit, we considered all the affected source files one by one. As indicated in Section 3.2, there is at least one Java file in every analyzed commits. For every source file, we calculated how many different developers commited on that file at least once from the beginning of the available history, including the actual commit as well. Therefore this value will be at least 1.

At this point we have a positive integer number for every affected source files of the commit in question. But for further analysis we need a value describing the ownership of the actual commit. For this we chose to calculate the geometric mean of the collected values for files. This expresses well the overall ownership of the file based actual ownership values.

For example, consider a small artificial project with 4 java files: A.java, B.java, C.java and D.java. This project have been developed by the following developers: sulley, mike, randall and celia. In Table 1 the rows represent commits. The first column contains the revision number, while the second one contains the author of that commit. Then the odd columns indicate if the actual file was affected by the commit in question, and the even columns contain the number of different developers of that file up to the current commit. The last column contains the calculated ownership value.

**Table 1.** A simple example

| Rev. | Author | A | | B | | C | | D | | Own. |
|------|--------|---|---|---|---|---|---|---|---|------|
| 1 | sulley | + | 1 | + | 1 | | | | | 1.00 |
| 2 | mike | + | 2 | | | + | 1 | | | 1.41 |
| 3 | sulley | + | 2 | + | 1 | | | | | 1.41 |
| 4 | randall | | | | | | | + | 1 | 1.00 |
| 5 | celia | + | 3 | | | | | + | 2 | 2.45 |
| 6 | randall | + | 4 | | | | | + | 2 | 2.83 |
| 7 | sulley | + | 4 | + | 1 | | | | | 2.00 |
| 8 | mike | + | 4 | | | | | | | 4.00 |

In the first revision sulley added files A.java and B.java. The ownership values are initialized to 1 for both of the files, and the geometric mean of 1 and 1 is 1.

In the second revision `mike` modified file `A.java` and added file `C.java`. At this point the ownership value of source file `A.java` has been increased to 2 (`sulley` and `mike`), and the value of file `B.java` was initiated to 1. The ownership value of the second revision is $\sqrt{1 \cdot 2} \approx 1.41$.

With these 2 examples the other 6 revisions are easy to understand.

From this scenario the following can be seen:

- File `A.java` is the "hot area" of the "project", modified by every developer.
- File `B.java` is an example of intensive modification by a certain developer (`sulley` in this case), therefore having a clear ownership.
- File `C.java` is an example of adding a source file once and never modifying later.
- File `D.java` is an example of a common code of two developers (`randall` and `celia` in this case).

As a result of the above described method, we get an ownership value for every revision.

## 3.5  Independence of the Values

At this point we have a maintainability change sign and an ownership value for each commit. Before going further to the statistical tests performed on these data, we argue that the calculated values are totally independent.

On one hand, the maintainability value of the system is calculated solely from the source code, no version control data is considered. The sign of the maintainability change of the actual commit depends (mainly) on the code delta, neither the code history nor the author are taken into account.

On the other hand, the ownership value is calculated solely from version control historical data, particularly the author of the past and the present commits is considered. Therefore we can state that the data series are independent form each other.

## 3.6  Comparison Tests

We divided the commits into 3 subsets based on the sign of maintainability changes, and analyzed their calculated ownership values. We omitted the neutral maintainability changes (i.e. no change in maintainability values), therefore 2 set of numbers remained:

- ownership values of the commits with positive maintainability change, i.e. code quality increase, and
- ownership values of the commits with negative maintainability change, i.e. code quality decrease.

The null hypothesis is that there is no significant difference between these values. The alternative hypothesis is that the ownership values related to commits with positive maintainability changes are significantly lower than those related to negative maintainability changes.

Considering the limitations of the data (e.g. it is not normally distributed) we chose the Wilcoxon rank correlation test (also known as Mann-Whitney U test) for comparison. This test compares all the elements of the first data set with all the elements of the other one, taking all the possible combinations into consideration. The null hypothesis is that the number of "greater" elements is the same as the number of "less" elements. The alternative hypothesis is that the elements of one of the sets are significantly higher than the elements of the other.

By default the Wilcoxon rank correlation test performs a two-tailed analysis. This means that it tells if the values in the checked subsets differ significantly, but does not tell the direction of the deflection. In this case we were not satisfied with the information if the elements of one subset significantly differ from the elements of the other one, we were also interested in which values were higher. For this reason we executed the one-tailed Wilcoxon-rank test with the alternative of "less". Practically it means that our alternative hypothesis is that in case of the above comparison the number of "less" elements are significantly higher than the number of "greater" elements.

The most important result of this test is the well-known p-value, indicating the probability of the result being at least as extreme as the observed, provided that the null-hypothesis is true. In the results section we present these p-values for the analyzed systems.

We interpret the p-values as follows:

— below 0.01: very strong significance,
— between 0.01 and 0.05: strong significance,
— between 0.05 and 0.1: significant,
— between 0.1 and 0.5: not significant,
  between 0.5 and 0.9: contradiction, and
— above 0.9: the opposite statement is true.

We performed the test using the `wilcox.test()` function in R [18]. R runs on multiple platforms; we performed the evaluation on Windows 7. As a result, we get p-values for all software systems the test was performed on.

Going back to our running example, Table 2 shows how the maintainability changed in each revision. For a better overview we repeated the ownership values in this table as well.

Now we have the following ownership value sets:

— Ownership values related to positive maintainability changes: {1.00, 1.00, 2.83}
— Ownership values related to negative maintainability changes: {1.41, 2.45, 4.00}

Considering all the comparison combinations (there are $3 * 3 = 9$ cases) we get the following. In 7 cases (the two 1.00 in all comparisons and comparing 2.83 with 4.00) the elements in the first data set are less than the elements in

Table 2. Example Maintainability Changes

| Revision | Ownership | Maintainability change |
|---|---|---|
| 1 | 1.00 | positive |
| 2 | 1.41 | neutral |
| 3 | 1.41 | negative |
| 4 | 1.00 | positive |
| 5 | 2.45 | negative |
| 6 | 2.83 | positive |
| 7 | 2.00 | neutral |
| 8 | 4.00 | negative |

the second one, and in 2 cases (comparing 2.83 from the first data set with 1.41 and 2.45 from the second one) the result of the comparison is just the opposite. The p-value in this example is about 0.19, indicating that the elements in the first subset is less, but not significantly, than the elements in the second set. The obvious reason for this is the small number of observation.

## 4    Results

### 4.1    Analyzed Systems

We executed the tests on four software systems. These are four independent systems, i.e. we executed four independent tests. The initial selection criteria for the subject systems were the following: availability of at least 1,000 commit and at least 200% code increase during the analyzed period. The analysis was – as earlier – performed on the following 4 systems:

- **Ant** – a command line tool for building Java applications.[1]
- **Gremon** – a proprietary greenhouse work-flow monitoring system.[2]
- **Struts 2** – a framework for creating enterprise-ready Java web applications.[3]
- **Tomcat** – an implementation of the Java Servlet and Java Server Pages technologies.[4]

Table 3 shows some basic properties of these systems.

To provide an overview about some interesting aspects of the analyzed systems we present a couple of diagrams. First let us consider Figure 1. The small empty circles on this strip chart represent the commits in a system. On the y-coordinate the developers are listed in a decreasing order according to their number of contributions. The topmost developer is always the one with the largest contribution. On the left of the diagrams the user IDs of the developers are displayed. In case of Gremon – as it is an industrial project – the real user IDs are masked. On the right of the diagrams the portions of the total contributions are displayed. The x-coordinate represents the revisions of a system.

---

[1] http://ant.apache.org
[2] http://www.gremonsystems.com
[3] http://struts.apache.org/2.x
[4] http://tomcat.apache.org

**Table 3.** Analyzed Systems

|  | Ant | Gremon | Struts 2 | Tomcat |
|---|---|---|---|---|
| Number of developers | 37 | 13 | 26 | 15 |
| Maximum logical lines of code | 106,413 | 55,282 | 152,081 | 46,606 |
| Number of commits | 6,102 | 1,158 | 1,749 | 1,292 |
| Maintainability increases | 1,482 | 456 | 498 | 269 |
| Maintainability no change | 3,051 | 365 | 710 | 704 |
| Maintainability decreases | 1,569 | 337 | 541 | 319 |

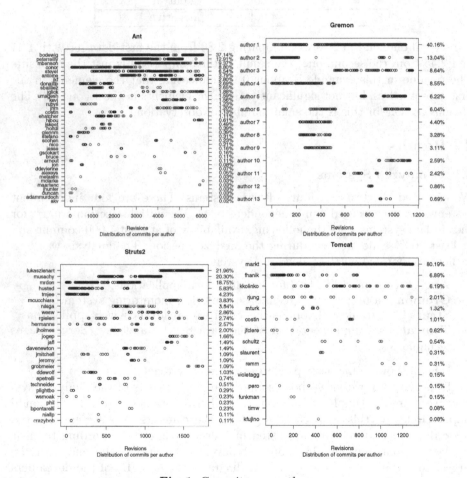

**Fig. 1.** Commits per authors

The black lines are actually several empty circles over one another; those are the periods when the developer in question was the most active.

Figure 2 illustrates the contributions of the authors from two aspects: the number of commits per developer, and the number of touched files per developer.

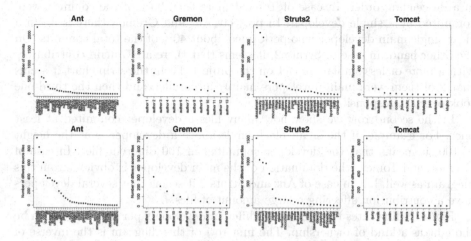

**Fig. 2.** Contributions per author

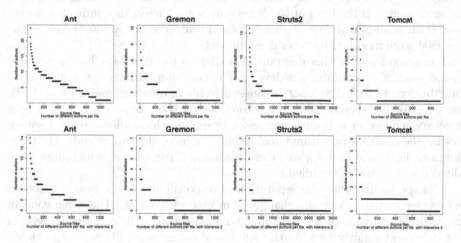

**Fig. 3.** Number of authors per file

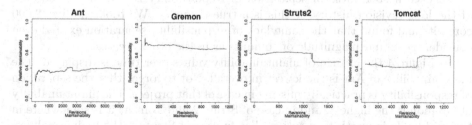

**Fig. 4.** Maintainability change

The diagrams in the first row contain the number of commits per author in a descending order. In case of Tomcat more then 80% of the commits were commited by a single developer. In projects Ant and Gremon there is again a clear single main developer who performed about 40% of the total commits. On the other hand, in case of Struts 2, it seems that there are 3 main contributors with a more or less similar impact on the project. From the strip chart it seems that 2 of them were main developers mainly in parallel, and then the third one took over the responsibility.

In the second row we show how many files a developer commited at least once. For example, if the small black circle above a developer is at the height of 100, it means that the developer commited in 100 different files. In case of Gremon and Tomcat the domination of the main developer is obvious from this diagram as well, but in case of Ant and Struts 2 it seems that several developers have a contribution affecting a large amount of files.

Figure 3 illustrates the number of different developers per files, which can be thought as a kind of ownership. The first row of this diagram is the inverse of the second row of Figure 2, namely how many different developers commited to a single file. The black circles seem to be lines here as there are many files with the same values. If the lines at the lower values are longer, that indicates clearer separation of responsibility. Higher values indicate the hot areas: these are the files that were modified by several developers.

The second row of this diagram is similar to the first one, but it contains a relaxation: the commits of a developer is not counted here if the number of contributions of that developer on the source file in question was at most 2. We applied this rule because we wanted to eliminate the possible bias caused by a directory rename or a branch merge for example, which affects several source files by the contributor without real modifications of the source code. On these diagrams lines at 0 also appear, e.g. containing those sources which have been added once but never modified.

It is spectacular that the separation of responsibility is the best – based on the earlier statistics not surprisingly – in case of Tomcat. The separation of responsibility in case of Struts 2 and Gremon seems sufficient, but in case of Ant it is spectacularly bad. As the number of commits in this project is higher than the total number of commits in the rest of the 3 projects all together, we checked if such mess in separation of responsibility is caused just because of the long revision history or this is a true tendency. We took the first 2,000 commits and found that the same lack of responsibility separation existed even considering similar magnitude of commits as in the other 3 cases.

In Figure 4 the change of maintainability values over time is displayed. The maintainability of Ant is the lowest in overall. Not to forget that the separation of responsibility is practically missing in case of that project. The maintainability of Gremon is the highest, and in case of Struts 2 and Tomcat it is somewhere in the middle. It is worth to mention that in case of Struts 2 the maintainability decreased and in case of Tomcat it increased over time.

## 4.2    Results of the Statistical Tests

Table 4 shows the results of the Wilcoxon rank test described in Section 3.6. The results vary from very strong to not significant, but neither of them contradict.

Table 4. Results

| System | p-value | Significance |
|--------|---------|--------------|
| Ant | 0.033728 | strong |
| Gremon | 0.059604 | significant |
| Struts 2 | 0.000014 | very strong |
| Tomcat | 0.213841 | not significant |

The result for project Ant is solid, above 0.01 but below 0.05. Indeed, the result for the first 2,000 commits is 0.002897, which is even below 0.01. It seems that the results weaken in the later phase of the project, when already too many sources have been contributed by too many developers.

The results for project Gremon is somewhat above 0.05, but still significant. On the other hand, results for Struts 2 is very strong.

The tests for Tomcat show absolutely no significance; however, the results are not contradicting either. The reason for this might be the fact that more than 80% of the commits were performed by the same author, which causes a huge bias compared to the other projects. Therefore Tomcat is an atypical project from this respect.

## 4.3    Discussion

For proper interpretation of the above results we address some possible misinterpretations.

One could simplify the method as follows: the more people work on a system, the more complex it will be, and the more complex a systems is, the harder to maintain it. We consider this relationship as already known and did not even check it. On the other hand, we state that the effect of the *future* modifications on source files changed by more developers in the *past* is more likely to lower the maintainability compared with modifications on files that have been changed by less number of developers earlier. Note that this is not trivial: a source file with several earlier contributor is likely to be more complex than those with clear ownership, and our statement is that the already low quality source code is more likely to become even worse than that of higher quality.

For the sake of better understanding we interpret our result as follows: source files which have been modified by more developers in the past is more likely to become more complex in the future than those with less number of contributors. An even more precise statement would be the following, which is harder to conceptualize: the number of earlier contributors of modifications resulting in code quality increase is more likely to be lower than those resulting in code quality decrease.

Now we highlight the limitations of the results. It would be an inappropriate interpretation that the quality of the code with clear ownership increases, and the quality of common code decreases. There are quality decreases in the sources with clear ownership, and quality increases even in the hottest code areas. The presented results are much more modest, but significant.

Regarding the strength of the results, we stress that the 4 analyzed systems have been fixed before the case study. These are the same as we used in our earlier studies ([9], [8], [7], and others under development).

## 5    Threats to Validity

The factors which might threaten the validity of the results are the following.

The significance of the results are varying from very strong to not significant. We consider the system with a not significant (but not contradictory) result as an extraordinary one, as more than 80% of the commits were performed by a single developer.

We eliminated the commits related to non-traceable maintainability changes. The cardinality of these operations is between one third and one half of the total number of commits. This is a relatively large amount of data to be excluded. An enhanced model considering also the commits with no maintainability change could provide other results. However, we do not expect a different final conclusion even in the case of such a model.

There are several quality models, and there is no such unique model which is accepted by the whole industry. Using another model could provide a different result. We know that no quality model (including the one we used) is perfect. However, as most product quality models rely on a similar source of information (i.e. source code metrics) we do not expect that the results are so much dependent on the actual quality model used.

The ownership values are based on files, but the maintainability data are calculated for commits. Therefore we needed a conversion of ownership values from files to commits. We found the geometric mean to be an adequate approach; however, one could argue that we should use another aggregating method (e.g. taking the mean, the median, the maximum element, or using a more sophisticated approach). We run some tests using another aggregation method, but found no significant difference in the results of the statistical tests.

For the calculation of ownership values of a file we considered the most straightforward approach: the number of developers modifying the file so far. More sophisticated methods like introducing a tolerance (considering a contributor only above a certain threshold), or using relative basis instead of an absolute one could further enhance the methodology and the results could be more adequate.

## 6    Conclusions and Future Work

In this study we investigated if different number of developers of source code have an impact on the maintainability change of future commits.

We estimated the relative maintainability of every revision of four analyzed software systems, using the Columbus Quality Model, which compares the metrics of the analyzed system with the metrics of software systems found in a benchmark, and with the help of these results we identified if the net quality change caused by the actual commit was positive, neutral or negative. Additionally, we defined the source code related ownership value to be the number of different contributors so far. We aggregated the file related values to estimate a commit based ownership value by calculating the geometric mean of the file related values affected by the actual commit. Then we divided the commits into 3 sets based on the sign of maintainability change caused by the commit, therefore getting sets of commits related to positive, neutral and negative maintainability change. We omitted the commits related to neutral maintainability changes, then we took the already calculated ownership values of the two remaining subsets.

We tested the straightforward null-hypothesis regarding these ownership values that any difference in the distribution of them is casual, as the ownership value was calculated using historical data from version control system, and the maintainability change is affected solely by the actual commit. Furthermore, it considers the source code only and no other information like the developer itself. We, on the other hand, by executing Wilcoxon rank correlation test, found the following significant tendency: the ownership values related to positive maintainability changes are more likely to be lower than those related to negative maintainability changes.

We executed the test on 4 different systems selected in advance. The test for one of them resulted a very strong (p-values 0.000014), one a strong (0.034), the third one a significant (0.060) and finally the fourth one a not significant (0.21) outcome. We are convinced that the system with low significance (but not contradicting) result is an exceptional one, as more than 80% of the commits were performed by a single developer. Considering the results we can conclude that common code is more likely to erode further than code with clear ownership.

A practical use of this result could be the following. The efforts which can be spent on code quality is typically very limited. This result (and other results of similar research) can help in prevention, and in more efficient allocation of these efforts. As a prevention it is proposed that the boundaries of responsibilities within a software system should be as clear as possible. If this rule has been already somehow broken, it is recommended to pay special attention on source files of common code in case of any modifications, e.g. by mandating more strict code review rules.

As this study is part of a longer term research, we have concrete plans regarding the future steps. In short term, we plan to investigate the effect of other information found in the version control system on maintainability, like file name, date, comment or the size of the files.

As a final step, we plan to aggregate the results, and then implement a software which identifies the hot areas of the source code of the analyzed system. An IDE plug-in, which visually marks these areas could be useful for architects, and it could automatically warn the developers.

In longer term, we plan to take other information into consideration, like information found in issue tracking systems or developer environment interactions. An even more accurate result could be obtained by considering the type of the software systems as well (e.g. standalone or client-server applications).

Moving to other languages and comparing them with Java could also be an interesting future research direction. For example, the quality model used for this research was adopted to C# by Hegedűs [13].

Our long term goal is to fine-tune the formula of code erosion as much as possible in order to understand why it happens, and with this information in hand we could give hints how to avoid it with the least additional effort.

**Acknowledgments.** The publication is partially supported by the European Union projects titled "Telemedicine-focused research activities on the field of Mathematics, Informatics and Medical sciences", project number: TÁMOP-4.2.2.A-11/1/KONV-2012-0073 and "REPARA – Reengineering and Enabling Performance And poweR of Applications", project number: 609666.

# References

1. Bakota, T., Hegedűs, P., Körtvélyesi, P., Ferenc, R., Gyimóthy, T.: A probabilistic software quality model. In: 2011 27th IEEE International Conference on Software Maintenance (ICSM), pp. 243–252. IEEE (2011)
2. Bakota, T., Hegedűs, P., Ladányi, G., Körtvélyesi, P., Ferenc, R., Gyimóthy, T.: A cost model based on software maintainability. In: 2012 28th IEEE International Conference on Software Maintenance (ICSM), pp. 316–325. IEEE (2012)
3. Bird, C., Nagappan, N., Devanbu, P., Gall, H., Murphy, B.: Does distributed development affect software quality?: an empirical case study of windows vista. Communications of the ACM **52**(8), 85–93 (2009)
4. Bird, C., Nagappan, N., Gall, H., Murphy, B., Devanbu, P.: Putting it all together: using socio-technical networks to predict failures. In: 20th International Symposium on Software Reliability Engineering, ISSRE 2009, pp. 109–119. IEEE (2009)
5. Bird, C., Nagappan, N., Murphy, B., Gall, H., Devanbu, P.: Don't touch my code!: examining the effects of ownership on software quality. In: Proceedings of the 19th ACM SIGSOFT Symposium and the 13th European Conference on Foundations of Software Engineering, pp. 4–14. ACM (2011)
6. Chidamber, S.R., Kemerer, C.F.: A metrics suite for object oriented design. IEEE Transactions on Software Engineering **20**(6), 476–493 (1994)
7. Faragó, C.: Variance of the quality change of the source code caused by version control operations. In: The 9th Conference of PhD Students in Computer Science (CSCS) (2014)
8. Faragó, C., Hegedűs, P., Ferenc, R.: The impact of version control operations on the quality change of the source code. In: Murgante, B., Misra, S., Rocha, A.M.A.C., Torre, C., Rocha, J.G., Falcão, M.I., Taniar, D., Apduhan, B.O., Gervasi, O. (eds.) ICCSA 2014, Part V. LNCS, vol. 8583, pp. 353–369. Springer, Heidelberg (2014)
9. Faragó, C., Hegedűs, P., Végh, Á.Z., Ferenc, R.: Connection between version control operations and quality change of the source code. Acta Cybernetica **21**, 585–607 (2014)

10. Fritz, T., Murphy, G.C., Hill, E.: Does a programmer's activity indicate knowledge of code? In: Proceedings of the 6th Joint Meeting of the European Software Engineering Conference and the ACM SIGSOFT Symposium on The Foundations of Software Engineering, pp. 341–350. ACM (2007)
11. Gyimóthy, T., Ferenc, R., Siket, I.: Empirical validation of object-oriented metrics on open source software for fault prediction. IEEE Transactions on Software Engineering **31**(10), 897–910 (2005)
12. Hattori, L., Lanza, M.: Mining the history of synchronous changes to refine code ownership. In: 6th IEEE International Working Conference on Mining Software Repositories, MSR 2009, pp. 141–150. IEEE (2009)
13. Hegedűs, P.: A probabilistic quality model for C# - an industrial case study. Acta Cybernetica **21**(1), 135–147 (2013)
14. ISO/IEC: ISO/IEC 9126. Software Engineering - Product quality 6.5. ISO/IEC (2001)
15. LaToza, T.D., Venolia, G., DeLine, R.: Maintaining mental models: a study of developer work habits. In: Proceedings of the 28th International Conference on Software Engineering, pp. 492–501. ACM (2006)
16. Mockus, A., Fielding, R.T., Herbsleb, J.: A case study of open source software development: the apache server. In: Proceedings of the 22nd International Conference on Software Engineering, pp. 263–272. ACM (2000)
17. Nordberg III, M.E.: Managing code ownership. IEEE Software **20**(2), 26–33 (2003)
18. R Core Team: R: A Language and Environment for Statistical Computing. R Foundation for Statistical Computing, Vienna, Austria (2013). http://www.R-project.org/
19. Rahman, F., Devanbu, P.: Ownership, experience and defects: a fine-grained study of authorship. In: Proceedings of the 33rd International Conference on Software Engineering, pp. 491–500. ACM (2011)
20. Weyuker, E.J., Ostrand, T.J., Bell, R.M.: Do too many cooks spoil the broth? using the number of developers to enhance defect prediction models. Empirical Software Engineering **13**(5), 539–559 (2008)

# Adding Constraint Building Mechanisms to a Symbolic Execution Engine Developed for Detecting Runtime Errors

István Kádár, Péter Hegedűs[(✉)], and Rudolf Ferenc

Department of Software Engineering, University of Szeged, Szeged, Hungary
{ikadar,hpeter,ferenc}@inf.u-szeged.hu

**Abstract.** Most of the runtime failures of a software system can be revealed during test execution only, which has a very high cost. The symbolic execution engine developed at the Software Engineering Department of University of Szeged is able to detect runtime errors (such as null pointer dereference, bad array indexing, division by zero) in Java programs without running the program in real-life environment.

In this paper we present a constraint system building mechanism which improves the accuracy of the runtime errors found by the symbolic execution engine mentioned above. We extend the original principles of symbolic execution by tracking the dependencies of the symbolic variables and substituting them with concrete values if the built constraint system unambiguously determines their value.

The extended symbolic execution checker was tested on real-life open-source systems as well.

**Keywords:** Software engineering · Symbolic execution · Java runtime errors · Constraint system building

## 1 Introduction

Nowadays, producing great, reliable and robust software systems is quite a big challenge in software engineering. About 40% of the total development costs go for testing and maintenance activities, moreover, bug fixing of the system also consumes a considerable amount of resources. The symbolic execution engine developed at the Software Engineering Department of University of Szeged supports this phase of the software engineering lifecycle by detecting runtime errors (such as null pointer dereference, bad array indexing, division by zero) in Java programs without running the program in real-life environment.

According to the theory of symbolic execution [1] the program does not run with specific input data, but the inputs are handled as symbolic variables. When the execution of the program reaches a branching condition containing a symbolic variable, the execution continues on both branches. At each branching point both the affected logical expression and its negation are accumulated on the true and false branches, thus all of the execution paths will be linked to a unique formula over the symbolic variables.

© Springer International Publishing Switzerland 2015
O. Gervasi et al. (Eds.): ICCSA 2015, Part V, LNCS 9159, pp. 20–35, 2015.
DOI: 10.1007/978-3-319-21413-9_2

The paper describes a constraint system construction mechanism, which improves the accuracy of the runtime errors found by the symbolic execution engine mentioned above by treating the assignments in the program as conditions too. Thus we can track the dependencies of the symbolic variables extending the original principles of symbolic execution. The presented method also substitutes the symbolic variables with concrete values if the built constraint system unambiguously determines their value. To build and satisfy the constraint systems we used the open-source Gecode constraint satisfaction tool-set [2].

The paper explains in detail how the algorithm is implemented that builds the constraint system for each execution path, how it is integrated into the symbolic execution engine of the Department of Software Engineering, and how the algorithm enhanced the effectiveness of the engine. The extended symbolic execution checker was tested on real-life open-source systems as well and we compared it with our previous tool [3] based on SymbolicPathFinder.

## 2   Background

### 2.1   Symbolic Execution

During its execution, every program performs operations on the input data in a defined order. Symbolic execution [1] is based on the idea that the program is operated on symbolic variables instead of specific input data, and the output will be a function of these symbolic variables. A symbolic variable is a set of the possible values of a concrete variable in the program, thus a symbolic state is a set of concrete states. When the execution reaches a selection control structure (e.g. an if statement) where the logical expression contains a symbolic variable, it cannot be evaluated, its value might be also true and false. The execution continues on both branches accordingly. This way we can simulate all the possible execution branches of the program.

During symbolic execution we maintain a so-called *path condition (PC)*. The path condition is a quantifier-free logical formula with the initial value of true, and its variables are the symbolic variables of the program. If the execution reaches a branching condition that depends on one or more symbolic variables, the condition will be appended to the current PC with the logical operator *AND* to indicate the true branch, and the negation of the condition to indicate the false branch. With such an extension of the PC, each execution branch will be linked to a unique formula over the symbolic variables. In addition to maintaining the path condition, symbolic execution engines make use of so called *constraint solver* programs. Constraint solvers are used to solve the path condition by assigning values to the symbolic variables that satisfy the logical formula. Path condition can be solved at any point of the symbolic execution. Practically, the solutions serve as test inputs that can be used to run the program in such a way that the concrete execution follows the execution path for which the PC was solved.

SymbolicChecker, the symbolic execution engine developed at the Software Engineering Department does not aim to generate test inputs, but to find as many true positive runtime errors in the program as possible. In accordance

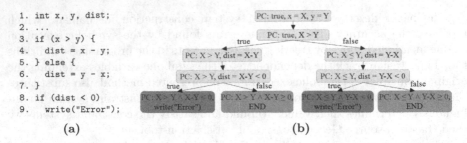

```
1. int x, y, dist;
2. ...
3. if (x > y) {
4.     dist = x - y;
5. } else {
6.     dist = y - x;
7. }
8. if (dist < 0)
9.     write("Error");
```

(a)                     (b)

**Fig. 1.** (a) Sample code that determines the distance of two integers on the number line (b) Symbolic execution tree of the sample code handling variable x and y symbolically

with this goal we changed and extended the standard path condition building method described above.

Figure 1 (a) shows a sample code that determines the distance of two integers x and y. The symbolic execution of this code is illustrated on Figure 1 (b) with the corresponding symbolic execution tree. We handle x and y symbolically, their symbols are X and Y respectively. The initial value of the path condition is true. Reaching the first if statement in line 3, there are two possibilities: the logical expression can be true or false; thus the execution branches and the logical expression and its negation is added to the PC as follows:

$$true \land X > Y \Rightarrow X > Y, \quad and \quad true \land \neg(X > Y) \Rightarrow X \leq Y.$$

The value of variable *dist* will be a symbolic expression, X-Y on the true branch and Y-X on the false one. As a result of the second if statement (line 8) the execution branches, and the appropriate PCs are appended again. On the true branches we get the following PCs:

$$X > Y \land X - Y < 0 \Rightarrow X > Y \land X < Y,$$

$$X \leq Y \land Y - X < 0 \Rightarrow X \leq Y \land X > Y.$$

It is clear that these formulas are unsolvable, we cannot specify such X and Y that satisfy the conditions. This means that there are no such x and y inputs with which the program reaches the *write("Error")* statement. As long as the PC is unsatisfiable at a state, the sub-tree starting from that state can be pruned, there is no sense to continue the controversial execution.

## 2.2   SymbolicChecker, the Symbolic Execution Engine

The goal of SymbolicChecker is to detect those real runtime errors that other audit tools cannot detect and those which mostly can be discovered by a large amount of testing only. It is important for us that the detected errors be as accurate as possible, so we can eliminate the false positive hits and find more numerous true positives that helps the software developers to create a higher-quality product and makes the maintenance tasks easier. Generating test cases which lead to the errors is not a goal here, much more to produce a descriptive designation of the execution path that led to a fault.

In the present paper we do not give a detailed description of the Symbolic-Checker analysis tool, but in order to understand our new constraint building concept, its basic understanding is needed. SymbolicChecker is written in C++, the development is still in progress. Currently the detection of null pointer dereferences, array over-indexing, division by zero, and type cast errors are implemented.

SymbolicChecker performs the analysis by symbolically executing each method of the system one-by-one. The parameters of the method under execution and the referred but not initialized variables are handled as symbols at the beginning of the analysis. It is important that we only report an error if it is guaranteed that during the execution the value that causes the problem can be determined by constant propagation. For example, if a method call is guaranteed to pass a null value, and it is guaranteed that the called method dereferences this parameter, we will fire an error, but if the dereferenced variable is a symbol we will not, because its value is unknown or uncertain. To limit the size of the symbolic execution tree, its maximum depth and the maximum number of states can be specified.

The symbolic execution is performed using the language-dependent abstract semantic graph (ASG) [4] of the program by interpreting the the nodes of this graph in a defined order. The order is defined by the language-independent control flow graph (CFG) [5]. The output of the SymbolicChecker contains the detected errors indicating their type, the execution path from the entry point to the exact location where the error occurred and a probability that estimates how likely the analyzed method runs onto the detected fault.

In SymbolicChecker *Definition* is a comprehensive name for all the data that appears during the symbolic execution. For example, the concrete or symbolic variables, constants, parameters of methods, their return value, or the result of sub-expressions are also Definitions. Actually, the symbolic execution of the program is the propagation of these Definition objects. Basically there are two types of Definitions: ValueDefinition and SymbolDefinition. ValueDefinition objects store specific, concrete values and SymbolDefinition instances represent the symbolic variables.

### 2.3 JPF Checker

In one of our previous works [3] we used SymbolicPathFinder [6] to create a tool named JPF Checker with the same goal as with SymbolicChecker: to detect runtime errors in Java programs without modifying the source code and without having to run it in a real-life environment. This SymbolicPathFinder based tool used the conventional constraint building mechanism. In Section 4 we compare this approach to our new concept which is implemented in SymbolicChecker.

## 3    Constraint Building

### 3.1    Principles

In this work we developed a constraint building mechanism and integrated it into SymbolicChecker which allows us to detect runtime errors that a conventional symbolic execution system cannot.

As we described in Section 2.2, SymbolicChecker reports errors only if the value causing the problem becomes concrete. The tool does not fire for symbolic variables because if a variable is a symbol it actually means that it's value is doubtful, not known. It may occur that during the symbolic execution of a program most of the variables turn into symbols, which makes finding runtime errors rather difficult.

The main idea behind the developed constraint building mechanism is that if during the analysis the program sets up conditions (constraints) that unambiguously determines the value of one or more symbolic variables, then we can convert these symbols into concrete values and the symbolic execution can be continued on the actual path using the concreted variables. Since these variables handled like concrete data, it is possible to detect errors that otherwise SymbolicChecker could not find. The conditions we mentioned above are determined by the conditional control structures (if, switch, while, etc.) and expressed by the assignments of the program, including the impacts of the increment and decrement operators $(++, --)$ of the Java language.

Overall, the goal of the implemented constraint building mechanism is the concretion of as many symbols as possible, which helps to find more runtime errors. In order to achieve this (1) it is necessary to build a special path condition (PC), that contains the dependencies of the symbolic variables too determined by the assignments of the program, and (2) if the constraints in the PC determine the values of some symbols unambiguously, the execution has to be continued using these concrete values on the actual path. Since such an extended path condition includes – in some form – also those conditions that are in the PC built the conventional way, as long as it is infeasible, the code parts that are unreachable can also be skipped. Therefore, false positive defects can be eliminated.

To demonstrate the basic idea of extending the PC, consider the code snippet in Figure 1a. In this example, the conventional path condition of the path which passes through the true branch of the if statement in line 3, and the false branch of if in line 8 is the following:
$$X > Y \ \wedge \ \neg(X - Y < 0) \ \Rightarrow \ X > Y \ \wedge \ X - Y \geq 0.$$

According to our concept the extended PC of the same path is the following:
$$X > Y \ \wedge \ \neg(dist < 0) \ \wedge \ dist = X - Y.$$

It can be seen that variable $dist$ is also included in the constraint system as a symbol, on the one hand in the negation of condition in line 8 where X-Y was not substituted, on the other hand the constraint that expresses the assignment in line 4. As a result, the constraint system contains information about variable dist as well that could be useful in the later stages of the execution.

In this example, the extended PC does not contain constraints which could unambiguously determine any variable, thus the benefit of the extension is not obvious here. The code snippet in Figure 2a shows an example where the extended PC indeed has some gains.

Executing the code symbolically in Figure 2a handling variable c as a symbol, the following constraint system will be built in the program state at line 7:

```
1. // c is an int symbol
2. double b = 2*c + 4;
3. int a = b + 9;
4. if (a > 8) {
5.    ...
6.    if (a < 10) {
7.        // concretion of b
8.        int p = 1/b;
9.    }
10. }
```

```
1. // a and b are symbols
2. if (b > 0) {
3.    ...
4.    if (a == 0) {
5.        // concreting a{
6.        ...
7.    }
8. }
```

(a) Sample code that provides symbol concretion, which helps to find runtime errors that a conventional symbolic execution tool cannot.

(b) Code snippet which points out a path condition that has more solutions, but concreted symbol a

**Fig. 2**

$$a > 8 \ \wedge \ a = b + 9 \ \wedge \ b = 2 \cdot c + 4 \ \wedge \ a < 10.$$

The constraint system above includes constraints that are introduced by the if statements of the code and the dependencies of symbol $a$, i.e. those constraints that are given by the assignments that defines variable $a$. After satisfying this constraint system it can be obtained that $a$ can only be 9, which implies that the value of b and c symbols are unambiguous too: $b = 0.0$ and $c = -2$. In such a situation the execution continues on that path for which the extended PC was satisfied. In case of the considered example if the execution continues with the b=0.0 value, at line 8 a division by zero error can be detected. As long as symbol $b$ would not be included in the PC, and if its unambiguous value would not be used, the detection of division by zero would fail.

In real-life programs, quite big constraint systems are be built, which contain lots of symbols. Easy to see that satisfying such a large set of constraints as a whole, it has a low probability that there is only one possible solution.

Figure 2 shows a code snippet that highlights the problem in question. Considering the path that passes along the true branches of both if statements, the path condition is $b > 0 \ \wedge \ a = 0$. Although there are infinite number of solutions of this formula, because the $b > 0$ constraint can be satisfied by any positive integer, the formula determines the value of symbol $a$ unambiguously, which would be preferred to use in later stages of the execution.

To overcome the problem we decompose the path condition into connected components that is, to constraint sets that are independent i.e. does not contain the same variables. The connected components can be satisfied individually and if some of them determines a variable unambiguously then the obtained values can be used later in the execution. Two constraints are in the same component if they contain at least one common variable. After such a decomposition the path condition becomes a set of constraint sets.

The essential steps of the algorithm of our constraint system building is shown in Figure 3. This algorithm will be executed after each branching point in the symbolic execution tree.

First of all, it is determined if the accumulation of the PC happens on the true or on the false branch then dependent upon this the created logical

```
1. Constraint constraint;
2. set<Constraint> actualConstraints;
3. if (onTrueBranch()) {
4.    constraint = constraintBuilder.createConstraint();
5. } else if (onFalseBranch()) {
6.    constraint = constraintBuilder.createNegatedConstraint();
7. }
8. actualConstraints.insert(constraint);
9. actualConstraints.union(dependenciesOfSymbolsInConstraint);
10. pathCondition.insert(actualConstraints);
11. decomposedPC = decompose(pathCondition);
12. foreach (set<Constraint> s : decomposedPC) {
13.    constraintSolver.solve(s);
14.    if (s.hasSolution) {
15.      if (!s.hasMoreSolution) {
16.        buildBackSolutions(s);
17.      }
18.    } else {
19.      weight = 0.0;
20.      break;
21.    }
22. }
```

**Fig. 3.** Pseudo code of the algorithm of constraint system building

expression or its negation is stored in variable *constraint* (lines 3-7) (the handling of switch statement of the Java programming language is not shown in the pseudo code). It is important to note that we build constraint exactly from that logical expression that is determined in the source code, there is no substitution of variables like in case of variable *dist* in Section 2.1, in example 1. Next, the created constraint is added to the *actualConstraints* constraint set (line 8), and also the dependencies of the symbols included in this constraint are inserted (line 9). These dependencies are defined by the assignments of the code, later we will discuss how they are created. The next step is that the path condition of the current execution path is extended by the *actualConstraints* constraint set (line 10), then the PC will be decomposed into connected components in line 11. As long as one of the connected components cannot be satisfied the weight of the current path is set to 0.0 indicating that there is no sense to continue the execution because of the contradictory conditions (line 19). On the other hand, if there is at least one solution, the algorithm examines its uniqueness (line 13), if the solution is unique, the concrete values are built back into the current symbolic state (line 16).

It have to be emphasized that a concreted symbol is built back into the a state only once and only into that state for which the constraint system concreted it.

## 3.2   Implementation

We used the Gecode constraint solver tool-set [2] for building and satisfying our constraint systems. Basically, we can differentiate two kinds of constraints: (1) conditions in the conditional control structures of the program (including the loops too) and (2) the dependencies of symbols which are included in these conditions. In the following, we describe how did we implement the building of the constraint system integrated into SymbolicChecker.

As we described in Section 2.2, for every kind of data that appears in a program during the symbolic execution (e.g. variables, literals, sub-expressions, etc.) a *Definition* object is created. In fact, the execution of the program is nothing else than the proper propagation of these Definition objects. The task is to achieve the tracking that determines what other Definitions a Definition object is created from and what operations it uses. This is how the relations between symbolic variables are described.

For the implementation we added a so-called *constraintSolverExpression* data member to the class Definition and a dependency set too, which is a set of constraints. These attributes are propagated with the Definitions along the program by the symbolic execution.

The constraintSolverExpression represents an expression object created using the Gecode constraint solver. With such an expression object Gecode can represent the inner structure of expressions that is, which operands are they created from using which operators. The constraintSolverExpressions is propagated in the following way: when an operation is performed on Definition objects we take the constraintSolverExpressions of the operands and perform the operation on the expressions too, and the resulting compound constraintSolverExpression will be set in the resulting Definition object. In case of operations performed on ValueDefinitions for efficiency reasons the operation on the constraintSolverExpressions is not performed, instead we simply create a new Gecode expression which stores the calculated value.

The dependency set contains the dependencies of those symbols which are in the constraintSolverExpression which are defined by the assignments of the program. In case of those assignments where the right side is a SymbolDefinition, for the variable which is on the left we create a new symbol. This new symbol will not take over the constraintSolverExpression of the right side, but the relation between left and right side Definitions is expressed by an equality constraint between them. The dependency set is propagated in the following manner. After performing an operation the dependency set of the resulting Definition will be the union of the dependency sets of the operands. In case of an assignment which has SymbolDefinition on the right side, the dependency set of the newly created SymbolDefinition on the left will be the dependency set of the right side symbol extended by the constraint which defines equality between the two sides.

```
1. // d is a symbol
2. int b = d + 3;
3. int c = 2*d;
4. int a = b { c;
5. if (42 == a) {
6.   ...
7. }
```

**Fig. 4.** Sample code for demonstrating the propagation of dependency sets

The branching conditions in selection control structures which defines the branching points of the symbolic execution tree are also expressions in the program, thus they appear as Definition objects (actually as SymbolDefinitions) in SymbolicChecker. Variable *constraint* in the algorithm shown in Figure 3 is created from the constraitnSolverExpression of such a Definition object, and constraint set *dependenciesOfSymbolsInConstraint* is the dependency set of this Definition too. Constraint set *actualConstraints* by which the path condition will be extended is the union of the above mentioned *constraint* and *dependenciesOfSymbolsInConstraint*.

In the followings, we demonstrate the building of the constraint system and the propagation of dependency sets for the example code in Figure 4. Variable $d$ is handled as symbol, it is a SymbolDefinition which dependency set is empty and its constraintSolverExpression is a Gecode expression which contains only a simple unknown variable. Firstly, in line 2 a ValueDefinition is created for literal 3, which dependency set is empty, then operation $+$ creates the $d+3$ SymbolDefinition. The dependency set of $d+3$ is the dependency set of the left and of the right side, which is also an empty set:

$$SymbolDef(d+3).depset = SymbolDef(d).depset \cup ValueDef(3).depset = \emptyset.$$

After the execution of the assignment the dependency set of $b$ is:

$$SymbolDef(b).depset = SymbolDef(d+3).depset \cup \{b = d+3\} = \{b{=}d{+}3\}.$$

Dependency set of symbol $c$ created at line 3 is quite similar:

$$SymbolDef(c).depset = SymbolDef(2*d).depset \cup \{c = 2 \cdot d\} = \{c = 2 \cdot d\}.$$

At the left hand side of assignment at line 4, dependency set of SymbolDefinition $b$-$c$ is the union of dependency set of $b$ and $c$:

$$SymbolDef(b-c).depset = SymbolDef(c).depset \cup SymboldDef(b).depset$$
$$= \{b = d+3,\ c = 2 \cdot d\}.$$

Then the dependency set of $a$:

$$SymbolDef(a).depset = SymbolDef(b-c).depset \cup \{a = b - c\}$$
$$= \{b = d+3,\ c = 2 \cdot d,\ a = b - c\}.$$

Dependency set of SymbolDefinition created from expression $42 == a$ at line 5 is the same as the dependency set of $a$, thus the path condition on the true branch of if statement is the following:

$$PC = \{42 = a\} \cup \{b = d+3,\ c = 2 \cdot d,\ a = b - c\}$$
$$= \{42 = a,\ b = d+3,\ c = 2 \cdot d,\ a = b - c\}.$$

## 4    Evaluation

SymbolicChecker with our constraint building mechanism was tested in a variety of ways. This section contains the results of these tests. First of all, we demonstrate the advantages of our algorithm through two examples emphasizing the difference of JPF Checker and SymbolicChecker without using the constraint building mechanism. After that, we write about the experiences got form the tests we have performed on large, real-life systems.

In *run()* method of the example code shown in Figure 5 SymbolicChecker with constraint building detects an array over-indexing fault. First of all, we follow what is the cause of the runtime error, then we look at how the new approach helps detecting it. Line 5 defines an array called *arr* with size of *max*. As long as parameter $n$ is grater than 0 (line 6), a sequence of operations will be performed which aims to calculate two sums based on the content of the array. This sequence of operations fills the array at first (lines 7-9), then starting from $n$, summarizes the *member* data members of objects on every second index

```
1. class Example {                              27.    public int getCharPos(char c) {
2.                                              28.       return c { 'a' + 1;
3.    public void run(int n) {                  29. }
4.       int max = getCharPos('w');             30.
5.       A[] arr = new A[max];                  31.    private int gcd(int x, int y) {
6.       if (n > 0) {                           32.       while (y != 0) {
7.          for (int i = 0; i < max; ++i) {     33.          int m = x % y;
8.             arr[i] = new A(max - i);          34.          x = y;
9.          }                                   35.          y = m;
10.         int sum1 = 0;                        36.       }
11.         while (n < max) {                    37.       return x;
12.            if (n % 2 == 0) {                 38.    }
13.               sum1 += arr[n].getMember();    39.
14.            }                                 40. }
15.            n++;                              41.
16.         }                                    42. class A {
17.         System.out.println("Sum1: " + sum1); 43.    private int member;
18.         int negOfGcd = -gcd(n, arr[0]);      44.
19.         int sum2 = 0;                        45.    public A(int member) {
20.         while (negOfGcd < max) {             46.       this.member = member;
21.            sum2 += arr[negOfGcd++].getMember(); 47. }
22.         }                                    48.
23.         System.out.println("Sum2: " + sum2); 49.    public int getMember() {
24.      }                                       50.       return member;
25.   }                                          51.    }
26. }                                            52. }
```

**Fig. 5.** Example code with the analysis of method run().

into the variable *sum1* (lines 11-16). Next, the code calls method *gcd()* with arguments $n$ and the 0th element of array *arr* (line 18) and summarizes the elements of the array starting from the negation of the return value of *gcd()* (lines 20-22). Method *gcd()* calculates the greatest common divisor of the numbers and its return value must be a positive integer if the arguments are $n$ and *arr[0]*. Because of this, variable negOfGcd guaranteed to be negative which causes an ArrayIndexOutOfBoundsException runtime error that results in the halt of the program.

When starting the analysis with method run(), variable $n$ is the only symbol. Variable *max* is concrete, array arr is also instantiated concretely and all of its elements are concrete values too. However, the execution of the loop in line 11 depends on $n$. On the false branch the execution continues from line 16, on the true branch we enter into the body of the loop, and after executing it we will branch again depending on the condition at line 11. This operation will continue until it reaches the maximum depth of the symbolic execution tree. If the execution paths entered into the loop at least once and then exited, the following constraints must be part of the extended path condition:

$$n_{prev} < max \ \wedge \ n = n_{prev} + 1 \ \wedge \ \neg(n < max) \Rightarrow$$

$$n_{prev} < max \ \wedge \ n = n_{prev} + 1 \ \wedge \ n >= max.$$

In this formula $n_{prev}$ means the instance of symbol $n$ when the execution just entered the loop. At this state the $n_{prev} < max$ constraint can be defined. In line 15 as the result of the incrementation a new symbol is created and the $n = n_{prev} + 1$ constraint is built. Since in the next iteration the execution do not enters into the loop, but continues on the false branch it is necessary to create the $\neg(n < max)$ constraint too. After satisfying the constraint set above, symbol $n$ will be determined unambiguously, and its value is equal to the value of variable $max$. This means that if the execution exits the while loop, the value of $n$ must be $max$.

Building back the unambiguous value of $n$ into the current symbolic state, the arguments of method call $gcd()$ are both concrete values, thus it will be executed concretely and its return value will also be a concrete number. As we assumed the return value must be a positive integer, this leads to a bad array indexing in line 21.

The example detailed above highlights that a concreted symbolic variable can make a significant part of the execution concrete. The spread of symbols can be reduced, thus fewer variables have to be handled as unknown and uncertain data. As a result, the analysis becomes faster, because fewer execution paths have to be examined. In the shown example without concreting variable $n$ we should have explored the whole symbolic execution tree of method gcd(), which is rather expensive because of the loop inside.

The demonstrated ArrayIndexOutOfBoundException cannot be detected nor by the JPF Checker, nor by SymbolicChecker without constraint building.

In the second example we show a real code part from the log4j logging system. Consider method $org.apache.log4j.net.SMTPAppender.sendBuffer()$ in Figure 6 from log4j version 1.2.11, in which we point out that our new approach can also eliminate false positive faults as the conventional path condition construction.

In line 228 of method $sendBuffer()$, $get()$ method of class $CyclicBuffer$ is called, which returns a $LoggingEvent$ reference. First of all, method get() initializes the reference $r$ to $null$ (line 103), then if the $numElems$ data member is greater than 0, $r$ gets a new value. However, on the false branch it returns the null-initialized $r$ reference. Following this false branch, in method $SMTPAppender.sendBuffer()$ variable event is initialized to null in line 227, this null value will be propagated into method $SimpleLayout.format()$, which dereferences it in line 60.

However this null dereference would be a false positive error, because in line 60 the null value never occurs. In line 224 we get the $numElems$ member of object $cb$ for which the first iteration of the for loop at line 225 defines a constraint. The PC looks like this:

$$0 < len \ \wedge \ len = cb.numElems.$$

Nevertheless, method $get()$ called at line 227 returns null only on the false branch where the $numElems > 0$ constraint is not satisfied, thus the path condition is the following:

$$0 < len \ \wedge \ len = cb.numElems \ \wedge \ \neg(numElems > 0).$$

This formula, however, unsatisfiable, which means that the execution can not continue on this path. Variable $event$ will not get the null value in line 227, so the method $format()$ of class $SimpleLayout$ cannot dereference it. This actually means that the execution enters the for loop in line 225 only if the value of variable $len$ is at least 1, but in this case method $get()$ cannot return null on the false branch of the if statement in line 104.

```
                                 public class CyclicBuffer {
                                   int numElems;
                                   ...
                                 101.   public
                                 102.   LoggingEvent get() {
// SMTPAppender.java             103.     LoggingEvent r = null;
public class SMTPAppender extends 104.    if(numElems > 0) {
                 AppenderSkeleton { 105.    numElems--;
                                 106.       r = ea[first];
  ...                            107.       ea[first] = null;
  protected Layout layout;       108.       if(++first == maxSize)
  protected CyclicBuffer cb =    109.         first = 0;
          new CyclicBuffer(bufferSize); 110.   }
  ...                            111.     return r;
  protected                      112.   }
  void sendBuffer() {
    ...                          ...
224.     int len = cb.length();  119.   public
225.     for(int i = 0; i < len; i++) { 120.   int length() {
226.                             121.     return numElems;
227.       LoggingEvent event = cb.get(); 122.  }
228.       sbuf.append(layout.format(event)); ...
229.       if(layout.ignoresThrowable()) { }
230.         String[] s =
                 event.getThrowableStrRep(); public class SimpleLayout extends Layout {
231.         if (s != null) {    ...
232.           for(int j = 0; j < s.length; j++) { 56.  public
233.             sbuf.append(s[j]); 57.   String format(LoggingEvent event) {
234.           }                 58.
235.         }                   59.     sbuf.setLength(0);
236.       }                     60.     sbuf.append(event
237.     }                                    .getLevel().toString());
  ...                            61.     sbuf.append(" - ");
}                                62.     sbuf.append(event
  ...                                        .getRenderedMessage());
}                                63.     sbuf.append(LINE_SEP);
                                 64.     return sbuf.toString();
                                 65.   }
                                   ...
                                 }
```

**Fig. 6.** Method org.apache.log4j.net.SMTPAppender.sendBuffer() and its environment

The elimination of the discussed false positive error would fail using SymbolicChecker without the constraint building mechanism, but the JPF Checker would also eliminate it, because in this case no symbols are concreted and the unsatisfiability of the path condition is also tested by the JPF/SPF based tool.

We have run SymbolicChecker with the presented constraint building mechanism on large Java systems too, however the evaluation of the results is not entirely finished yet. Manually reviewing the reported errors is rather time-consuming because of the difficulty of interpreting the long execution paths from the entry point to the point where the error was detected in the source code. Which can be seen in the results so far is that there are significantly fewer runtime errors in the resultant report obtained by SymbolicChecker that uses the constraint building mechanism compared to the ones that JPF Checker produces. This does not mean that the report of SymbolicChecker does not contain false positive results, but most of them draw attention to real errors and potential sources of errors.

Considering the duration of the analyzes, the run-time of SymbolicChecker using the constraint building stays below the run-time of JPF Checker, but this duration is about twice longer then the run-time of SymbolicChecker running it without the constraint building mechanism. The analysis of the log4j logging library took slightly less than half an hour without constraint building, and the duration is about an hour using the new approach. Of course, we expected such a time requirement of our constraint building algorithm because the building of the constraint system, decomposing it to connected components and especially its satisfaction are rather computation intensive tasks.

## 5   Related Work

In this section we present works that are related to our research. First, we introduce some existing tools and technicques for runtime error detection mainly in Java programs, then we show the possible applications of the symbolic execution. We also summarize the problems that have been solved successfully by SymbolicPathFinder that we used for implementing our approach. Finally, we present works that completed or modified the symbolic execution technique.

The work of Weimer and Necula [7] focuses on proving safe exception handling in safety critical systems. They generate test cases that lead to an exception by violating one of the rules of the language. Unlike they do not generate test inputs based on symbolic execution but solving a global optimization problem on the control flow graph (CFG) of the program.

The JCrasher tool [8] by Csallner and Smaragdakis takes a set of Java classes as input. After checking the class types it creates a Java program which instantiates the given classes and calls each of their public methods with random parameters. This algorithm might detect failures that cause the termination of the system such as runtime exceptions. The tool is capable of generating JUnit test cases and can be integrated to the Eclipse IDE. JCrasher creates a driver environment but it can analyze public methods only and instead of symbolic execution it generates random data which is obviously not feasible for examining all possible execution branches.

The DART [9] (Directed Automata Random Testing) by Godefroid et al. tries to eliminate the shortcomings of the symbolic execution e.g. when it is unable to handle a condition due to its unlinear nature. DART executes the program with random or predefined input data and records the constraints defined by the conditions on the input variables when it reaches a conditional statement. In the next iteration taking into account the recorded constraints it runs the program with input data that causes a different execution branch of the program. The goal is to execute all the reachable branches of the program by generating appropriate input data.

The idea of symbolic execution is not new, the first publications and execution engines appeared in the 1970's. One of the earliest work is by King that lays down the fundamentals of symbolic execution [1] and presents the EFFIGY system that is able to execute PL/I programs symbolically. Even though EFFIGY handles only integers symbolically, it is an interactive system with which the user is able to examine the process of symbolic execution by placing breakpoints and saving and restoring states.

Starting from the last decade the interest about the technique is constantly growing, numerous programs have been developed that aim at dynamic test input generation using symbolic execution. The EXE (EXecution generated Executions) [10] presented by Cadar et al. at the Stanford University is an error checking tool made for generating input data on which the program terminates with failure. The input generation is done by the STP built-in constraint solver that solves the path condition of the path causing the failure. The basic difference between Symbolic Checker and EXE is that for running EXE one needs to declare the variables to be handled symbolically.

Further description and comparison of the above mentioned and other tools can be found in the work of Coward [11] and Cadar[12].

Song et al. applied the symbolic execution to the verification of networking protocol implementations [13]. The SymNV tool creates network packages with which a high coverage can be achieved in the source code of the daemon, therefore potential rule violations can be revealed according to the protocol specifications.

The main application of the Java PathFinder [14] and its symbolic execution extension, the SymbolicPathFinder [6] is the verification of the internal projects in NASA. Bushnell et al. describes the application of Symbolic PathFinder in TSAFE (Tactical Separation Assisted Flight Environment) [15] that verifies the software components of an air control and collision detection system. The primary target is to generate useful test cases for TSAFE that simulates different wind conditions, radar images, flight schedules, etc.

In our previous work [3] we used Symbolic PathFinder to create a tool named JPF Checker with the same goal as we have in case of Symbolic Checker: to detect runtime errors in Java programs without modifying the source code and without running it in a real-lif environment. This Symbolic PathFinder based tool used the conventional constraint building mechanism. In section 4 we compare this approach to our new concept which implemented in Symbolic Checker.

System MIX [16] combines symbolic execution with static type checking based techniques. It designates type and symbolic blocks in the program, which determines which code-part should be analysed using symbolic execution and wich one using static type checking. In the border of these blocks so-called mix-rules are used to convey the neccessary information. MIX is intended to provide a compromise between the precise but resource intensive symbolic execution and the less precise but faster type checking.

Shannon and others [17] built an abstraction layer above the Java string handling using finite state automatas. In addition to the implementation of the java.lang.String class StringBuilder and StringBuffer classes are included. As a result, the system is able to handle constraints that contains strings and string operations, thus it can be applied to programs that are working on more complex strings, such as SQL queries. Currently Symbolic Checker does not handle string constraints, we plan to deliver this development in the future.

Durring symbolic execution it may occur that the built path condition contains function calls, e. g. $if(y >= f(x))$. The so-called concolic (concrete-symbolic) [18] execution provides a possible solution for this problem using a special constraint builden mechanism. The main idea of this approach is that two path conditions are maintained at the same time. One of them contains those conditions witch do not includes function calls, and the other is the so-called complex PC, in which there are conditions that includes function calls too. First, the algorithm satisfies the simple PC and assnigs values to the included symbols, then these values are used to execute those included functions concretely which execution depended on those symbols which values have been determined int the first step. This method also capitalize on turning symbols int to concrete values, like the approach we present in this paper.

## 6    Summary and Future Work

The basic principles of symbolic execution has been known for decades, and several tools were made that utilizes the possibilities offered by this technique. SymbolicChecker which we developed at the Software Engineering Department of University of Szeged is differing from the most of these tools because it does not aim to generate test inputs, but to detect execution paths that lead to runtime errors and dangerous code parts as accurately as possible. In order to reach this goal we developed a constraint building mechanism and integrated it into SymbolicChecker, which differs from the ones which are used in other systems. The presented approach builds a constraint system for each execution path, which includes constraints over the variables too that depends on the inputs handled as symbolic variables and in case of unambiguity the concreted values are used in the later stages of the analysis. As a result, runtime errors can be detected that would not be possible using a conventional symbolic execution tool. For example the demonstrated ArrayIndexOutOfBoundsException in Section 4 cannot be detected nor by the JPF Checker, nor by SymbolicChecker without constraint building. By concreting symbolic variables the size of the symbolic execution tree can be reduced as well, which implies improvements in performance also. The ability to eliminate false positive results is achieved by ignoring those paths that carries contradictory constraints.

The results so far are promising and we continue the development of our tool. First, the review and evaluation of the results we get on large systems will take place, which will determine the future tasks. We plan to optimize the whole symbolic execution engine included the constraint building mechanism, as well as develop new methods and techniques that make the detection of runtime errors even more accurate.

**Acknowledgment.** The publication is partially supported by the European Union FP7 project "REPARA – Reengineering and Enabling Performance And poweR of Applications", project number: 609666.

## References

1. King, J.C.: Symbolic Execution and Program Testing. Communications of the ACM **19**(7), 385–394 (1976)

2. Gecode Tool-set. http://www.gecode.org/
3. Kádár, I., Hegedűs, P., Ferenc, R.: Runtime exception detection in java programs using symbolic execution. Acta Cybernetica **21**(3), 331–352 (2014)
4. Ferenc, R., Beszédes, Á., Tarkiainen, M., Gyimóthy, T.: Columbus – reverse engineering tool and schema for C++. In: Proceedings of the 18th International Conference on Software Maintenance (ICSM 2002), pp. 172–181. IEEE Computer Society, October 2002
5. Allen, F.E.: Control flow analysis. SIGPLAN Not. **5**(7), 1–19 (1970)
6. Păsăreanu, C.S., Rungta, N.: Symbolic pathfinder: symbolic execution of Java bytecode. In: Proceedings of the IEEE/ACM International Conference on Automated Software Engineering, ASE 2010, pp. 179–180. ACM, New York (2010)
7. Weimer, W., Necula, G.C.: Finding and preventing run-time error handling mistakes. In: Proceedings of the 19th Annual ACM SIGPLAN Conference on Object-oriented Programming, Systems, Languages, and Applications, OOPSLA 2004, pp. 419–431. ACM, New York (2004)
8. Csallner, C., Smaragdakis, Y.: JCrasher: an Automatic Robustness Tester for Java. Software Practice and Experience **34**(11), 1025–1050 (2004)
9. Godefroid, P., Klarlund, N., Sen, K.: DART: directed automated random testing. In: Proceedings of the 2005 ACM SIGPLAN Conference on Programming Language Design and Implementation, PLDI 2005, pp. 213–223. ACM, New York (2005)
10. Cadar, C., Ganesh, V., Pawlowski, P.M., Dill, D.L., Engler, D.R.: EXE: automatically generating inputs of death. In: Proceedings of the 13th ACM Conference on Computer and Communications Security, CCS 2006, pp. 322–335. ACM, New York (2006)
11. Coward, P.D.: Symbolic Execution Systems - a Review. Software Engineering Journal **3**(6), 229–239 (1988)
12. Cadar, C., Godefroid, P., Khurshid, S., Păsăreanu, C.S., Sen, K., Tillmann, N., Visser, W.: Symbolic execution for software testing in practice: preliminary assessment. In: Proceedings of the 33rd International Conference on Software Engineering, ICSE 2011, pp. 1066–1071. ACM, New York (2011)
13. Song, J., Ma, T., Cadar, C., Pietzuch, P.: Rule-based verification of network protocol implementations using symbolic execution. In: Proceedings of the 20th IEEE International Conference on Computer Communications and Networks (ICCCN 2011), pp. 1 8 (2011)
14. Java PathFinder Tool-set. http://babelfish.arc.nasa.gov/trac/jpf
15. Bushnell, D., Giannakopoulou, D., Mehlitz, P., Paielli, R., Păsăreanu, C.S.: Verification and validation of air traffic systems: tactical separation assurance. In: 2009 IEEE Aerospace Conference, pp. 1–10 (2009)
16. Khoo, Y.P., Chang, B.Y.E., Foster, J.S.: Mixing type checking and symbolic execution. In: Zorn, B.G., Aiken, A. (eds.) PLDI, pp. 436–447. ACM (2010)
17. Shannon, D., Zhan, D., Hajra, S., Lee, A., Khurshid, S.: Abstracting symbolic execution with string analysis testing. In: Academic and Industrial Conference Practice and Research TechniquesMUTATION, 2007. Taicpart-Mutation (2007)
18. Păsăreanu, C.S., Rungta, N., Visser, W.: Symbolic execution with mixed concrete-symbolic solving. In: Proceedings of the 2011 International Symposium on Software Testing and Analysis, ISSTA 2011, pp. 34–44. ACM, New York (2011)

# Comparison of Software Quality in the Work of Children and Professional Developers Based on Their Classroom Exercises

Gergő Balogh[(✉)]

Department of Software Engineering, University of Szeged, Szeged, Hungary
geryxyz@inf.u-szeged.hu

**Abstract.** There is a widely accepted belief that education has positive impact on the improvement of expertise in software development. The studies in this topic mainly focus on the product, more closely the functional requirements of the software. Besides these, they often pay attention to the individual so-called basic skills like abstract and logical thinking. We could not find any references where the final products of classroom exercises were compared by using non-functional properties like software quality. In this paper, we introduce a case study where several children's work is compared to works created by professional developers and not qualified adults. These numerical properties are difficult to measure and compare objectively. The model used to measure the various aspects of software quality also known in the industrial sector, hence it provides a well established base for our research. Finally, we analyse and evaluate the results and briefly introduce further research plans.

## 1 Introduction

IT experts tend to feel that in their field the gap between the industrial and the educational world are wider then ever. The first time students get a well presented, real life picture about software engineering is at the university. By this time, they have already made their choice about their further studies without prior knowledge. There are some articles or case studies on each side of the chasm. Charlie McDowell et al. [3] suggests that pair programming produces more proficient, confident programmers and may help increase female representation in the field. Basili et al. [2] studied the fault-proneness in software programs using eight student projects. They observed that the various numerical properties of the source code were correlated with defects while others was not. We can only find a few researchers who try to build a bridge. For example, Mengel & Yerramilli (1999) [4] studied the quality of 90 C++ novice student programs through the static analysis of the source code. The study results are encouraging and show that definite correlations exist, so static analysis is a viable methodology for assessing student work. Brian M. Stecher et al. [6] describe an accountability system that comprises explicit educational goals, assessments for measuring the attainment of goals and judging success, and consequences

© Springer International Publishing Switzerland 2015
O. Gervasi et al. (Eds.): ICCSA 2015, Part V, LNCS 9159, pp. 36–46, 2015.
DOI: 10.1007/978-3-319-21413-9_3

(rewards or sanctions). This model focuses on the achieved goals and results, and does not measure the internal quality of the implementation.

Our opinion is that there is a growing need for a well defined model which is able to evaluate the performance of students in such a way that it is acceptable for the industrial sector as well. In this particular experiment, we seek for similarities or contrasts between the implementations of students and experts, which can be measured in an objective way. During the research, a software quality model was used to evaluate the high level properties of the solutions. This model serves as a bridge to make a reliable comparison between the educational and the industrial sector.

## 2    Background

To answer our question we used data gathered on special classroom sessions to compare the software quality of students and experts. The sessions took place in a single afternoon in three distinct parts, each with the duration of one and a half hour. The high-school student used the provided development environment in groups of two or three. After they had finished their tasks, we collected all solutions and analysed them with an automated software quality model. Both the development environment and the quality model will be introduced in this section. We will briefly describe the accomplished tasks and exercises as well.

### 2.1    Original and Reduced Quality Model

We use a modified version of the source code quality model implemented by FrontEndART Ltd. The original model which is based on the researches at the University of Szeged conforms to the ISO/IEC 25010 standard and is capable of qualifying the source code of a software system. Figure 1. shows the original quality model. The computation of the ISO/IEC 25010 high level quality characteristics, together with the maintainability of the system, is based on a directed acyclic graph whose nodes correspond to quality properties that can either be internal (low-level) or external (high-level). Internal quality properties characterize the software product from an internal (developer) view and are usually estimated by using source code metrics. External quality properties characterize the software product from an external (end user) view and are usually aggregated somehow by using internal and other external quality properties. The nodes representing internal quality properties are called sensor nodes as they measure internal quality directly. The other nodes are called aggregate nodes as they acquire their measures through aggregation. The edges of the graph represent dependencies between an internal and an external or two external properties. [1]

We modified the previously described model and used this reduced model in our experiment. During the modification we only deleted existing nodes. In particular, reuseability, documentation and all of their children was removed, to create a reduced quality model, which provide better measurement capabilities. In general, these nodes were deleted, because their value was irrelevant, in the

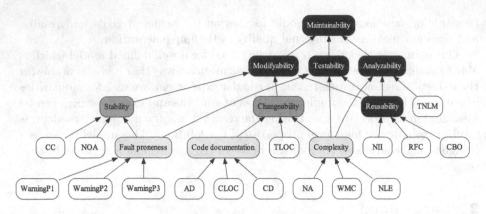

**Fig. 1.** Original quality model

context of the classroom exercises analysed in the following sections. We close
the introduction of reduced model by listing all eliminated nodes. The reason of
elimination and the meaning of the metric are also given.

**Number of Incoming Invocations (NII)** *Reason of exclusion:* The children
do not use inter-class calls and rarely use inter-method calls.
 **Method:** number of other methods and attribute initializations which
  directly call the method.
 **Class:** number of other methods and attribute initializations which directly
  call the local methods of the class.
**Response set For Class (RFC)** *Reason of exclusion:* The students only need
to implement methods in the same class and do not modify other classes.
 **Class:** number of local (i.e. not inherited) methods in the class (NLM) plus
  the number of directly invoked other methods by its methods or attribute
  initializations (NOI).
**Coupling Between Object classes (CBO)** *Reason of exclusion:* During
solving the tasks, only the provided API classes were used.
 **Class:** number of directly used other classes (e.g. by inheritance, function
  call, type reference, attribute reference).
**API Documentation (AD)** *Reason of exclusion:* Students do not write any
comments or documentation.
 **Class:** ratio of the number of documented public methods in the class +1 if
  the class itself is documented to the number of all public methods in the
  class + 1 (the class itself); however, the nested, anonymous, and local
  classes are not included.
 **Package** ratio of the number of documented public classes and methods
  in the package to the number of all of its public classes and methods;
  however, the classes and methods of its subpackages are not included
**Comment Lines of Code (CLOC)** *Reason of exclusion:* Students do not
write any comments or documentation.

**Method:** number of comment and documentation code lines of the method; however, its anonymous and local classes are not included.

**Class:** number of comment and documentation code lines of the class including its local methods and attributes; however, its nested, anonymous, and local classes are not included.

**Package** number of comment and documentation code lines of the package; however, its subpackages are not included.

**Comment Density (CD)** *Reason of exclusion:* Students do not write any comments or documentation.

**Method:** ratio of the comment lines of the method (CLOC) to the sum of its comment (CLOC) and logical lines of code (LLOC).

**Class:** ratio of the comment lines of the class (CLOC) to the sum of its comment (CLOC) and logical lines of code (LLOC).

**Package** ratio of the comment lines of the package (CLOC) to the sum of its comment (CLOC) and logical lines of code (LLOC).

## 2.2  Development Environment

We use a special integrated development environment called Greenfoot. Greenfoot is a project in the Programming Education Tools Group, part of the Computing Education Research Group at the School of Computing, University of Kent in Canterbury, UK. The Greenfoot Team currently includes Michael Kölling, Ian Utting, Davin McCall, Neil Brown, Philip Stevens and Michael Berry. [5]

Its main goal is to provide a simple and easy to use user interface for students to acquire basic programming skills. The users are able to interact with various elements of an object oriented program via an intuitive and simple user interface. The interface is a full IDE which includes project management, auto-completion, syntax highlighting, and other tools common to most IDEs. A couple of these features are shown in Figure 2a.

The graphical elements do not hide the underlying source code, so the users have to use the mouse, which is more natural for young children, and the keyboard together to accomplish their tasks. Its main concepts are *actors* who live in *worlds* to build games, simulations, and other graphical programs.

The main window is shown by Figure 2b. On the left side you can see the visual representation of a world object which acts as a canvas or scene for the whole project. The classes and their relations are shown on the right.

The creators also provide a basic class library for Greenfoot, which is highly customizable by the teachers to their needs. The objects are programmed in standard textual Java code providing a combination of programming experience in a traditional text-based language (Figure 2c.) with visual execution.

## 2.3  Implemented Classroom Exercises

In Hungary, teachers tend to use the *Logo programming language* in primary and secondary schools. Logo is an educational programming language designed

**(a)** Greenfoot integrated development environment features

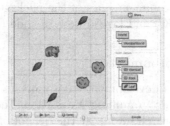

**(b)** Greenfoot main window

**(c)** Greenfoot source code editor

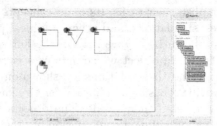

**(d)** A solution made by an expert

**Fig. 2.** Greenfoot integrated development environment for students

in 1967 by Daniel G. Bobrow, Wally Feurzeig, Seymour Papert and Cynthia Solomon. Today, the language is mainly remembered for its use of *turtle graphics,* in which commands for movement and drawing produced line graphics either on the screen or with a small robot called a turtle. The language was originally conceived to teach concepts of programming related to LISP and later to enable what Papert called body-syntonic reasoning where students could understand (predict and reason about) the turtle's motion by imagining what they would do if they were the turtle. There are substantial differences between the many dialects of Logo, and the situation is confused by the regular appearance of turtle graphics programs that mistakenly call themselves Logo. From the many implementations and IDE-s our teachers use IMAGINE.

To ease the transition from a toy language (Logo) to a programming language used in real world (Java), we reimplemented the basic logic of turtle graphics in Java and integrated it into the Greenfoot development environment. With this API students can create new Java classes to control the turtle and draw some simple graphics. We renamed turtles to ladybugs, because the original word *'teknos'* contains a diacritics on the second-to-last letter, while *'katica'* does not.

Two base classes were provided, namely `Ladybug` and `Katica`. The latter is a subclass of the first, it wraps the original English instruction with their Hungarian equivalent. `Ladybug` implements the following methods and properties.

`turn()` Turns the ladybug.
`moveTo()` Moves the ladybug to the given position.
`move()` Moves the ladybug to the given distance.
`penUp()` Takes the pen up.
`penDown()` Puts the pen down.
`setColor()` Sets the color of the pen.
`setLocation()` Sets the location of the ladybug.

The students and the experts accomplished the same classroom exercises. Each tasks implemented in a unique class inherited either the `Ladybug` or the `Katica` classes. The following four exercises were solved and their results were analysed.

1. Create a ladybug which is able to draw a rectangle with a specific size.
2. Create a ladybug which is able to draw a rectangle with a specific size.
3. Create a ladybug which is able to draw a triangle with a specific size.
4. Create a ladybug which is able to draw a polygon with the given number and length of sides.

Figure 2d. shows a solution made by an expert developer. In this example, each ladybug was placed on the world and the program was already executed.

## 3   Evaluation

After each student and expert had solved the above mentioned tasks, we measured the quality of their source code with the previously introduced modified model. We collected altogether 37 solutions from the students and from 3 experts. During the initialization of the measurement phase, all solutions which does not solve the task, i.e. it do not contain any source code, were eliminated. We measure the quality of the code of each user as it was a separate system. These results were aggregated into the four well known descriptive metrics: minimum, maximum, average, and the median of a given quality or source code metric. In the following sections, we will compare the high and low level metrics and we will also inspect some corner case solutions.

## 3.1 Comparison of High Level Metrics

We measured the following high level software quality metrics.

**Maintainability** Maintenance cost of the software system due to its source code

**Testability** Resources needed to test and verify the modifications made in the software

**Fault proneness** The probability that a failure occurs during the operation of the system

**Complexity** The general complexity of the software source code

**Modifiability** Risk of altering the source code without causing side effects

**Stability** Probability of operational failures caused by modifications of the software

**Comprehensibility** How difficult it is to understand the source code

**Changeability** Resources needed to alter the behavior of the software

**Analyzability** Expected cost of detecting faults and their casuses during operation

**Minor rule violations** Minor issues in the code that e.g. decrease the code readibility.

**Major rule violations** Major issues in the code that can cause e.g. performance issues.

**Critical rule violations** Critical issues in the code that can cause bugs and unintended behaviour.

Figure 3b. and 3a. show the statistical descriptors of the high level metrics. The average value represented on the bar-charts. The minimum, the maximum and the median were also displayed as points. The metric of top level – called maintainability – was highlighted to emphasize the fact that it aggregates all other metrics.

We cannot find any significant differences among the average performance of developers and students, however the lower and upper limits shows greater differences. The ranges of code quality are much wider in the case of students than in the case of experts. To emphasize these differences we connected the lower and upper limits with dashed lines. However these lines solely added to make the previously noted differences more easy to see. The order of these high level metrics is irrelevant, and these data points represent discreet values.

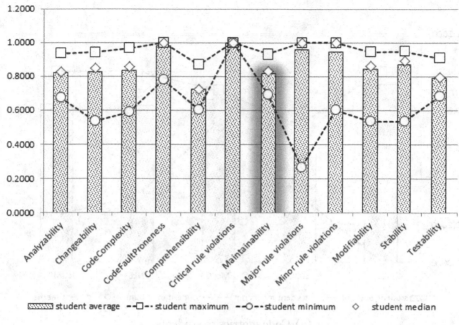

**(a)** Code quality metrics of students

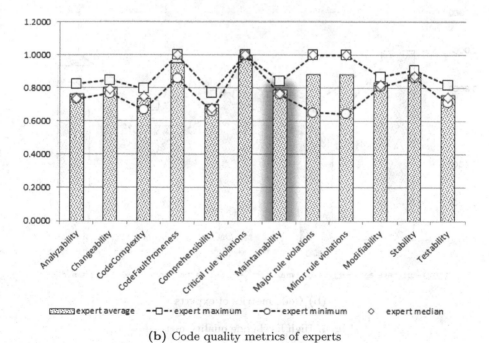

**(b)** Code quality metrics of experts

**Fig. 3.** High level code quality metrics

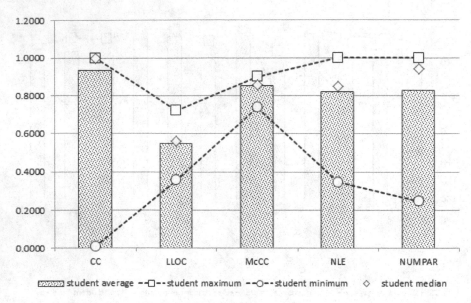

(a) Code metrics of students

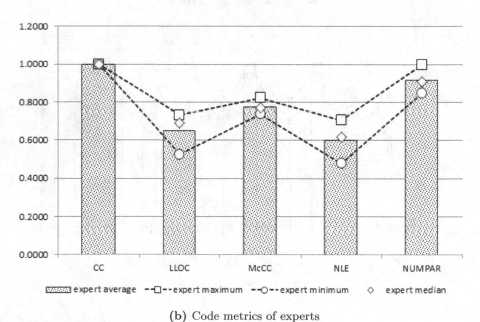

(b) Code metrics of experts

**Fig. 4.** High level code quality metrics

## 3.2   Comparison of Low Level Metrics

The following low level metrics were measured and evaluated.

**CC** The real value between 0 and 1 expresses which amount of the item is covered by code duplication.

**LLOC** The metric counts all non-empty, non-comment lines. Lines of nested classes or packages are not counted.

**McCC** The number of decisions within the specified method plus 1, where each if, for, while, do...while and ?: (conditional operator) counts once, each N-way (switch) counts N+1 and each try block with N catch counts N+1.

**NLE** NLE for a method is the maximum of the control structure depth. Only if, switch, for, foreach, while and do...while instructions are taken into account but if...else if does not increase the value. NLE for a class is the maximum of the NLE values of its methods.

**NUMPAR** The number of parameters of a method (the ellipsis is counted as one parameter).

Low level metrics follow the same pattern as seen with high level metrics. There are not any significant differences between the average and median values of students and developers, but the performance of the first ones tend to fluctuate more wildly. The values of LLOC and NUMPAR are closer to the optimal value calculated from the benchmark of the quality model. However, this does not mean that developers implemented the task with more code, but their solutions were more similar to the systems in the benchmark.

## 4   Conclusion

In this paper we used a simplified version of quality model based on the researches at the University of Szeged that conforms to the ISO/IEC 25010 standard and is capable of qualifying the source code of a software system to measure and compare the quality of source code created by students and experts. The subjects of our analysis were distinct solutions of predefined classroom exercises.

The results suggest that there are not any significant differences between the average performance of the two groups. These similarities can be explained with the fact that students were guided by an expert i.e. the teacher. On the other hand, the quality of source code produced by experts have less fluctuation. They tend to provide more stable performance.

Outliers can be found either direction form the average or median among the solutions of the students. We suggest that these represent the children who have more or less affinity for abstract thinking and logical problem solving.

In general, we conclude that these data and the results of their analysis suggests some interesting ideas. However, we are aware that this is just a stepping stone for further researches.

# References

1. Bakota, T., Hegedus, P., Kortvelyesi, P., Ferenc, R., Gyimothy, T.: A probabilistic software quality model. In: 2011 27th IEEE International Conference on Software Maintenance (ICSM), pp. 243–252. IEEE, September 2011. http://ieeexplore.ieee.org/lpdocs/epic03/wrapper.htm?arnumber=6080791
2. Basili, V., Briand, L., Melo, W.: A validation of object-oriented design metrics as quality indicators. IEEE Transactions on Software Engineering **22**(10), 751–761 (1996). http://ieeexplore.ieee.org/lpdocs/epic03/wrapper.htm?arnumber=544352
3. McDowell, C., Werner, L., Bullock, H.E., Fernald, J.: Pair programming improves student retention, confidence, and program quality. Communications of the ACM **49**(8), 90–95 (2006). http://dl.acm.org/ft_gateway.cfm?id=1145293&type=html
4. Mengel, S.A., Yerramilli, V.: A case study of the static analysis of the quality of novice student programs. ACM SIGCSE Bulletin **31**(1), 78–82 (1999). http://www.dl.acm.org/citation.cfm?id=384266.299689
5. Kölling, M., Utting, I., McCall, D., Brown, N., Stevens, P., Berry, M.: Greenfoot homepage (2014). http://www.greenfoot.org/home
6. Stecher, B.M., Kirby, S.N., Barney, H., Pearson, M.L., Chow, M., Hamilton, L.S.: Organizational Improvement and Accountability (2004). http://www.rand.org/pubs/monographs/MG136.html

# Characterization of Source Code Defects by Data Mining Conducted on GitHub

Péter Gyimesi, Gábor Gyimesi, Zoltán Tóth[✉], and Rudolf Ferenc

Department of Software Engineering, University of Szeged, Szeged, Hungary
{Gyimesi.Peter,Gyimesi.Gabor}@stud.u-szeged.hu,
{zizo,ferenc}@inf.u-szeged.hu

**Abstract.** In software systems the coding errors are unavoidable due to the frequent source changes, the tight deadlines and the inaccurate specifications. Therefore, it is important to have tools that help us in finding these errors. One way of supporting bug prediction is to analyze the characteristics of the previous errors and identify the unknown ones based on these characteristics. This paper aims to characterize the known coding errors.

Nowadays, the popularity of the source code hosting services like GitHub are increasing rapidly. They provide a variety of services, among which the most important ones are the version and bug tracking systems. Version control systems store all versions of the source code, and bug tracking systems provide a unified interface for reporting errors. Bug reports can be used to identify the wrong and the previously fixed source code parts, thus the bugs can be characterized by static source code metrics or by other quantitatively measured properties using the gathered data.

We chose GitHub for the base of data collection and we selected 13 Java projects for analysis. As a result, a database was constructed, which characterizes the bugs of the examined projects, thus can be used, inter alia, to improve the automatic detection of software defects.

**Keywords:** Bug database · GitHub · Data mining

## 1 Introduction

The characterization of source code defects is a popular research area these days. Programmers tend to make mistakes despite the assistance provided by the development environments, and also errors may occur due to the frequent changes and not appropriate specifications, therefore, it is important to get more tools to help the automatic detection of errors. For automatic recognition of defects, it is required to characterize the known ones.

One possible way of characterization is to try retrieving useful information from defective code parts. This requires the knowledge whether a given source code contains bugs or not. Defective sections of code can be characterized in different aspects after locating them.

© Springer International Publishing Switzerland 2015
O. Gervasi et al. (Eds.): ICCSA 2015, Part V, LNCS 9159, pp. 47–62, 2015.
DOI: 10.1007/978-3-319-21413-9_4

During the software development cycle, programmers use a wide variety of tools, including bug tracking, task management, and version control systems. There are numerous commercial and open source software systems available for these purposes. Furthermore, different web services are built to meet these needs. The most popular ones like SourceForge, Bitbucket, Google Code and GitHub fulfill the above mentioned functionalities. They usually provide more services, such as source code hosting and user management. Different APIs make it possible to retrieve various data properties, thus can be used as data sources. For example, they can be used to examine the behavior or the co-operation of users or even to analyze the source code itself. Since most of these services include bug tracking, it raises the idea to use this information in the characterization of source code defects [16]. To do this, the bug reports managed by these source code hosting providers must be connected to the appropriate source code parts [14]. A common practice in version control systems is to describe the changes in a comment belonging to a commit and often provide an ID for the associated bug report which the commit is supposed to fix [10]. This can be used to identify the faulty versions of the source code. Processing diff files can help us to obtain the code sections affected by the bug [15]. We can use textual similarity between faulty code parts [2], if we have such a database. We can also use static source code metrics [5], for which we only need one tool that is able to produce them. For the sake of completeness, other information extracted from the services can be involved in a database, such as user statistics.

To build a database containing useful bug characterization information, we have chosen GitHub since it has several regularly maintained projects and also a well defined API that makes it possible to integrate an automatic data retrieval mechanism in our own project. We selected 13 Java projects, which are suitable for such examination and serve as a base for creating a bug database with different characterizations. We have taken into consideration all reported bugs stored in the bug tracking system. With attached diff files we located the affected source parts. For the characterization we used static source code metrics and some other ones we defined based on the set of data retrieved from GitHub. The set of these metrics describe the projects from many aspects that can be a good starting point to execute different bug prediction techniques (e.g building models for prediction).

## 2   Related Work

Many approaches have been presented dealing with bug characterization and localization. Zhou et al. published a study describing BugLocator [16], a tool that detects the relevant source code files that need to be changed in order to fix a bug. BugLocator uses textual similarities (between initial bug report and the source code) in order to rank potential fault-prone files. Prior information about former bug reports is stored in a bug database. Ranking is based on the idea that descriptions with high similarity assume that the related files are highly similar too. A similar ranking is done by Rebug-Detector [12] a tool made by

Wang et al. for detecting related bugs from source code using bug information. The tool focuses on overridden and overloaded method similarities.

ReLink[14] is developed to explore missing links between changes committed in version control systems and fixed bugs. This tool could be helpful for software engineering research that are based on the linkage data, such as software defect prediction. ReLink mines and analyzes information like bug reporter, description, comments, date from bug database and then try to pair the bug with the appropriate source code files based on the set of source code information extracted from a version control system.

The history of version control systems shows us the concerned files and their changed lines only, but software engineers are also interested in which source code elements (e.g. classes or methods) are affected by a change or a bug [13]. Tóth et al. presented a method for tracking low level source code elements' (class, method) positions in files by processing version control system log information [15]. This method helps to keep source code positions up-to-date.

Kalliamvakou et al. mined GitHub repositories to investigate their characteristics and their qualities [10]. They presented a detailed study discussing different project characteristics, such as (in)activity. Further research questions were involved – whether a project is standalone or a part of a more massive system. Results have shown that the extracted data set can serve as a good input for various investigations, however one must use them with mistrust and always verify the usefulness and reliability of the mined data. It is a good practice to choose projects with many developers and commits, moreover should keep in mind that the most important point is to choose projects that fit well for your own purpose. In our case we have tried to create a database that is reliable (some manual validation is performed) and general enough for testing different bug prediction techniques.

Bird et al. presented a study on distributed version control systems, thus the paper focuses mainly on Git [3]. They examined the usage of version control systems and the available set of data (such as whether the commits are removable, modifiable, movable) gathered by the use of them (with respect of differentiate central and distributed systems). The main purpose of this paper was to draw attention on pitfalls and help researchers to avoid such pitfalls during the processing and analysis of mined Git information set.

Many research papers have shown that using a bug tracking system improves the quality of the developed software system. Bangcharoensap et al. introduced a method to locate the buggy files in a software system very quickly using the bug reports managed by the bug tracking system [2]. The presented method contains three different approaches to rank the fault-prone files, namely:

- Text mining: ranks files based on the textual similarity between a bug report and the source code itself.
- Code mining: ranks files based on prediction of the potential buggy module using source code product metrics.
- Change history: ranks files based on prediction of the fault-prone module using change process metrics.

They used the gathered project data collected on Eclipse platform to investigate the efficiency of the proposed approaches. Finally, they showed that these three ways are suitable to locate buggy files. Furthermore, bug reports with short description and many specific words greatly increase the effectiveness of finding the weak points (the files) of the system.

Not only the above presented method can be used to predict the occurrence of a new bug, but a significant change in source code metrics can be also a clue that the relevant source code files contain a potential bug or bugs [9]. Couto et al. presented a paper that shows the possible relationship between changed source metrics (used as predictors) and bugs [5]. They described an experiment to discover more robust evidences towards causality between software metrics and the occurrence of bugs.

Previously mentioned approaches use a self-made database for their own purpose as we could seen this advice in the work of Kalliamvakou et al. too [10]. Bug prediction techniques and approaches can be presented and compared in different ways; however, there are some basic points that can serve as common components. One common element can be a database used for evaluation of the various approaches. PROMISE [11] is a database that contains many bugs gathered from open source and also from industrial software systems. The main purpose of PROMISE is to support prediction methods and summarize a bunch of bugs and their characterization extracted from various projects. A similar database for bug prediction was presented, and commonly known as *Bug prediction dataset* [8]. The reason for creating this data set was mainly inspired by the idea of measuring the performance of the different prediction models and also comparing them to each other. This database handles the bugs and the relevant source code parts at class level, in other words the bugs are assigned to classes located in the source code.

At last but not least, the iBUGS database is presented [7] that contains a large amount of information describing projects from the aspect of testing different automatic defect localization methods. Bug describing information comes from version control systems and from bug tracking systems too. iBUGS used the following open source projects to extract the bugs from (in parentheses the number of extracted bugs are shown):

- AspectJ – an extension for the Java programming language to support aspect oriented programming (223).
- Rhino – a JavaSript interpreter written in Java (32).
- Joda-Time – provides a quality replacement (extension) for the Java date and time classes.

An attempt was performed on generating the iBUGS database in an automatic way and the generated set was compared with the manually validated set of bugs [6]. iBUGS is a very promising database since the set of validated bugs is considerable (263), although three projects only cannot guarantee the generality sufficiently. Our database includes several various projects from GitHub which are available from a public API. The given dataset extends the previously shown databases by including more metrics and storing more entries. The shown works successfully made use of their narrowed datasets, thus an extended database can serve as a base for further investigations. Besides 52 static source code metrics, our dataset contains additional metrics extracted from version control and user management systems. We used the diff files from the version control system to automatically identify the faulty source elements (classes).

## 3   Approach

In this section we will introduce some prerequisites and the process of creating a database containing bug characterization. At first, we will show some collected information dealing with GiHub, then define different metrics to characterize the reported bugs. Later in this section we will present the data mining process including data collection, processing raw data, analysis of source code versions, and extracting the characteristics of reported bugs.

### 3.1   GitHub

GitHub is one of today's most popular source code hosting services. It is used by several major open source projects for managing their project, among others Node.js, Ruby on Rails, Spring Framework, Zend Framework, and Jenkins. GitHub offers public and private Git repositories for its users, with some collaborative services, for example built-in bug and issue tracking systems. Since this set of abilities are supported by GitHub, we decided to use this source code hosting service (the well-defined API supports extracting these characteristics). This system can be used for bug reporting, since any GitHub user can add an issue. Issues can be labeled by the collaborators. The system provide some basic labels, such as "bug", "duplicate" and "enhancement", but anybody can customize these tags if required. In an optimal case, the collaborators review these reports and label them with the proper labels, for instance, the bug reports with "bug" label. For us, the most important feature of bug tracking is that we can refer to an issue from the comment of the commit, thereby we can identify a connection between the source code and the reported bug. GitHub has an API[1] that can be used for managing repositories from other systems, or query information about them. This information include events, feeds, notifications, gists, issues, commits, statistics, and user data.

With the GitHub Archive[2] project that also uses this API, we can get up-to-date statistics about the public repositories. For instance, Table 1 presents the number of created repositories in 2014 using the top 10 languages.

As can be seen, this is a large amount of information, and since this is public, it can be useful for mining different properties of the projects stored in GitHub. It means we can obtain the list of commits related to bug reports.

---

[1] https://developer.github.com/v3/
[2] http://www.githubarchive.org/

**Table 1.** The number of created repositories in 2014 by the top 10 languages

| Language | Number of repositories |
|---|---|
| JavaScript | 792 613 |
| Java | 562 142 |
| Ruby | 480 181 |
| CSS | 354 845 |
| PHP | 347 113 |
| Python | 317 525 |
| C | 290 113 |
| C++ | 164 936 |
| C# | 123 707 |
| Objective-C | 119 454 |

### 3.2 Metrics

The characterization of developed software systems by certain aspects is a difficult task, because a lot of subjective factors also play roles in them. With metrics we can measure the properties of a project objectively. These properties can describe the whole system itself from various points of view. Metrics can be obtained from the source code, from the project management data or from the execution traces of the source code. There are several different ways to measure them. From software metrics we can deduce higher-level metrics, such as the quality of source code or the distribution of defects, but they can be used to build a cost estimation model, apply performance optimization or to improve activities supporting software quality. In our case, the static source code metrics and the metrics obtained from GitHub are taken into consideration. These can be used to characterize the defective code sections on file level or even on source code element (class) level.

**Source Code Metrics.** The area of object-oriented source code metrics has been researched for many years [4], so no wonder that several tools have been developed for measuring them. These tools are suitable for detailed examination of systems written in various programming languages.

The static object-oriented source code metrics can be divided into several types or groups: size, inheritance, coupling, cohesion and complexity. Calculated product and process metrics can be used for different quality assurance methods as well. One such example can be a development of a quality rating model [1] or another application can be the determination of the correlation between the distribution of the bugs and the calculated metrics [9]. For such purposes a database containing readily extracted software metrics and located bugs provides a great opportunity. The list of used software metrics in the characterization is shown in Table 2.

**Table 2.** Used metrics for characterization

| Abbreviation | Full name |
| --- | --- |
| LCOM5 | Lack of Cohesion in Methods 5 |
| NOA | Number of Ancestors |
| NOC | Number of Children |
| NOD | Number of Descendants |
| NOP | Number of Parents |
| NOI | Number of Outgoing Invocations |
| NOS | Number of Statements |
| CBOI | Coupling Between Object classes Inverse |
| NPA | Number of Public Attributes |
| TCLOC | Total Comment Lines of Code |
| TNLM | Total Number of Local Methods |
| TNLG | Total Number of Local Getters |
| TNLA | Total Number of Local Attributes |
| NPM | Number of Public Methods |
| CLOC | Comment Lines of Code |
| NLPM | Number of Local Public Methods |
| AD | API Documentation |
| TNLS | Total Number of Local Setters |
| NLPA | Number of Local Public Attributes |
| TNPM | Total Number of Public Methods |
| TNPA | Total Number of Public Attributes |
| NLG | Number of Local Getters |
| NLM | Number of Local Methods |
| DIT | Depth of Inheritance Tree |
| NLA | Number of Local Attributes |
| NLE | Nesting Level Else-If |
| TNOS | Total Number of Statements |
| CD | Comment Density |
| NLS | Number of Local Setters |
| LOC | Lines of Code |
| LLOC | Logical Lines of Code |
| TCD | Total Comment Density |
| RFC | Response set For Class |
| NG | Number of Getters |
| NL | Nesting Level |
| NM | Number of Methods |
| NA | Number of Attributes |
| NS | Number of Setters |
| TNLPM | Total Number of Local Public Methods |
| DLOC | Documentation Lines of Code |
| TNLPA | Total Number of Local Public Attributes |
| NII | Number of Incoming Invocations |
| WMC | Weighted Methods per Class |
| TNG | Total Number of Getters |
| TLLOC | Total Logical Lines of Code |
| TNA | Total Number of Attributes |
| PUA | Public Undocumented API |
| TLOC | Total Lines of Code |
| TNS | Total Number of Setters |
| TNM | Total Number of Methods |
| PDA | Public Documented API |
| CBO | Coupling Between Object classes |

**Metrics Extracted from Version Control System.** In addition to the static source code metrics we gathered also other metrics from the available data. From the version control system the number of modifications and fixes on a file can be easily determined; moreover, committer identity can be mapped to the changed files. Furthermore, GitHub provides statistics about the users, that includes the number of commits per user on a project. From these data we determined to create the following metrics on file level:

- Number of modifications
- Number of fixes
- Number of opened issues
- Number of modifications the committer performed on the project

### 3.3    Data Mining

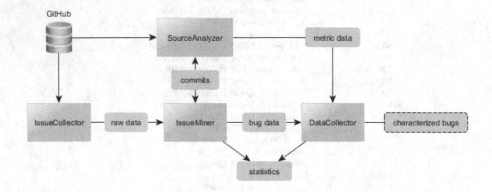

**Fig. 1.** The components of the process

We carried out the data processing in multiple steps. First we collected the data from GitHub by our IssueCollector program. Then we processed the raw data and created statistics, for which we have developed a program component called IssueMiner. For the next step we applied the SourceAnalyzer component, which downloads the necessary source code versions from GitHub and analyzes them. After this, we connected the results of the analysis with the data downloaded from GitHub, and located the defective sections of code and characterized them with the calculated metrics. For this purpose, the DataCollector tool was developed. The process and components are illustrated in Figure 1.

**The Criteria for Choosing Projects.** We considered a number of criteria when searching for appropriate projects on GitHub. First of all, we searched for Java language projects, especially larger ones, because these are more suitable for this kind of analysis. It was also important to have an adequate number of commits which use bug labels in reports to separate them clearly from other reports, moreover to refer to the appropriate bug report from the description of the commits. In addition, we preferred the currently active projects. We found many projects during the search, which would have fulfilled most aspects but in many cases developers used an external bug tracker system, so it would have been difficult to process them.

**The List of Selected Projects.** We have selected 13 projects based on the previously described aspects. Some properties of the projects will be presented, but first we introduce the set of selected software systems. The following projects were considered adequate for selection:

- **JUnit**[3]: A Java framework for writing unit tests.
- **Mission Control Technologies**[4]: Originally developed by NASA for the space flight operations. It is a real-time monitoring and visualization platform that can be used for monitoring any other data as well.
- **OrientDB**[5]: A popular document-based NoSQL graph database. Mainly famous for its speed and scalability.
- **Neo4j**[6]: The world's leading graph database with high performance.
- **MapDB**[7]: A versatile, fast and easy to use database engine in Java.
- **mcMMO**[8]: An RPG game based on Minecraft.
- **Titan**[9]: A high-performance, highly scalable graph database.
- **Oryx**[10]: It is an open source software with machine learning algorithms that allows the processing of huge data sets.
- **jHispter**[11]: A versatile software for generating Java Web applications.
- **Universal Image Loader**[12]: An Android library that assists the loading of images.
- **Netty**[13]: It is an asynchronous event-driven networking framework.
- **ANTLR v4**[14]: A popular software in the field of language processing. It is a powerful parser generator for reading, processing, executing, or translating structured text or binary files.
- **Elasticsearch**[15]: A popular RESTful search engine.

Table 3 provides a more accurate picture of the projects. This table shows the number of bug reports and commits of the projects. Explanation of used abbreviations are described in the following:

---

[3] https://github.com/junit-team/junit
[4] https://github.com/nasa/mct
[5] https://github.com/orientechnologies/orientdb
[6] https://github.com/neo4j/neo4j
[7] https://github.com/jankotek/MapDB
[8] https://github.com/mcMMO-Dev/mcMMO
[9] https://github.com/thinkaurelius/titan
[10] https://github.com/cloudera/oryx
[11] https://github.com/jhipster/generator-jhipster
[12] https://github.com/nostra13/Android-Universal-Image-Loader
[13] https://github.com/netty/netty
[14] https://github.com/antlr/antlr4
[15] https://github.com/elasticsearch/elasticsearch

Table 3. Statistics about the selected projects

| | NC | NCBR | NBR | NOBR | NCLBR | ANCBR |
|---|---|---|---|---|---|---|
| Android Universal I. L. | 914 | 52 | 80 | 5 | 75 | 0,69 |
| ANTLR v4 | 2 941 | 109 | 146 | 16 | 130 | 0,84 |
| Elasticsearch | 9 764 | 979 | 1 331 | 91 | 1 240 | 0,79 |
| jHipster | 1 436 | 52 | 68 | 0 | 68 | 0,76 |
| jUnit | 1 942 | 66 | 75 | 4 | 71 | 0,93 |
| MapDB | 1 052 | 80 | 109 | 18 | 91 | 0,88 |
| mcMMO | 4 476 | 251 | 635 | 11 | 624 | 0,40 |
| Mission Control T. | 975 | 15 | 37 | 9 | 28 | 0,54 |
| Neo4j | 29 208 | 76 | 268 | 112 | 156 | 0,49 |
| Netty | 6 254 | 567 | 747 | 28 | 719 | 0,79 |
| OrientDB | 8 404 | 362 | 710 | 212 | 498 | 0,73 |
| Oryx | 295 | 29 | 27 | 0 | 27 | 1,07 |
| Titan | 1 690 | 50 | 88 | 6 | 82 | 0,61 |

**NC** Number of Commits
**NCBR** Number of Commits per Bug Reports
**NBR** Number of Bug Reports
**NOBR** Number of Open Bug Reports
**NCLBR** Number of CLosed Bug Reports
**ANCBR** Average Number of Commits per Bug Reports

Figure 2 shows the number of commits for closed bug reports. This shows that there is a relatively large number of cases without a single commit. There are possible causes, for example, bug report is not referred from the commit description, the error has already been fixed, or a commit was not made with the purpose to fix the problem.

Figure 3 shows the ratio of the number of commits per projects, illustrating the activity and the size of the projects. Neo4J is dominant if we only consider the number of commits, however bug report related activities are slight.

### 3.4 Data Collection

At the beginning we saved the data for the selected projects via the GitHub API. It was necessary, because the data is continuously changing on GitHub due to the activity of the projects and we need a consistent data source for the analysis.

The saved data set includes the users assigned to the repository (Contributors), the open and the closed bug reports (Issues), and all of the commits. About users we stored the user id and the number of commits on the repository they formerly have applied to. From open issues we only stored the date of their creation. For closed issues we stored the creation date, closing date and the commit identifiers with their creation dates. The data we stored for the commits includes the id of the contributor, the id of the development branch, the parent(s) of the commit and the affected files with the diff files.

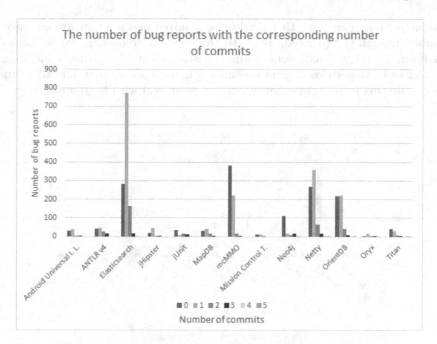

**Fig. 2.** The number of bug reports with the corresponding number of commits

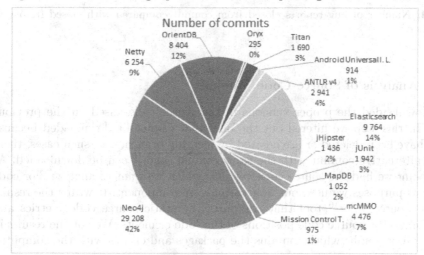

**Fig. 3.** The number of commits per projects

## 3.5 Processing of Raw Data

Data saved from GitHub is only a raw data set that includes all commits. We only need the ones that are relevant to the bug reports. These are the commits with a reference to a bug issue (fix) and the commits applied after the submission of the report but before the commit referenced the issue. Between these commits some further ones can occur that need to be removed because they are no longer available through Git (deleted, merged). During data processing, we performed such filtering and we made some more statistics about the projects including the number of issues closed from commits. Figure 4 shows that not all of the projects use this feature. In other words, we must deal with both options.

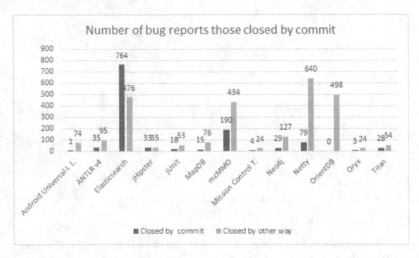

**Fig. 4.** Number of bug reports closed from commit compared with closed from web interface

## 3.6  Analysis of Source Code Versions

We downloaded the proper versions of the source code based on the previous step. In this step we filtered out the ones that cannot be downloaded because they have been deleted or are created by merging branches. In such cases, there is an alternative commit with the same content, and it can be downloaded. At this point we had the source code versions which we want to analyze. For code analysis purposes we used our SourceAnalyzer component. It wraps the results of the SourceMeter[16] tool that computes the static source code metrics and determines the source code positions of the code elements. We got the results in the form of graphs which contains the packages and classes with the computed data.

## 3.7  Extracting the Characteristics of Bugs

The next step is to link the two data sets – the graphs of analysis and the data gathered from GitHub – and extract the characteristics of the bugs. In this step we determine the followings:

---

[16] https://www.sourcemeter.com/

- the source code elements affected by the commits
- the static source code metrics of the affected source code elements
- the number of modifications and fixes of the files in each commit
- the last user modified a file in relevant commits
- the number of open bug reports in relevant commits

This was carried out by our DataCollector program. To determine the affected source code parts, we used diff files. These files contain the differences between two source code versions in a unified diff format. In the following, a unified diff file snippet is shown.

```
--- /path/to/original   ''timestamp''
+++ /path/to/new   ''timestamp''
@@ -1,4 +1,4 @@
+Added line
-Deleted line
 This part of the
 document has stayed the
 same
```

Each difference contains a header information specifying the starting line number and the number of affected lines. Using this prior information, we can get the range of the modification. To obtain a more accurate result, we subtracted the unmodified code lines from this range. The diff files generated by GitHub contain additional information about which method is affected. For us, it does not carry any extra information because the difference can affect multiple source code elements. Thus, there is no further task to do but to examine the source code elements in every modified file and identify which ones of them are affected by the changes (method uses the source code element positions). We identified the source code elements by their fully qualified names. With this algorithm we got the affected source elements as a result set.

Now we have enough information to calculate some additional metrics. The program calculates the number of open bug reports for each commit. This is done by counting the issues whose creation date is earlier than the creation of the commit, and the closing time is later. Furthermore, it calculates the number of modifications of each file. At first, it arranges the commits by the order of creation time. Starting with the earliest one, it increases the counter on a file if it is affected by a commit. During this process it is also counting the number of fixes. A modification is considered as a fix if it is in the last commit for a closed issue. Lastly, it determines the user for each file who most recently has modified it. It is used to connect the user statistics with the modifications. The user statistics means the number of modifications on a project applied by the user. The number of modifications is collected at the time of downloading the data from GitHub and not at the time the commit was made.

Next, our program determines the commits that were performed after the creation time of an issue and before the first commit for fixing that issue. These are essential because these versions presumably contained the buggy source code

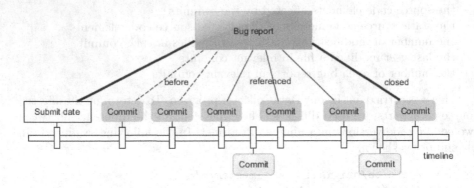

**Fig. 5.** The relationship between the bug reports and commits

parts. The relation between the bug report and the commits is shown in Figure 5. To mark the code sections affected by the bug in these commits, the program accumulates the modifications on issue level. It is done by collecting the fully qualified name of these elements. Then these metrics are exported into a CSV document. This is done for each bug report due to resource saving purposes because the graphs can be very large in size. The metrics for files and classes are exported to different files. One file specifies whether a source code was buggy or not, the other one contains assignment of source code elements and the number of bugs related to them. Thus, four types of output is generated in this manner. Finally, it concatenates these files resulting in a large set of data. The first line of this CSV file contains the header with the metric names.

Once our dataset is created it can serve as an input for building a model for fault prediction that one can use to forecast fault-prone spots in any developed software system. The database can be easily updated since only a filtered analysis should be performed (from a given date) that can extend the previous version of the dataset.

## 4   Conclusion and Future Work

In this study, we developed a method and performed its implementation, which generates a bug related database mined from GitHub project hosting service using static source code metrics of the relevant code parts. It identifies the faulty source code elements from the past automatically by using diff files from the version control system. This way it allows the simultaneous processing of several publicly available projects located on GitHub, thereby resulting in the production of a large database. Previous studies have dealt with only few larger data sets created under strict management, as opposed to our way. Additionally, our dataset contains new static source metrics compared to the other databases, allowing the examination of the relationship between these metrics and software bugs. Furthermore, in result of examining the projects on GitHub, we selected 13 suitable Java projects, which we used to build the database.

We are planning to expand the database with additional projects and additional data sources, such as SourceForge and Bitbucket. We also plan to refine the metrics and define new features. The calculated file level metrics – with further analysis – can be determined for lower level source elements, i.e. for classes and methods, and data gathered from GitHub also can be used do define more metrics. Our ultimate goal is to use the data to examine the correlation between the bugs and the source code metrics, and to apply the results to facilitate the automatic recognition of source code defects.

**Acknowledgments.** The publication is partially supported by the European Union FP7 project "REPARA – Reengineering and Enabling Performance And poweR of Applications", project number: 609666.

We also thank Zsuzsanna Fehér and László Szoboszlai for their valuable assistance.

# References

1. Bakota, T., Hegedus, P., Kortvelyesi, P., Ferenc, R., Gyimothy, T.: A probabilistic software quality model. In: 2011 27th IEEE International Conference on Software Maintenance (ICSM), pp. 243–252 (September 2011)
2. Bangcharoensap, P., Ihara, A., Kamei, Y., Matsumoto, K.: Locating source code to be fixed based on initial bug reports - a case study on the eclipse project. In: 2012 Fourth International Workshop on Empirical Software Engineering in Practice (IWESEP), pp. 10–15 (October 2012)
3. Bird, C., Rigby, P.C., Barr, E.T., Hamilton, D.J., German, D.M., Devanbu, P.: The promises and perils of mining git. In: 6th IEEE International Working Conference on Mining Software Repositories, MSR 2009, pp. 1–10 (May 2009)
4. Shyam, R.: Chidamber and Chris F Kemerer. A metrics suite for object oriented design. IEEE Transactions on Software Engineering **20**(6), 476–493 (1994)
5. Couto, C., Silva, C., Valente, M.T., Bigonha, R., Anquetil, N.: Uncovering causal relationships between software metrics and bugs. In: 2012 16th European Conference on Software Maintenance and Reengineering (CSMR), pp. 223–232 (March 2012)
6. Dallmeier, V., Zimmermann, T.: Automatic extraction of bug localization benchmarks from history. Technical report, Universitat des Saarlandes and Saarbrücken and Germany (2007)
7. Dallmeier, V., Zimmermann, T.: Extraction of bug localization benchmarks from history. In: Proceedings of the twenty-second IEEE/ACM international conference on Automated software engineering, pp. 433–436. ACM (2007)
8. D'Ambros, M., Lanza, M., Robbes, R.: An extensive comparison of bug prediction approaches. In: Proceedings of MSR 2010 (7th IEEE Working Conference on Mining Software Repositories), pp. 31–41 (2010)
9. Gyimothy, Tibor, Ferenc, Rudolf, Siket, Istvan: Empirical validation of object-oriented metrics on open source software for fault prediction. IEEE Transactions on Software Engineering **31**(10), 897–910 (2005)
10. Kalliamvakou, E., Gousios, G., Blincoe, K., Singer, L., German, D.M., Damian, D.: The promises and perils of mining github. In: MSR 2014 Proceedings of the 11th Working Conference on Mining Software Repositories, pp. 92–101 (2014)
11. Menzies, T., Caglayan, B., He, Z., Kocaguneli, E., Krall, J., Peters, F., Turhan, B.: The promise repository of empirical software engineering data (June 2012)

12. Wang, D., Lin, M., Zhang, H., Hu, H.: Detect related bugs from source code using bug information. Computer Software and Applications Conference (COMPSAC) (2010)
13. Chadd, C.: Williams and Jeffrey K Hollingsworth. Automatic mining of source code repositories to improve bug finding techniques. IEEE Transactions on Software Engineering **31**(6), 466–480 (2005)
14. Wu, R., Zhang, H., Kim, S., Cheung, S.-C.: Relink: recovering links between bugs and changes. In: Proceedings of the 19th ACM SIGSOFT Symposium and the 13th European Conference on Foundations of Software Engineering, pp. 15–25. ACM (2011)
15. Toth, Z., Novak, G., Ferenc, R., Siket, I.: Using version control history to follow the changes of source code elements. Software Maintenance and Reengineering (CSMR) (2013)
16. Zhou, J., Zhang, H., Lo, D.: Where should the bugs be fixed? more accurate information retrieval-based bug localization based on bug reports. In: 2012 34th International Conference on Software Engineering (ICSE) (2012)

# Software Component Score: Measuring Software Component Quality Using Static Code Analysis

Berkhan Deniz[✉]

Aselsan Electronics Inc., Defense System Technologies (SST) Group, Ankara, Turkey
berkhand@aselsan.com.tr

**Abstract.** Static code analysis is a software verification method which analyzes software source code in terms of quality, security and reliability. Unlike other verification activities, static analysis can be automated; thus it can be applied without running the software or creating special test cases. Software metrics are widely practiced by many companies and researchers in order to evaluate their software. In this study, the software component quality measurement method which is developed in an embedded software team will be described. The method is based on automatically collected metrics and predetermined set of rules. First, the measured and calculated metrics under this method will be defined and the reasons for selecting these metrics will be described. Then, the software quality score calculation method using these metrics will be explained. Finally, the gains obtained with this method and the future plans will be related.

**Keywords:** Static code analysis · Quality measurement · Code based metrics · Software component quality · Coding standards

## 1 Introduction

Static analysis (also called static code analysis, source code analysis, and static program analysis), is a software verification method which analyzes software source code in terms of quality, security, and reliability. Unlike other verification activities, static analysis can be automated; thus it can be applied without running the software or creating special test cases.

Software developers and software testers can detect various run-time errors such as overflow, divide by zero, memory and pointer errors via static analysis. These analyses not only detect some software errors but also provide various format checks (i.e. indentation check), architectural evaluations, and coding standard conformity [1].

In this study, the software component score assignment approach based on automatically collected metrics will be explained. It is intended not to use the components under a certain score for future projects without making the necessary improvements.

The rest of the study is organized in this way: In Section 2 the literature about static code analysis and its relationship with software quality is presented; in Section 3 the so called quality measurement model, used metrics, their selection criteria and gains obtained are explained. Section 4 discusses the factors affecting the validity of the suggested method. Finally, in Section 5 the results and the suggestions for future studies are provided.

© Springer International Publishing Switzerland 2015
O. Gervasi et al. (Eds.): ICCSA 2015, Part V, LNCS 9159, pp. 63–72, 2015.
DOI: 10.1007/978-3-319-21413-9_5

## 2    Literature

### 2.1    Measuring Quality

Fault detection metric, which is related to reliability of the software, is the most reported quality metrics in the literature. This metric focuses on the total number of defects detected in the software product. The total number of changes (improvement or repair) in the software product during the maintenance period is also used for measuring quality [4]. In addition, rework effort (the entire work spent for correcting problems) is also used for measuring the quality.

In [5], isolation and ease of fixing of problems and strength of errors are used as quality indicators. The change density (number of changes per SLOC) and the modified code ratio between software releases are other indicators of software quality.

Furthermore, fault-proneness is designed and used as a quality metric which is defined as the probability of defect detection in a software system as a function of structural characteristics of the system [6].

### 2.2    Using Code Based Metrics for Quality Measurement

In previous section, we discussed about the general software product characteristics for quality measurement. In this section, we will focus on code-based metrics, especially on object-oriented metrics.

There are different OO metrics defined in the literature. However, Chidamber and Kemerer's metrics suite for OO design is the deepest research in OO metrics investigation. These metrics are known as CK metrics, and by far, these are the most popular OO metrics [7]:

- Weighted Methods per Class (WMC)
- Depth of Inheritance Tree (DIT)
- Number of Children (NOC)
- Between Object Classes (CBO)
- Response for a Class (RFC) Lack of Cohesion in Methods (LCOM)

Apart from the object-oriented metrics (i.e. CK metrics), functional metrics are also widely used. Cyclomatic complexity, source line of code, and line of comments is examples of these metrics.

Software developers and managers can use code-based metrics for different purposes: system level forecasting, prior determination of unsafe components through early measures, and the development of safety design and programming instructions [7, 2]. Furthermore, selecting appropriate metrics from all the alternative metrics support the software developers and managers to identify the quality and structure of the software design and code [12]. In a previously published research, whether a software module would be fault-prone was successfully predicted by linking metrics and earlier software data [12]. Likewise, component defects were predicted using object-oriented metrics obtained from design, code, and requirements [8]. Other authors empirically investigated the usability of object-oriented metrics in forecasting fault-proneness while considering the severity of defects [10].

Furthermore, object oriented metrics support software developers and managers conduct assessments of the necessary development and testing efforts [2]. Moreover, developers analyze and collect metrics in order to validate the software design quality, and hence, help developers improve software quality and productivity [3]. Likewise, metrics are essential especially when the developers decide on a new technology because metrics material provides rapid response in the new feature for software designers and managers [9]. Hence, if metrics are properly used, the costs of the implementation and maintenance reduce, and software product's quality improves significantly. The advantages of software metrics are summarized in Table 1.

**Table 1.** Summary of advantages of software metrics

| |
|---|
| System level forecasting |
| Prior identification of unsafe components |
| Development of safety design and programming instructions |
| Identify the quality and structure of the software design and code |
| Prediction of fault-proneness |
| Prediction of development and testing efforts |
| Validation of the software design quality |
| Improve software quality and productivity |
| Rapid response when a new technology is adopted |
| Reduction of implementation and maintenance cost |

# 3   Suggested Software Component Quality Measurement Method

As it was mentioned in previous sections, the software assessment method by using metrics is widely used by many software companies and researchers. In this section, the suggested software component quality measurement method which uses the auto collected component metrics and predefined rules will be explained.

## 3.1   The Development Environment and Infrastructure of the Suggested Method

This method was developed in Aselsan's (Turkey's largest defense systems company) Defense System Technologies Group by the Air Defense Weapon Systems and Image Processing team. The team develops embedded software products for weapon systems by using C++ language.

The team has created coding style and naming rules guides which are actively used during the software development process. Additionally, the team uses a subset of MISRA C++ [8] coding standard that is modifies according to the team's needs. By utilizing the programs developed by the team, the appropriateness of the software components to the coding style guide, naming rules guide and subset of the coding standard are measured automatically. The details of the measured metrics will be given in the next section.

The team develops the products by using a UML based software development tool. The metrics automatically collected from the UML models of the generated software components are also utilized in the suggested method and their details will also be given in the next section.

Finally, structural metrics of the components, such as source line of codes, line of comments, blank line of codes, are also used in the suggested quality method and collected automatically.

## 3.2    Collected Metrics

The metrics collected beneath the suggested component quality measurement method are separated into 5 groups:

**Structural Metrics.** Line of codes – LOC, source line of codes – SLOC, comment line of codes – CLOC, blank line of codes – BLOC.

**Number of Structural Component Elements.** Number of packages, number of classes, number of interface methods, number of virtual interface methods, number of static interface methods, number of attributes, number of arguments, number of types.

**Coding Standard Metrics.** Number of coding style violations, number of naming rules processed items, number of naming rules violations, number of MISRA C++ standard violations, number of compilation warnings.

**UML Model Metrics.** Number of packages with no description, number of classes with no descriptions, number of interface methods with no descriptions, number of structures defined manually (i.e. not through the UML model), number of enumerations defined manually (i.e. not through the UML model), number of types selected manually (i.e. not through the UML model), number of uninitialized attributes.

**Unit Test Metrics.** Number of unit tests, number of unit test warnings, number of unit test errors.

By combining the metrics listed above, 17 combined metrics are defined and the component quality measurement method is based upon these metrics (see Table 2).

**Table 2.** The metrics used

| M1: Comment line of codes per line of codes |
| --- |
| M2: Blank line of codes per line of codes |
| M3: Compilation warnings per source line of codes |
| M4: Coding style violations per number of methods |
| M5: The rate of naming rules violations (The number of naming rules violations / number of naming rules processed items) |
| M6: MISRA C++ standard violations per source line of codes |
| M7: The rate of packages with no descriptions (The number of packages with no description / the number of packages) |
| M8: The rate of classes with no descriptions (The number of classes with no description / the number of classes) |
| M9: The rate of interface methods with no descriptions (The number of interface methods with no description / the number of interface methods) |
| M10: The rate of non abstract interface methods ((The number of virtual interface methods + The number of static interface methods) / the number of interface methods) |
| M11: The rate of manually defined structure types (The number of manually defined structure types / the number of types) |
| M12: The rate of manually defined enumeration types (The number of manually defined enumeration types / the number of types) |
| M13: The rate of manually selected arguments and attributes (the number of manually selected arguments and attributes / the sum of arguments and attributes) |
| M14: The rate of uninitialized attributes (The number of uninitialized attributes / the number of attributes) |
| M15: The rate of unit tests (The number of unit tests / (The total number of methods – the number of interface methods)) |
| M16: The rate of unit test warnings (The number of unit test warnings / the number of unit tests) |
| M17: The rate of unit test errors (The number of unit test errors / the number of unit tests) |

### 3.3    The Reasons of the Selection of These Metrics

The metrics used in suggested method are given in Table 2. In this section, it will be discussed why these metrics are appropriate in terms of measuring software component quality.

M1 and M2 are line of codes metrics. They are used to assess the understandability of the software components' source code.

M3, M4, M5, and M6 are metrics which are used to assess the appropriateness of the components' source code with respect to various standards. By providing the components to be convenient to these standards, it is aimed the components to gain the related quality attributes which these standards bring out [1]. It is widely accepted that developing software with respect to previously determined rules ensures to improve the understandability and reliability; therefore these metrics are used in the suggested method.

M7, M8, and M9 are metrics extracted from the UML model. They measure whether related descriptions about the units in the components, the missions of these units and their intended usages are provided in the source code of the components. For the sake of measuring understandability and maintainability of the components, these metrics are used in the suggested method.

M10 is a metric about the interface methods of the component. As a good programming practice, it is aimed to provide all interface methods of the developed components to be abstract; hence to make all the classes inherited from these interfaces to implement these methods. Therefore, it is intended to make the developer of the component be aware of the changes in the signature of the interface methods as a result of updates just during the compile time. This metric is important in terms of the adaptability of the components.

M11, M12, and M13 are the kind of metrics collected from the UML model. The used UML based development tool provides interface for the developers to create structure and enumeration types; and to select attribute and argument types both manually or by using the model. When the structure and enumeration types are defined by using the model, they become accessible through the application programming interfaces that the development tool provides outside and additionally the changes in the definitions of these types are automatically put in use all over the source code. When the attribute and argument types are selected by using the model, the compilation errors which can happen due to the signature changes of these types are prevented and the related dependency changes are applied automatically into the component. By making the definitions using the model but not manually, the development times of the components are reduced; and maintainability and adaptability of the components are improved. As a consequence these metrics are put into the suggested method.

M14 measures the number of uninitialized attributes in the component. Uninitialized attributes are among the most important sources of errors and security gaps in the components, especially in embedded components [1]. This metric is thought to be critical in terms of component security and reliability; and hence used in the suggested method.

Finally; M15, M16, and M17 are metrics related to the components' unit tests. During the development of the components, the team planned to make unit tests for every method (except the interface methods) in the components for the sake of robustness of the components. Therefore, these metrics are put into use in the suggested method.

## 3.4    Component Quality Measurement Method

In this section, the component quality measurement method developed by using the metrics in Table 2 will be explained.

Firstly, it will be expressed how the metrics in Table 2 are assessed. For each metric, a three-level success criterion (in terms of percentages) is defined. These levels are "Success", "Warning", and "Fail".

The success criteria of the metrics are shown in Table 3. When the table is examined in detail, it can be seen which metrics are more important and which metrics have less tolerance in the suggested method, can be identified. For instance, in some metrics the success criteria is 0-1% (i.e. M3, M9, M11, M12, and M17); for M14 it is strictly 0%.

The suggested method is comprised of 4 phases:

1. Automatically measurement of the metrics in Section 3.2 for the component of which the quality measurement will be made
2. By using the metrics measured, calculation of the metrics in Table 2
3. The assessment of these metrics according to the success criteria in Table 3
4. Giving points to metrics as a result (for successful metrics 100, for metrics with warning 50, for failed metrics 0) and the average of these points is taken as the "component score"

**Table 3.** Metrics' success criteria

|  | Success | | Warning | | Fail | |
|---|---|---|---|---|---|---|
|  | Min | Max | Min | Max | Min | Max |
| M1 | 20% | 100% | 10% | 20% | 0% | 10% |
| M2 | 0% | 30% | 30% | 50% | 50% | 100% |
| M3 | 0% | 1% | 1% | 2% | 2% | 100% |
| M4 | 0% | 10% | 10% | 20% | 20% | 100% |
| M5 | 0% | 10% | 10% | 20% | 20% | 100% |
| M6 | 0% | 10% | 10% | 20% | 20% | 100% |
| M7 | 0% | 5% | 5% | 20% | 20% | 100% |
| M8 | 0% | 10% | 10% | 50% | 50% | 100% |
| M9 | 0% | 1% | 1% | 2% | 2% | 100% |
| M10 | 0% | 20% | 20% | 40% | 40% | 100% |
| M11 | 0% | 1% | 1% | 2% | 2% | 100% |
| M12 | 0% | 1% | 1% | 2% | 2% | 100% |
| M13 | 0% | 5% | 5% | 10% | 10% | 100% |
| M14 | 0% | 0% | 0% | 0% | 0% | 100% |
| M15 | 50% | 100% | 10% | 50% | 0% | 10% |
| M16 | 0% | 10% | 10% | 20% | 20% | 100% |
| M17 | 0% | 1% | 1% | 2% | 2% | 100% |

In Table 4, the measured metrics for a sample component is shown. The metrics calculated using these metrics are shown in Table 5. Additionally, in this table, the assessment of the calculated metrics, grading them and by taking average calculation of the component score is also shown.

**Table 4.** Measured metrics for a sample component

| | | |
|---|---|---|
| **Structural metrics** | Line of codes: | 28175 |
| | Blank line of codes: | 4968 |
| | Source line of codes: | 13795 |
| | Comment line of codes: | 9303 |
| **Structural element counts** | Packages: | 14 |
| | Classes: | 67 |
| | Methods: | 356 |
| | Interface methods: | 71 |
| | Virtual interface methods: | 34 |
| | Static interface methods: | 35 |
| | Attributes: | 860 |
| | Arguments: | 456 |
| | Types: | 75 |
| **Coding standard metrics** | Coding style violation count: | 0 |
| | Naming rules processed items count: | 2282 |
| | Naming rules violations count: | 130 |
| | MISRA C++ violations count: | 227 |
| | Compilation warning count: | 4 |
| **UML Model Metrics** | Packages with no description: | 1 |
| | Classes with no description: | 5 |
| | Interface methods with no descriptions: | 0 |
| | Manually defined structure types count: | 0 |
| | Manually defined enumeration types count: | 0 |
| | Manually selected types count: | 0 |
| | Uninitialized attributes: | 0 |
| **Unit Test Metrics** | Unit tests count: | 0 |
| | Unit tests warning count: | 0 |
| | Unit tests error count: | 0 |

**Table 5.** Calculated metrics for the sample component and component score

| Metrics | Calculation (%) | Assessment | Metric Point |
|---|---|---|---|
| M1: Comment line of codes per line of codes | 33,02% | Success | 100 |
| M2: Blank line of codes per line of codes | 17,63% | Success | 100 |
| M3: Compilation warnings per source line of codes | 0,03% | Success | 100 |
| M4: Coding style violations per number of methods | 0,00% | Success | 100 |
| M5: The rate of naming rules violations (The number of naming rules violations / number of naming rules processed items) | 5,70% | Success | 100 |
| M6: MISRA C++ standard violations per source line of codes | 1,65% | Success | 100 |
| M7: The rate of packages with no descriptions (The number of packages with no description / the number of packages) | 7,14% | Warning | 50 |
| M8: The rate of classes with no descriptions (The number of classes with no description / the number of classes) | 7,46% | Success | 100 |
| M9: The rate of interface methods with no descriptions (The number of interface methods with no description / the number of interface methods) | 0,00% | Success | 100 |
| M10: The rate of non abstract interface methods ((The number of virtual interface methods + The number of static interface methods) / the number of interface methods) | 50,70% | Fail | 0 |

**Table 6.** (*Continued.*)

| Metrics | Calculation (%) | Assessment | Metric Point |
|---|---|---|---|
| M11: The rate of manually defined structure types (The number of manually defined structure types / the number of types) | 0,00% | Success | 100 |
| M12: The rate of manually defined enumeration types (The number of manually defined enumeration types / the number of types) | 0,00% | Success | 100 |
| M13: The rate of manually selected arguments and attributes (the number of manually selected arguments and attributes / the sum of arguments and attributes) | 4,94% | Success | 100 |
| M14: The rate of uninitialized attributes (The number of uninitialized attributes / the number of attributes) | 1,05% | Fail | 0 |
| M15: The rate of unit tests (The number of unit tests / (The total number of methods – the number of interface methods)) | 0,00% | Fail | 0 |
| M16: The rate of unit test warnings (The number of unit test warnings / the number of unit tests) | 100,00% | Fail | 0 |
| M17: The rate of unit test errors (The number of unit test errors / the number of unit tests) | 100,00% | Fail | 0 |
| | | Component score: | 67,65% |

### 3.5    The Gains Made by Component Quality Score Measurement

By the suggested method which is actively used during the software development process, firstly, it is aimed to standardize the component assessments. The developers are forced to calculate the component score for each version of the components released. Additionally, for each version of the components which will be used in released products, it is accepted not to have failed metrics. Therefore, until having every metrics in the accepted levels, it is expected to continue the components' development.

Additionally, it is ensured that the out-of-company developers (i.e. subcontractors) and the new team members will develop components convenient with the used standards (naming, MISRA, coding style) thanks to the suggested method. Hence, it is provided that the components are developed convenient with the used rules and standards, metrics are measured automatically, and the developers take feedback to see where they have made mistakes.

## 4    Assessment

The method explained in this study is created by a team developing safety-critical embedded software by weapon systems. The measured and calculated metrics are thought to be appropriate and sufficient for the software development field. Additionally, the success criteria of the calculated metrics are determined by the team objectively. It is expected to update the suggested assessment and grading methods when it is to be used in different software development fields.

## 5    Conclusion and the Future Plans

It is provided that, by the suggest method, the developed components' quality will at least improve among sequent releases. Since the team forces the components to be over a threshold quality score in order to be used in the released products, it is assured that the used component versions will at least have a minimum quality in terms of the suggested method.

Future plans include measurement of components' some other code based metrics (i.e. CK metrics) and to find out their relationship with the metrics measured and calculated in this method. Additionally, it is in our future plans to find the relationship of the components' post release errors and their calculated quality score.

## References

1. Louridas, P.: Static code analysis. Software, IEEE **23**(4), 58–61 (2006)
2. Rosenberg, L.H.: Applying and interpreting object oriented metrics. In: Software Technology Conference (1988)
3. Stamelos, I., Angelis, L., Oikonomou, A., Bleris, G.L.: Code quality analysis in open source software development. Information Systems Journal **12**(1), 43–60 (2002)
4. Briand, L.C., Wust, J., Daly, J.W., Porter, D.V.: Exploring the relationships between design measures and software quality in object-oriented systems. The Journal of Systems and Software **51**(3), 245–273 (2000)
5. El-Emam, K.: Object-oriented metrics: a review of theory and practice. In: Advances in Software Engineering, pp. 23–50. Springer-Verlag New York, Inc., New York (2002)
6. Ural, E., Tekin, U., Buzluca, F.: Nesneye Dayalı Yazılım Metrikleri ve Yazılım Kalitesi. Yazılım Kalitesi ve Yazılım Geliştirme Araçları Sempozyumu (2008)
7. Subramanyam, R., Krishnan, M.S.: Empirical analysis of CK metrics for object-oriented design complexity: implications for software defects. IEEE Transactions on Software Engineering **29**(4), 297–310 (2003)
8. MISRA-C++: 2008: Guidelines for the Use of the C++ Language in Critical Systems. MIRA Limited (2008)
9. Nagappan, N., Ball, T., Zeller, A.: Mining metrics to predict component failures. In: ICSE 2006 Proceedings of the 28th International Conference on Software Engineering, Shanghai, China, pp. 452–461 (2006)
10. Jiang, Y., Cukic, B., Menzies, T., Bartlow, N.: Comparing design and code metrics for software quality prediction. In: PROMISE 2008 Proceedings of the 4th International Workshop on Predictor Models in Software Engineering, Leipzig, Germany, pp. 11–18 (2008)

# Validation of the City Metaphor in Software Visualization

Gergő Balogh(✉)

Department of Software Engineering, University of Szeged, Szeged, Hungary
geryxyz@inf.u-szeged.hu

**Abstract.** The rapid developments in computer technology has made it possible to handle a large amount of data. New algorithms have been invented to process data and new ways have emerged to store their results.

However, the final recipients of these are still the users themselves, so we have to present the information in such a way that it can be easily understood. One of the many possibilities is to express that data in a graphical form. This conversion is called visualization. Various kinds of method exist, beginning with simple charts through compound curves and splines to complex three-dimensional scene rendering. However, they all have one point in common; namely, all of these methods use some underlying model, a language to express its content.

The improved performance of graphical units and processors have made it possible and the data-processing technologies have made it necessary to renew and to reinvent these visualization methods. In this study, we focus on the so-called city metaphor which represents information as buildings, districts, and streets.

Our main goal is to find a way to map the data to the entities in the fictional city. To allow the users to navigate freely in the artificial environment and to understand the meaning of the objects, we have to learn the difference between a realistic and an unrealistic city. To do this, we have to measure how similar it is to reality or the city-likeness of our virtual creations. Here, we present three computable metrics which express various features of a city. These metrics are compactness for measuring space consumption, connectivity for showing the low-level coherence among the buildings, and homogeneity for expressing the smoothness of the landscape. These metrics will be defined in a formal and an informal way and illustrated by examples. The connections among the high-level city-likeness and these low-level metrics will be analyzed. Our preliminary assumptions about these metrics will be compared to the opinions of users collected by an on-line survey. Lastly, we will summarize our results and propose a way to compute the city-likeness metric.

**Keywords:** Software visualization · City-metaphor · Validation · Metric

## 1 Introduction

In theory software systems could became infinitely complex by their nature. In theory, there is no limit of control flow embedding, or the number of methods,

© Springer International Publishing Switzerland 2015
O. Gervasi et al. (Eds.): ICCSA 2015, Part V, LNCS 9159, pp. 73–85, 2015.
DOI: 10.1007/978-3-319-21413-9_6

attributes, and other source code elements. In practice, these are limited by the computational power, time and storage capacity. To comprehend these systems, developers have to construct a detailed mental image. These images are gradually built during the implementation of the software system.

Often these mental images are realized as physical graphics with the aid of data visualization software. For example, different kinds of charts are used that emphasize the difference among various measurable quantities of the source code, or UML diagrams which are able to visualize complex relations and connections among various entities in the system.

There are several visualization techniques which use various real-life entities to represent abstract concepts. We used the city as a metaphor for the software system itself. To make it easier to navigate in a virtual environment and to help interpret the underlying connections and concepts, the generated world has to be similar to the real world. In our case it means we have to generate realistic cities to represent the abstract concepts and relations among the properties of the source code. To create such a city without manual intervention, we need a way to connect low-level properties of the city with its degree of realism.

## 1.1  Research Questions

Our brains are hard-wired to grasp the meaning of real objects. To make navigation easier in a virtual environment and to help to interpret the underlying connections and concepts for the user, the generated world has to be quite similar to the real world. In our case it means we have to generate realistic cities to represent the abstract concepts and relations among the properties of the source code. To create such a city without human intervention, we need a way to connect low-level properties of the city with its degree of realism. These problems can be summarized in the following research questions.

**RQ. 1** Is it possible to define a set of low-level, directly measurable metrics of the generated city that is able to describe its properties in a meaningful way?
  – We will define a set of proposed metrics that fulfill these requirements.
  – To ensure the validity of these constructs, we will conduct a user survey.
**RQ. 2** Is there a way to determine the degree of realism of the generated city based on low-level metrics?
  – We will look for a way to combine the low-level metrics in order to create a high-level metric that is able to express the similarity between a real city and a generated city.

## 2  Background

### 2.1  Related Works

Everyone is different it seems, each person having his or her own points of views. We use various tools to comprehend the world. Some of us need numbers, others use abstract formulas; but most of us need to see the information as colors, shapes or figures. To fulfill the expectations of users, many data visualization techniques and tools have been designed and implemented. It is beyond the scope of this article to exhaustedly evaluate these techniques and tools, but in our opinion traditional visualization tools like Rigi [6], sv3D [3] and SHriMP Views [4] were built on innovative ideas. It is often difficult to interact with them, and they usually "fall behind" in graphics terms compared to those in today's computer games, for instance. To address these issues we implemented a tool which combines today's common interactive, three dimensional game interface with an elaborate static code analysis.

Closely related approaches to our tool are CodeCity [5] and EvoSpace [2], which use the analogy of skyscrapers in a city. CodeCity simplifies the design of the buildings to a box with height, width, and color. The quantitative properties of the source code – called metrics – are represented by these attributes. In particular, each building represents a class where the height indicates the number of methods, the width represents the number of attributes, and the color tells us the type of the class. The buildings are grouped into districts as classes are tied together into namespaces. The diagram itself resembles a 3D bar chart with grouping. EvoSpace uses this analogy in a more sophisticated way. The buildings have two states: closed – when the user can see the large-scale properties like width and height, and open – when we are able to examine the small-scale structure of the classes, see the developers and their connections. It also provides visual entity tagging and quick navigation via the connections and there is also a small overview map.

## 2.2 Code Visualization in CodeMetropolis

Our tool called CodeMetropolis is a collection of command line tools. It takes the output graph of SourceMeter [1] and creates a Minecraft world from it. SourceMeter are a collection of various programs that are able to analyze and measure static artifacts related to the source code. The output is given with a unique binary format. The world uses the metropolis metaphor, which means that the source code metrics are represented by the various properties of the different building types. Figure 1 shows an example world.

In our representation there are two levels. The data level contains the various objects and their data, which are directly related to the measured artifacts, such as classes. However, the metaphor level is built up from the visual representations of these like buildings and floors. At the data level, each entity has its own property set – for example metrics, which are displayed at metaphor level. The buildings in our metropolis are parts of this metaphor, and they have some attributes which control the visual appearance. The items which are highlighted in Figure 2 represent various source code entities, while the properties are mapped to the attributes in order to visualize the data. In this concrete case, we can see some namespaces visualized as stone plates. They contain two classes represented by buildings. Lastly these have several methods whose size and complexity are mapped to the width and height of the floors.

**Fig. 1.** An typical code metropolis

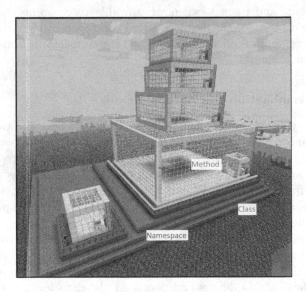

**Fig. 2.** Items of a city

## 3    Low-Level Metrics of Virtual Cities

In order to define low-level metrics we have to specify the exact model of a generated city. In this study we mainly focus on cities which only contain buildings like skyscrapers. These buildings could be represented by their bounding box, which is a box with the same width, length and height as the maximal width, length and height of the building itself. The current model does not represent the inner structure of the buildings. The buildings are grouped together into various types, which could be districts at the metaphor level or namespaces and packages at the data level. Buildings also have a position on the plain. Based on the above statements, the current model defines a building as a box with the following properties:

**Width** the maximal size of the building along the x-axis.
**Length** the maximal size of the building along the y-axis.
**Height** the maximal size of the building along the z-axis.
**Position** an ordered pair of numbers that represents the location of the pivot point of a building on the plain.
**Type** the unique identifier of the set that the building belongs to.

A city or metropolis is a set of the type of buildings defined here. The buildings cannot be rotated or have any intersecting region.

Three low-level metrics were constructed during the study. Our main design goal was to provide scalable, parameterizable and normalized values which can describe meaningful properties of a city. This means that these metrics have to be independent of the following:

– the area of the city
– the height of the city, i.e. the height of the tallest building
– the number of buildings in the city
– the size of the buildings

With the help of these metrics we can compare the generated metropolises against each other in a formal and automatic way. For example, a huge city like Las Vegas could be compared with a smaller city like Cambridge. Parametriz-ability is the ability to change the distribution of the metrics, which allows us to fine-tune the values between the lowest and highest possible values, namely zero and one. With these in mind the following metrics were constructed:

*Compactness.* This expresses the density of the buildings in the city. It is the ratio of the total area of the buildings over the convex hull of the city (Figure 3a).

*Homogeneity.* This expresses the smoothness of the small scale landscape. It is the distance between any two buildings weighted by the difference in their height. A gamma-correction is used in both parts (Figure 3b).

*Connectivity.* This describes the coherence among buildings. It is the distance between any two buildings guarded by the their type. A gamma-correction is used in the last parts (Figure 3c).

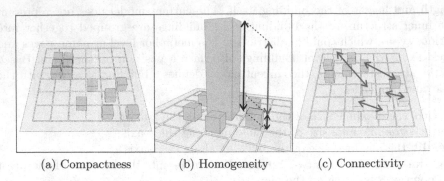

(a) Compactness          (b) Homogeneity          (c) Connectivity

**Fig. 3.** Low level metrics

## 3.1  Formal Definition

Let us define the buildings as a tuple with 6 items and the collection of these as an unordered set.

$$\mathbb{B} = \{\text{buildings}\} \tag{1}$$

$$b \in \mathbb{B} \tag{2}$$

$$b = \begin{pmatrix} x_b & y_b & z_b \\ |x|_b & |y|_b & |z|_b \end{pmatrix}, \tag{3}$$

where

$$(x_b, y_b, z_b) \in \mathbb{N}^3 \text{ is a predefined pivot point of the building} \tag{4}$$

$$|x|_b, |y|_b, |z|_b \in \mathbb{N} \text{ is the width, the length and the height of the building} \tag{5}$$

$$|\hat{z}|_b = \frac{|z|_b}{\max\limits_{d \in \mathbb{B}} |z|_d} \text{ is the normalized height of the building} \tag{6}$$

We will define the distance between any two buildings as the Euclidean distance between their pivot points.

$$b, d \in \mathbb{B} \tag{7}$$

$$\|b; d\| = \|(x_b; y_b); (x_d; y_d)\| \in \mathbb{R} \tag{8}$$

Furthermore, the following sets will be the corner points of a building:

$$b \in \mathbb{B} \tag{9}$$

$$P_b \subset \mathbb{N}^3 \text{ is the corner points of building } b \tag{10}$$

$$\underline{P}_b \subset P_b \text{ is the lower corner points of building } b \tag{11}$$

$$\overline{P}_b \subset P_b \text{ is the upper corner points of building } b \tag{12}$$

We will also use the convex hull of a building and a set of points:

$$R \subset \mathbb{N}^3 \tag{13}$$

$$\text{Conv } R \text{ is the convex hull of point set } R \tag{14}$$

$$D \subseteq \mathbb{B} \tag{15}$$

$$\text{Conv } D = \text{Conv} \left( \bigcup_{d \in D} \underline{P}_d \right) \tag{16}$$

The following notation will be used to denote some basic properties of the buildings.

$$b \in \mathbb{B} \tag{17}$$

$$R \subset \mathbb{N}^3 \tag{18}$$

$$A_b = |x|_b \cdot |y|_b \text{ is the area of the building} \tag{19}$$

$$A_R \in \mathbb{N} \text{ is the area of the convex hull of the point set } R \tag{20}$$

$$P_R \in \mathbb{N} \text{ is the perimeter of the convex hull of the point set } R \tag{21}$$

$$D_R = \max_{a,b \in R} \|a; b\| \tag{22}$$

We define a classification over the set of the buildings, this is the type of the building. The type of a building is given with the following relation. It is equal to 1 if and only if two buildings are of the same type:

$$b, d \in \mathbb{B} \tag{23}$$

$$t(b), t(d) \in \mathbb{N} \text{ the type of the building} \tag{24}$$

$$\delta(b; d) \in \{0; 1\} \tag{25}$$

$$\delta(b; d) = \delta(d; b) \tag{26}$$

$$\delta(b; d) = \begin{cases} 1 & \text{if } t(b) = t(d) \\ 0 & \text{if } else \end{cases} \tag{27}$$

**Compactness.** With these notations, the previously introduced compactness metric could be defined as the ratio of the area of the convex hull of a set of buildings (i.e. the convex hull of the set of points of the buildings) over the total area of these buildings. Because our model does not allow any intersecting buildings, the lower limit will be 0 and the upper limit will be 1.

$$C \subseteq \mathbb{B} \tag{28}$$

$$\text{Comp } C = \frac{\sum_{d \in C} A_d}{A_{\text{Conv } C}} \tag{29}$$

**Connectivity.** The third metric called connectivity is defined as the sum of normalized distances between each pair of buildings. We introduced a connection guard and a final normalization part. With this formula we only charge a fee for the buildings with the same type. The size of the fee is greater if the buildings are farther away from each other. This metric was used during a competition to rank the solutions, so gamma correction was applied to the distance part, to be able to fine tune the order.

$$D \subseteq \mathbb{B} \tag{30}$$

$$\operatorname*{Conn}_{\gamma} D = \frac{1}{\binom{|D|}{2}} \sum_{\substack{d,b \in D \\ d \neq b}} \left( \underbrace{\delta(d;b)}_{\text{connection guard}} \underbrace{\left( \frac{\|d;b\|}{\max_{e,f \in D} \|e;f\|} \right)^{\gamma}}_{\text{distance part}} \right) \tag{31}$$

**Homogeneity.** Next, homogeneity is specified as the arithmetic mean of buildings weighted with the difference of their normalized height. As in the case of connectivity, an overall normalization is added to ensure that the values are bounded. This metric was used during a competition to rank the solutions, so gamma correction was applied to the height difference and to the distance part as well, to be able to fine tune the order.

$$D \subseteq \mathbb{B} \tag{32}$$

$$\operatorname*{Hom}_{\gamma,\zeta} D = \frac{1}{\binom{|D|}{2}} \sum_{\substack{d,b \in D \\ d \neq b}} \left( \underbrace{\left| |\hat{z}|_d - |\hat{z}|_b \right|^{\zeta}}_{\text{height difference part}} \underbrace{\left( \frac{\|d;b\|}{\max_{e,f \in D} \|e;f\|} \right)^{\gamma}}_{\text{distance part}} \right) \tag{33}$$

## 4  Validation by a User Survey

To validate the previously defined low-level metrics and to create a new high-level metric which is able to express the similarity between a generated city and a real one, a user survey was used. This contained several questions with a predefined list of choices. We asked the users to rank the cities according to their degree of realism. Furthermore, they had to decide which one of the two given cities could be used as an example from a specific point of view. For example, they had to choose the more compact or round one. These examples represented the upper limits of the low-level metrics. To reduce the unintentional influence of the researchers, we used the data of the SEDCup contest, held in the spring of 2013. To win, the teams had to design and implement an algorithm to create an artificial layout of houses with various properties. Only the generated cities were used as trial data in the survey without any preselection. The target audience contained mainly students and coworkers from the IT field. Altogether, 51 complete and 20 partial surveys were processed.

## 4.1 Validation of Construct of Low-Level Metrics

To address the first research question, the answers from the users were compared with the values of low-level metrics. Because their distributions are unknown, only their directions were considered. It meant that we counted how many times they agreed with the user opinion in the sense that one city had a bigger value than the other. During the evaluation we defined three ranges. If the users gave a positive answer of 40% or less we said that the users disagreed with the metric. If it was more than 40% but less than 60%, it meant that the results were not significant. Lastly, if it was 60% or more it meant that the users agreed with the metric. To avoid any further bias, the users did not know the exact values of the metrics when we did the survey. We summarize our findings on the low-level metrics in Table 1.

**Table 1.** Agreement among the users and metrics we applied

| metric | compactness | connectivity | homogeneity |
|---|---|---|---|
| agreement level | 58.62% | 78.57% | 74.07% |
| conclusion | not significant | agree | agree |

Based on these results, we can answer the first research question presented earlier. It is possible to define low-level metrics for the generated cities that accord with everyday notions, hence appear meaningful. To demonstrate this, we proposed a set of these metrics from which two provide a positive answer. We assume that the meaning of compactness could be better defined so as to improve the level of agreement.

## 5 Construction of a High-Level Metric

To answer the second research question in the study, a high-level metric was defined. It is able to describe the similarity between a real and a generated city. This metric is the weighted sum of the above-defined compactness, connectivity and homogeneity metrics. We used the answers for the ranking questions of the survey mentioned above. Users had to rank several predefined cities, from the most realistic to the least realistic. The rankings were compared and processed by various methods and algorithms to determine the weights of the low-level metrics.

### 5.1 Methodology

We used the following methodology to determine the weights for the novel high-level metric called DoR. First of all, the rankings made by users were compared with each other and their correlation was encoded in a graph. The Kendall tau correlation was used to measure the correlation between each pair of rankings. Any pair of observations are said to be concordant if the ranks for both elements agree. With this definition, the degree of correlation is the ratio of the difference between the number of concordant and discordant pairs over the total. That is,

$$\tau = \frac{\text{(number of concordant pairs)} - \text{(number of discordant pairs)}}{\frac{1}{2}n(n-1)} \tag{34}$$

The rankings were used as the nodes. The edges connected any two nodes if and only if the Kendall tau correlation coefficient was higher than 0.8.

Next, a community detection algorithm was applied to find the biggest community in the graph. This represents the majority opinion of the users. We used the ranking with the highest number of edges of this community, hence it mirrors the opinions of most of the users. A graph of this is shown in Figure 4. Here, we chose node 48 in cluster 9.

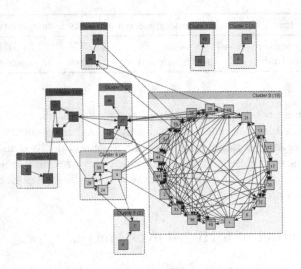

**Fig. 4.** Correlation graph of the ranking of generated cities

Based of this ranking and the values of the trial data, an inequality system was constructed. It has three free variables that correspond to the weights of the three low-level metrics. Any solution that satisfies these inequalities represents a weight distribution that respects the original ranking of the cities. To find a solution, we define a linear programming problem with a constant target function, hence the simplex method will stop at the first solution. The inequality system does not have any feasible solution, so we had to relax it by removing one of the constraints. This new linear programming problem yielded a solution that respects the original ranking with the exception of one switch. We could not find any better solution. Lastly, we normalized the weights.

## 5.2   Formal Definition

The trial data used was the city data given in Table 2. We listed the value of the low-level metrics for each with a unique identifier called name. The table also shows the order of cities according to our ranking.

**Table 2.** Trial data of generated cities along with the values of the low-level metrics

| name | compactness | connectivity | homogeneity |
|------|-------------|--------------|-------------|
| Epsilon | 0.460 | 0.969 | 0.871 |
| Eta | 0.123 | 0.819 | 0.733 |
| Gamma | 0.975 | 0.611 | 0.837 |
| Delta | 0.582 | 0.942 | 0.859 |
| Beta | 0.370 | 0.984 | 0.873 |
| Alpha | 0.799 | 0.987 | 0.889 |
| Zeta | 0.359 | 0.999 | 0.916 |

This data can be reexpressed as an incquality system, as we did below in matrix form. Using a trivial transformation, it can presented as a linear programing problem. To do this, we have to define the following matrices:

$$L = \begin{pmatrix} 0.46 & 0.969 & 0.871 \\ 0.123 & 0.819 & 0.733 \\ 0.975 & 0.611 & 0.837 \\ 0.582 & 0.942 & 0.859 \\ 0.37 & 0.984 & 0.873 \\ 0.799 & 0.987 & 0.889 \end{pmatrix} \tag{35}$$

$$R = \begin{pmatrix} 0.123 & 0.819 & 0.733 \\ 0.975 & 0.611 & 0.837 \\ 0.582 & 0.942 & 0.859 \\ 0.370 & 0.984 & 0.873 \\ 0.799 & 0.987 & 0.889 \\ 0.359 & 0.999 & 0.916 \end{pmatrix} \tag{36}$$

$$w = \begin{pmatrix} w_{comp} \\ w_{ecc} \\ w_{conn} \\ w_{homo} \end{pmatrix} \tag{37}$$

$$Lw < Rw \tag{38}$$

$$(L - R)w < 0 \tag{39}$$

$$A = (L - R) = \begin{pmatrix} 0.337 & 0.15 & 0.138 \\ -0.852 & 0.208 & -0.104 \\ 0.393 & -0.331 & -0.022 \\ 0.212 & -0.042 & -0.014 \\ -0.429 & -0.003 & -0.016 \\ 0.44 & -0.012 & -0.027 \end{pmatrix} \tag{40}$$

Because this problem does not have any feasible solution, we will relax it by removing the first inequality from the system. After solving it, a relaxed version with a simplified simplex algorithm, the calculated weights yield a set with an order shown in Table 3.

**Table 3.** Order of the cities according to the DoR metric

| name | the DoR metric |
| --- | --- |
| → Epsilon | 0.6094 |
| Eta | 0.5137 |
| Gamma | 0.5909 |
| Delta | 0.6009 |
| Beta | 0.6109 |
| Alfa | 0.6209 |
| Zeta | 0.6418 |

As a final step, we could define our new DoR metric using the following weights:

$$\mathrm{DoR}_{\zeta,\gamma_1,\gamma_2} B = -0.27 \, \mathrm{Comp}\, B + -0.73 \, \underset{\gamma_1}{\mathrm{Conn}}\, B + 4.43 \, \underset{\gamma_2,\zeta}{\mathrm{Hom}}\, B \tag{41}$$

As seen above, we are able to construct a sub-optimal, high-level metric; hence we can give a positive answer to our second research question. In the future, other methods and definition could be investigated and compared with the DoR metric.

## 6   Conclusions

We live in the age of information explosion where to grasp large amounts of data as quickly as possible is a basic requirement. One of the many possibilities is to convert the data into some clear graphical form, such as data that represents elements of a virtual city. In this study, we presented three computable metrics which express various features of such a city. These are compactness for measuring space consumption, connectivity for showing the low-level coherence among the buildings, and homogeneity for expressing the smoothness of the landscape. These metrics were defined in both a formal and informal way. We also constructed a high-level metric called DoR that is able to express the similarity between a generated metropolis and a real one. Both high- and low-level metrics were validated by a user survey. The opinions obtained in the survey were much as we had anticipated. The results show that it is possible to construct methods (using the DoR metric) which are able to estimate the degree of realism of a generated city. This method embodied as a software-systems could provide a full- or semi-automatic way for creating a life-like virtual environment within a reasonable time. In such a world we could use our everyday senses to perceive the data represented in a clear graphical way.

**Acknowledgments.** I would like to thank to András London, Norbert Hantos, Tamás Grósz for their help and suggestions during this study and for the evaluation of the results.

# References

1. FrontEndART Ltd.: SourceMeter homepage (2014)
2. Lalanne, D., Kohlas, J.: Human Machine Interaction: Research Results of the MMI Program
3. Marcus, A., Comorski, D., Sergeyev, A.: Supporting the evolution of a software visualization tool through usability studies. In: Proceedings of the 13th International Workshop on Program Comprehension, pp. 307–316, May 2005. http://ieeexplore.ieee.org/xpls/abs_all.jsp?arnumber=1421046 http://dl.acm.org/citation.cfm?id=1058432.1059368
4. Storey, M.A., Best, C., Michaud, J.: SHriMP views: an interactive environment for information visualization and navigation. In: CHI 2002 Extended Abstracts on Human Factors in Computing Systems - CHI 2002, p. 520, April 2002. http://dl.acm.org/citation.cfm?id=506459 http://dl.acm.org/citation.cfm?id=506443.506459
5. Wettel, R., Lanza, M.: CodeCity. In: Companion of the 13th International Conference on Software Engineering - ICSE Companion 2008, p. 921. ACM Press, New York, May 2008. http://scholar.godogle.com/scholar?hl=en&btnG=Search&q=intitle:codecity#1    http://scholar.google.com/scholar?hl=cn&btnG=Search&q=intitle:CodeCity#1 http://dl.acm.org/citation.cfm?id=1370175.1370188
6. Wong, K.: Rigi user's manual. Department of Computer Science, University of Victoria (1998).http://www.rigi.cs.uvic.ca/downloads/pdf/rigi-5_4_4-manual.pdf

# A Systematic Mapping on Agile UCD Across the Major Agile and HCI Conferences

Tiago Silva da Silva[1], Fábio Fagundes Silveira[1(✉)], Milene Selbach Silveira[2], Theodore Hellmann[3], and Frank Maurer[3]

[1] Universidade Federal de São Paulo – UNIFESP, São José dos Campos, Brazil
{silvadasilva,fsilveira}@unifesp.br
[2] Pontifícia Universidade Católica do Rio Grande do Sul, Porto Alegre, Brazil
milene.silveira@pucrs.br
[3] University of Calgary, Calgary, Canada
{tdhellma,frank.maurer}@ucalgary.ca

**Abstract.** Agile User-Centered Design is an emergent and extremely important theme, but what does it exactly mean? Agile User-Centered Design is the use of user-centered design (UCD) in Agile environments. We build on previous work to provide a systematic mapping of Agile UCD publications at the two major agile and human-computer interaction (HCI) conferences. The analysis presented in this paper allows us to answer primary research questions such as: what is agile UCD; what types of HCI techniques have been used to integrate agile and UCD; what types of studies on agile UCD have been published; what types of research methods have been used in Agile UCD studies; and what benefits do these publications offer? Furthermore, we explore topics such as: who are the major authors in this field; and is the field driven by academics, practitioners, or collaborations? This paper presents our analysis of these topics in order to better structure future work in the field of Agile UCD and to provide a better understanding of what this field actually entails.

**Keywords:** Agile software development · User-centered design · Usability · Systematic mapping · Empirical

## 1 Introduction

The combination of User-Centered Design (UCD) and agile development methodologies is a topic of increasing importance within both communities, but it is not entirely clear what this approach entails. In order to better understand Agile UCD – and provide an overview of the field – we decided to perform a systematic mapping study of Agile UCD in major agile and Human-Computer Interaction (HCI) conferences. Systematic mapping studies take a large number of papers as input and categorize them based on their titles and abstracts. The result of such a study is a framework for understanding what topics a given field encompasses. Systematic mapping studies take a large number of papers as input and categorize them based on their titles and abstracts [1]. The result of such a study is a framework

O. Gervasi et al. (Eds.): ICCSA 2015, Part V, LNCS 9159, pp. 86–100, 2015.
DOI: 10.1007/978-3-319-21413-9_7

for understanding a field of research at a high level – in other words to find out what topics a given field encompasses.

We restricted our initial search for papers to major agile conferences (Agile, XP, and the defunct XP/Agile Universe) – and HCI conferences – CHI (Conference on Human Factors in Computing Systems) and UIST (User Interface Software and Technology Symposium).

Our goal with this paper is to present a summary of what the field of Agile UCD is about. We accomplish this by asking research questions including: what has Agile UCD been used for; what types of research have been published; what types of HCI techniques have been used to integrate agile and UCD; what types of papers on agile UCD have been published; what types of research methods have been used in agile UCD studies; what benefits do publications propose their research will produce? As a secondary analysis, we also answer several additional questions: which authors lead the field; which countries produce the most research on this topic; and do academics, practitioners, or collaborations between the two groups contribute more publications to the field?

The remainder of this paper is organized as follows: Section 2 brings up a background of the Agile UCD field and explains the research method adopted for this work; Section 3 presents the primary results; Section 4 presents the answers for the secondary research questions; and Section 5 describes limitations, conclusions, and future work.

## 2    Background and Research Method

The main goal of a systematic mapping study is to provide an overview of a research field, and identify the quantity and type of research and results available within it. Often one wants to map the frequencies of publication over time to see trends and a secondary goal can be to identify the forums in which research in the area has been published [2].

This study is a first step for an update of the systematic literature review presented by Silva da Silva et al. [3] in 2011. In that paper, a total of 58 studies were retrieved. From these 58, 28 studies were published in conferences on agile methodologies – XP, Agile, and XP/Agile Universe. Since the 2011 publication, basic approaches to integrating agile and UCD have become robust enough for researchers and practitioners. This is different from 2011 when the authors used to say that integrating agile and UCD was important but not well addressed.

We strictly follow the process demonstrated by Hellmann et al. [1] which was in turn based on the guidelines provided by Petersen et al. [2]. This section provides details on how we conducted each step of the present systematic mapping. At least two of the authors participated in each step of the analysis of publications in order to allow us to immediately resolve disagreements as to how publications should be keyworded through discussion.

**Research Questions.** According to Petersen et al. [2], specific research questions are important for constraining the scope of a systematic mapping study.

For this study, we wanted to investigate what agile UCD means with respect to research papers published at two major agile software development and human-computer interaction conferences. We defined five primary research questions as follows:

- PRQ1. What is Agile UCD?
- PRQ2. What types of HCI techniques have been used to integrate agile and UCD?
- PRQ3. What types of papers on Agile UCD have been published?
- PRQ4. What types of research methods have been used in Agile UCD studies?
- PRQ5. What benefits do publications propose their research will produce?

We also investigate two secondary research questions. These questions are not less important than our primary research questions, but the analysis required to answer them is more superficial. The questions are:

- SRQ1. Which authors lead the field of Agile UCD?
- SRQ2. Is the field led by practitioners, academics, or collaborations between them?

**Conduct Search.** We decided to conduct the search in the last ten years by manually searching through the main Agile conferences – XP/Agile Universe, XP, and Agile – and through the main[1] HCI conferences – CHI and UIST – proceedings to identify papers on the topic of agile and user-centered design. The total number of papers in the main Agile conferences between the years 2003 and 2014 was 1108. The total number of papers in the main HCI conferences between the years 2003 and 2014 was 3723. The breakdown of these papers by conference can be seen in Fig. 1.

**Screening Papers.** After collecting this initial set of 4831 papers, the next step was to screen out those publications not relevant to our research questions. Different from Hellmann et al. [1], which did it in two steps (first eliminating papers on the basis of their titles and then based on their abstracts), we performed both in just one step.

For inclusion, papers titles and abstracts needed to make reference to UCD or to an implicitly UCD-related process, e.g., User eXperience (UX), Interaction Design (IxD), business analysis or user requirements. After this step, our paperset included 55 papers.

In a second step, papers were eliminated on the basis of their type. We excluded workshop proposals, tutorials, demos, and other non-research papers due to our interest in research. During this stage, a further nine papers were eliminated, leaving us with a final set of 46 papers, which can be seen in Fig. 1. These papers are listed at the following link:
http://www.silvadasilva.com/home/systematicmapping-iccsa15.

---

[1] According to Google Scholar Top publications in Human Computer Interaction.

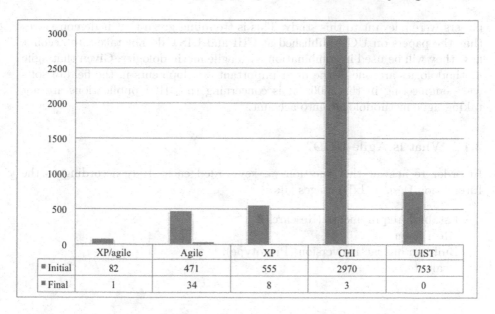

|         | XP/agile | Agile | XP  | CHI  | UIST |
|---------|----------|-------|-----|------|------|
| ■ Initial | 82       | 471   | 555 | 2970 | 753  |
| ■ Final   | 1        | 34    | 8   | 3    | 0    |

**Fig. 1.** Initial and final paperset by conference

**Keywording Using Abstracts.** After we finished the papers screening, we keyworded each paper based on its title and abstract in order to develop a framework for understanding the field of Agile UCD. This keywording was done using open and closed coding. In open coding, the keywords used to describe a paper are drawn out of the source material itself. This approach was used for PRQs 2 and 5. In closed coding, the list of keywords is pre-defined and keywords are simply picked from the list to apply to papers. This approach was used for PRQ 1, 3 and 4.

**Data Extraction and Mapping.** Data was extracted by the first author of this paper and then confirmed by the second author. We used Papers[2] to manage citations and keep track of the keywords applied to our paperset. The analysis and visualization preparation was performed in Microsoft Excel[3].

## 3   Primary Results

One of the most striking results of this research can be seen clearly in Fig. 1: the main HCI conferences, CHI and UIST, have hardly published any research on Agile UCD. Given that 3723 papers were published in CHI and UIST during the time period investigated, and given that CHI and UIST do contain a large number of publications on UCD, it is very surprising that only 3 of these

---

[2] http://www.papersapp.com
[3] http://products.office.com/en-us/excel

papers were relevant to this study. This is troubling given that it demonstrates that the papers on UCD published at CHI and UIST do not take into account how they will be used in combination with agile methodologies. Given that agile methodologies are one of the most important developments in the field of software engineering in the 2000s, it is concerning that HCI publications are not taking agile methodologies into account.

### 3.1   What Is Agile UCD?

In order to answer this question we keyworded each study according to the Interaction Design (IxD) stages [4]:

– Establish requirements (Research).
– (Re)Design.
– Build an interactive version (Prototype).
– Evaluate.

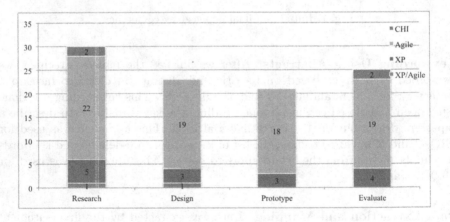

**Fig. 2.** Breakdown of IxD stages by conference

As we may observe in Fig. 2 and Fig. 3, there is a balance among the IxD stages. However, the main concern is with the Research stage.

As previously identified in Silva da Silva et al. [3], this is the main concern both for designers and agile practitioners. This topic is extremely important for the integration of agile and UCD. When speaking to cultural differences between designers and agile practitioners, it is widely known that Big Design – in this case, Design also encompasses Research – Up Front goes against the Agile principles.

The less focused stage is the Prototype one. Associated to our previous experience, this led us to conclude that this is the minor concern for agile practitioners. They usually do not care about how designers work in a day-to-day basis. For instance, no paper addressed how is a day-to-day of a UX designer. The

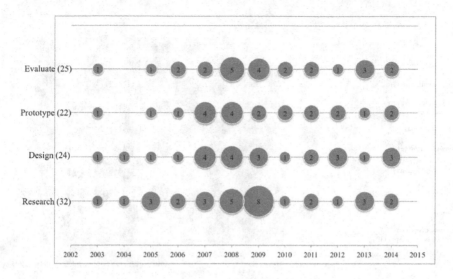

**Fig. 3.** Breakdown of IxD stages by year.

main issue is to find the right timing to research, design and evaluate the user experience of software products within agile iterations.

A breakdown of these results by conference can be seen in Fig. 2, while a breakdown of these topics by year can be seen in Fig. 3.

On the one hand, Fig. 3 shows us that the apex of agile UCD research was in 2009, and the main stage was Research. On the other hand, this figure also shows us that interest in all topics has dropped sharply after 2009, presenting a fair balance from 2011 to 2014.

## 3.2 What Types of HCI Techniques Have been Used in the Integration of Agile and UCD?

During the course of keywording our set of papers, we identified 16 different keywords related to the HCI techniques used to integrate agile and UCD. Due to space limitations, we limit the present discussion to the top 10 most frequently-occurring keywords.

The main HCI technique used for the integration of agile and UCD is usability testing on lightweight prototypes, as depicted in Fig. 4. By lightweight prototypes we mean low fidelity prototypes – either paper or digital. The following technique is Continuous Research, which means splitting the activities of research throughout the development process. Following that, we have Evolutionary Prototyping, Upfront Design, and Continuous Design that are all interconnected. Note that all the HCI techniques mentioned are adapted to or merged with an agile context. There are no examples of unmodified HCI techniques – like Heuristic Evaluation or Formal Usability Evaluations – in the papers we analyzed.

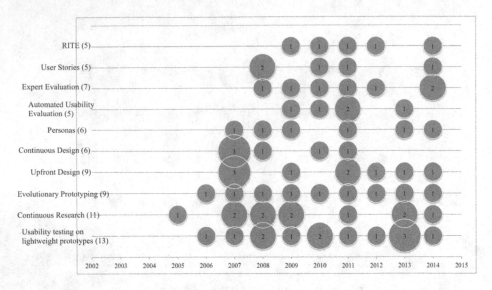

**Fig. 4.** Breakdown of HCI techniques by year. Due to space limitations, we limit the present discussion to the top 10 most frequently-occurring.

### 3.3  What Types of Papers on Agile UCD Have been Published?

In order to answer this question, we adopted the five types of papers identified by Wieringa et al. [5], they are: Solution, Validation, Philosophical, Opinion, Experience, and Evaluation.

We applied these paper types as keywords to each publication in our paperset in order to find out what kinds of publication were most common in this field. The results are broken down by year in Fig. 5.

Fig. 5 shows that Experience reports have traditionally been common in research publications since 2005 till 2014. At least one Solution paper per year has been published since 2005, except in 2012. Most of them propose a solution based on an experience report.

It is not clear why some Philosophical papers have been published just recently in 2011, 2012 and 2014, since a philosophical paper aims at providing a framework for understanding a field. This might has happen because nowadays the need of integrating agile and UCD is more than clear and based on a lot of experience reports, authors can draw conclusions and provide an understanding of the entire field.

### 3.4  What Types of Research Methods Have been Used in Agile UCD Studies?

Easterbrook et al. [6] describe a number of empirical methods available in the context of Software Engineering. The authors examine the goals of each empirical method and analyze the types of questions each best addresses. They culminate

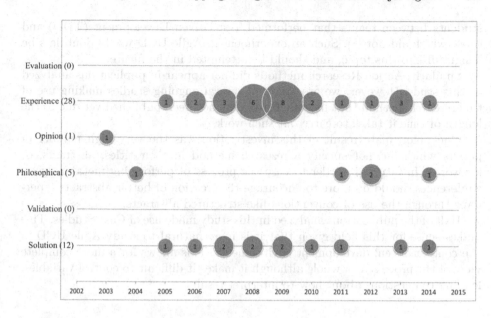

**Fig. 5.** Breakdown of research type by year

in five methods. Although Easterbrook et al. [6] did not identify Grounded Theory (GT) as one of the five classes of research method, they describe GT as a technique for developing theory iteratively from qualitative data. Therefore, we added GT as the sixth class of research method, as follows: Controlled Experiments (including Quasi-Experiments), Case Studies, Survey Research, Ethnographies, Action Research, and Grounded Theory.

Most of the studies cannot be considered a Case Study because they do not present Research Questions, which according to Easterbrook et al. [6] is precondition for conducting a case study. However, whenever the authors identify a paper that describes itself as a case study, we trust on the author's judgment to classify the paper.

The vast majority of the papers have no explicit research methods available in the abstracts, as depicted in Fig. 6. Thus, the most cited research method is Case Study, followed by Grounded Theory studies and Ethnographies and then Survey Research respectively.

No Controlled Experiments were reported in any publication included in this study. This might make sense given that the process of Agile UCD is complex and it would be difficult to design a rigorous, realistic experiment to conduct on this topic. However, as has been demonstrated in other subfields of software engineering, it is possible to conduct Controlled Experiments on small parts of a process. For example, the meta-analysis contained in Jeffries [7] lists a number of controlled and quasi-controlled experiments that were uncovered during their investigation of the process of test-driven development – including several experiments lasting weeks or months – that worked by, for example, dividing students between teams that performed Test-Driven Development (TDD) and those which did not [8]. Such an experiment in Agile UCD would doubtless be insightful into this topic, and should be attempted in the future.

Similarly, Action Research methods did not appear in publications analyzed in this study. However, we are aware of several ongoing studies making use of this research method that have unfortunately not been published yet due to the length of time it takes to carry out such work.

One unfortunate result of this investigation was the very high number of papers which did not specify a research method in their titles, abstracts, or keywords. In future, in order to ease the process of performing meta-analyses, conferences should do more to encourage the creation of better abstracts – perhaps through the use of conventions like structured abstracts.

Half of the publications analyzed in this study made use of Case Studies. This makes sense for this field given that it is more natural to study Agile UCD in the context of real development environments. This allows for a more complete view of the process as a whole although it makes it difficult to control variables, leading to questions about how to interpret results.

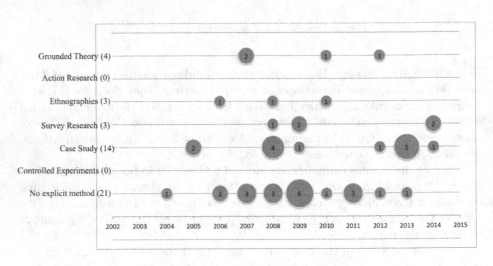

**Fig. 6.** Breakdown of research method by year

## 3.5    What Benefits do Publications Propose Their Research will Produce?

In order to answer this question, we open coded the abstracts and found the codes depicted in Fig. 8.

Improve Communication between Designers and developers or even with other members of the teams, with higher management levels is the main benefit cited.

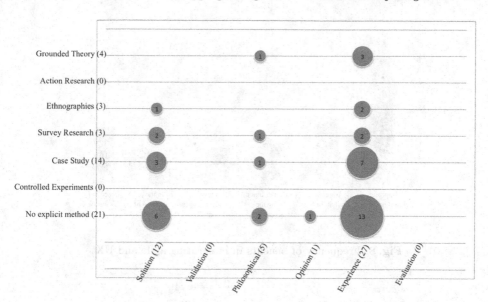

**Fig. 7.** Breakdown of research method by research type

Improve visibility, also seen as improving the big picture of the project is the second most cited benefit, followed by Improved Usability.

With regards to Improved Usability, since most of the papers are experience reports, we never have a baseline to check if the usability has improved. Papers never tell whether and how the usability of the product has improved (or not). The studies are carried out in real projects or products, therefore, we never have a version A of the product – developed without agile UCD – and a version B with an integrated process.

The Business Analysis improvement is a result of Business Analysts pairing with Interaction Designers. Designers have a special background that helps a lot in the business analysis. Moreover, designers improve their skills on business analysis and analysts improve their knowledge on UX, it is a two-way gain.

## 4   Secondary Research Questions

### 4.1   Which Authors Lead the Field of Agile UCD?

Fig. 9 shows us that Frank Maurer, with seven papers, is the leading author in the field, followed by Helen Sharp with five papers, Jennifer Ferreira and Robert Biddle with four papers published each.

### 4.2   Is the Field Led by Practitioners, Academics, or Collaborations?

We coded the author institutions of papers as one of: industry, academic, or both. This was done based on author affiliations at time of publication. The

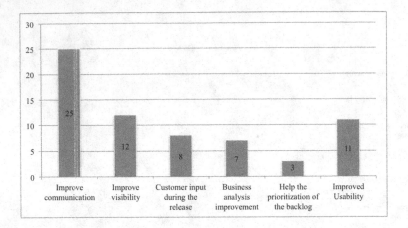

**Fig. 8.** Frequency of benefits in integrating agile and UX

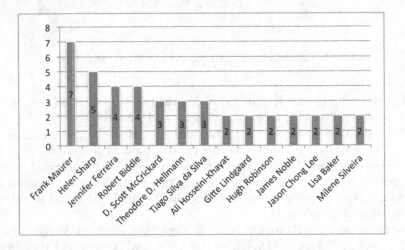

**Fig. 9.** Frequency of papers published by author

"both" keyword was used for papers with at least one author from each type of institution.

We feel that this is an important distinction as collaborations have the potential to combine the rigour of academic evaluations with the practicality of industry concerns. The distribution of papers across these three keywords can be seen in Fig. 10.

Fig. 10 provides an interesting look at the composition of each conference. An Industry author published the first paper on agile UCD at XP/Agile Universe conference. The XP conference looks the more academic one, having five publications of academic groups, though the Agile conference has more papers overall. Agile conference has a mix of both groups and also has the only three collaborations papers.

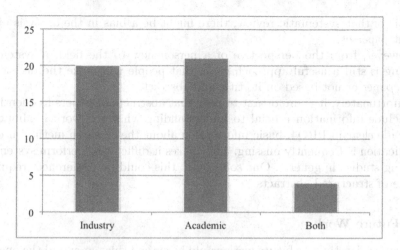

**Fig. 10.** Distribution of papers among Industry, Academic, and Collaborative sources.

# 5 Limitations, Future Work and Conclusion

## 5.1 Limitations

Out of five authors, only one participated on every step of this study. The paper search and selection process was done entirely by the first author, with a confirmation of these results by the third author, leading to a possibility of researcher bias.

With regards to the sources, there is a risk that, by restricting our sources to specific, hand-picked conferences, we are unreasonably biasing our results towards the idiosyncrasies of those conferences.

There seem to be two fundamental issues with systematic studies: keywording papers reliably given that different people will view papers in different ways (or even that different authors will use different terms or use terms in different ways); and finding a consistent initial paper set. If we search for papers manually, as in this study, the validity of our initial paperset suffers; however, if we search for papers using an automated system, we run a strong risk of missing relevant papers.

Additionally, when performing a manual search on specific conferences, there is an additional problem: how do researchers know which proceedings to search? This is a serious issue given that automated searches seem to be prone to missing relevant work.

A systematic mapping study is less reliable than a systematic review. Analyzing studies based only on the titles and abstracts is tricky and strongly limits our analysis. Our results could be different if we were considering the full text of each paper rather than the abstracts and titles alone.

This is another limitation of this study, due to the fact that the author that keyworded the set of papers of this study is the same that performed the main analysis of that systematic review, there might be a bias in the analysis of the current paperset.

However, from the perspective of a person new to the field, a systematic mapping is still a useful approach given that people will make the decision to read a paper or not based on its title and abstract.

Unfortunately, it is important to note that abstracts of papers frequently do not include information crucial to understanding what the work is about. As was made clear in PRQ4, basic information about the research method used in a publication is frequently missing. This makes it difficult to perform systematic mapping studies in general. One solution to this could be conferences requiring the use of structured abstracts.

## 5.2   Future Work

An obvious direction for future work would be extend this paper and the ongoing mapping into a full systematic review, enabling us to look in more depth to make sure that we have gathered all appropriate keywords for each paper.

Future work should also be done to attempt a Controlled Experiment on an aspect of Agile UCD. Along with the ongoing Action Research being done in this field, Controlled Experiment will ensure that all research methods are represented in this field.

## 5.3   Conclusion

In this paper we present the findings of a systematic mapping of the field of agile UCD wherein we manually searched the proceedings of the XP/Agile Universe, XP, Agile, CHI and UIST conferences for relevant papers. This allowed us to address a variety of research questions and provide insight into the distribution of topics, authors, etc. within this field.

As previously mentioned, a basic structure for the integration of agile and UCD is firm enough for researchers and practitioners. For instance, Sy [9], Ferreira et al. [10], Fox et al. [11], and Silva da Silva et al. [3] arrived at very similar proposals. Now we need details to confirm these proposals.

From the study of Silva da Silva et al. [3] till now, nine papers have been published in the current two main Agile and two main HCI conferences. Moreover, a workshop focused on Integrating Agile and User-Centered Design was carried in the XP conference in 2013.

Again, the first major discovery of this study was that HCI conferences do not publish a reasonable number of papers on Agile UCD – only 3 of the papers considered in this study came from HCI conferences, none of which were from UIST.

*PRQ1. What is Agile UCD?* We found an emphasis on the Research stage of the Interaction Design process. However, we also found that there is a balance among the stages of the design process, and the stage less concerned is the Prototype (Build and interactive version).

*PRQ2. What types of HCI techniques have been used to integrate agile and UCD?* The main usability technique used is performing usability tests on lightweight prototypes. Another important issue is to spread the research stage of Interaction Design cycle throughout the entire agile software development lifecycle – Continuous Research. Evolutionary prototyping, Upfront design, Continuous design and Personas respectively come afterwards. Automated usability evaluations emerged mainly from those papers that present tools for Graphical User Interface (GUI) testing.

*PRQ3. What types of papers on Agile UCD have been published?* The main type of paper identified were Experience Reports. This could be explained due to the fact that most of the studies were carried out in real agile environments, which is actually great for this context-dependent kind of research.

*PRQ4. What types of research methods have been used in Agile UCD studies?* As already mentioned, the majority of the studies did not provide the research method in the abstract. Case Study and Grounded Theory were the most common research methods in this field.

*PRQ5. What benefits do publications propose their research will produce?* The main benefit reported in the mapped studies is improved communication. The second benefit is improved usability. This benefit should be obvious; after all, this is the end goal of this integration. Other interesting benefits reported are the improved visibility and the improved business analysis.

*SRQ1. Which authors lead the field of agile UCD?* We enumerate the top 14 authors for the field of Agile UCD. This section may be of particular interest to those just getting started in the field of Agile UCD.

*SRQ2. Is the field led by practitioners, academics, or collaborations?* We found a good mix of academic and industry authors, though collaboration between the two groups could stand to be increased. We were also able to analyze the conferences individually to find that XP has a stronger academic showing while Agile has a good mix of industry and academic authors but far fewer collaborations. This knowledge could help authors decide which of the two major agile conferences to submit their work to.

Overall, this paper provides a detailed overview of the field of agile UCD with respect to the XP/Agile Universe, XP, Agile and CHI conferences. This paper provides details on the process of performing this kind of systematic mapping study that will be useful for making sure that future studies are more comprehensive.

Finally, most importantly, for newcomers, this study serves as an introduction to the field of agile UCD. For authors who have already published work on Agile UCD, this paper serves as a guide towards what could be done to improve the field as a whole.

**Acknowledgments.** The authors would like to thank FAPESP and CAPES for financial support.

# References

1. Hellmann, T., Chokshi, A., Abad, Z., Pratte, S., Maurer, F.: Agile testing: a systematic mapping across three conferences: understanding agile testing in the xp/agile universe, agile, and xp conferences. In: Agile Conference (AGILE), pp. 32–41, August 2013
2. Petersen, K., Feldt, R., Mujtaba, S., Mattsson, M.: Systematic mapping studies in software engineering. In: Proceedings of the 12th International Conference on Evaluation and Assessment in Software Engineering, EASE 2008, Swinton, UK, pp. 68–77. British Computer Society, UK (2008)
3. Silva da Silva, T., Martin, A., Maurer, F., Silveira, M.: User-centered design and agile methods: a systematic review. In: AGILE Conference (AGILE), pp. 77–86. IEEE (2011)
4. Preece, J., Rogers, Y., Sharp, H., Benyon, D., Holland, S., Carey, T.: Human-Computer Interaction. Addison-Wesley Longman Ltd., Essex (1994)
5. Wieringa, R., Maiden, N., Mead, N., Rolland, C.: Requirements engineering paper classification and evaluation criteria: A proposal and a discussion. Requir. Eng. **11**(1), 102–107 (2005)
6. Easterbrook, S., Singer, J., Storey, M.A., Damian, D.: Selecting empirical methods for software engineering research. In: Shull, F., Singer, J., SjÃÅberg, D. (eds.) Guide to Advanced Empirical Software Engineering, pp. 285–311. Springer, London (2008)
7. Jeffries, R.: Tdd: The art the art of fearless programming. IEEE Software **24**(3), 24–30 (2007)
8. Pancur, M., Ciglaric, M., Trampus, M., Vidmar, T.: Towards empirical evaluation of test-driven development in a university environment. In: IEEE Region 8 Proc. EUROCON. Computer as a Tool, vol. 2, pp. 83–6. IEEE (2003)
9. Sy, D.: Adapting usability investigations for agile User-Centered design - international journal of usability studies. Journal of Usability Studies **2**(3), May 2007
10. Ferreira, J., Noble, J., Biddle, R.: Agile development iterations and ui design. In: Agile Conference (AGILE), pp. 50–58, August 2007
11. Fox, D., Sillito, J., Maurer, F.: Agile methods and user-centered design: how these two methodologies are being successfully integrated in industry. In: Agile Conference, AGILE 2008, pp. 63–72, August 2008

# Boosting the Software Quality of Parallel Programming Using Logical Means

Mohamed A. El-Zawawy[1,2](✉)

[1] College of Computer and Information Sciences, Al Imam Mohammad Ibn
Saud Islamic University (IMSIU), Riyadh, Kingdom of Saudi Arabia
[2] Department of Mathematics, Faculty of Science, Cairo University,
Giza 12613, Egypt
maelzawawy@cu.edu.eg

**Abstract.** Parallel programming can be realized as a main tool to
improve performance. A main model for programming parallel machines
is the single instruction multiple data (SIMD) model of parallelism.

This paper presents an axiomatic semantics for SIMD programs. This
semantics is useful for designating and attesting partial correctness properties for SIMD programs and is a generalization of the separation's logical system (designed for sequential programs). The second contribution
of this paper is an operational framework to conventionally define the
semantics of SIMD programs. This framework has two sets of inference
rules; for running a program on a single machine and for running a program on many machines concurrently. A detailed correctness proof for
the presented logical system is presented using the proposed operational
semantics. Also the paper presents a detailed example of a specification
derivation in the proposed logical system.

**Keywords:** Concurrent separation logic (CSL) · Program logic · Program verification · Concurrency · Soundness · Distributed programs ·
Semantics of programming languages · Operational semantics

## 1 Introduction

The fact that power constraints have caused performance advances in sequential
computing to decline has led distributed programming [19] to become a main
tool to improve performance. One of the challenges of programming large-scale
machines is the hierarchical nature of machines. The hierarchy causes communication costs among different components of a machine. A main model to program
parallel machines is the the single instruction multiple data (SIMD) [21] archety-
pal of parallelism. Synchronization operations and global assembling communication are used in SIMD to integrate execution of independent threads.

SIMD has many advantages over other related models. The simplicity of
SIMD capacitates abstaining parallel errors, building creative programming, and

© Springer International Publishing Switzerland 2015
O. Gervasi et al. (Eds.): ICCSA 2015, Part V, LNCS 9159, pp. 101–116, 2015.
DOI: 10.1007/978-3-319-21413-9_8

implementing and amending program analysis and optimization. Also the local-view model (data locality) of SIMD execution results in scalability on massive machines and pleasant performance.

A main concern in distributed programming is resources. Usually, resources of the system are shared among a number of processes like processor time, network bandwidth, and memory. Hence efficiency of the system is strongly related to using resources correctly. Although the concept of sharing resources is simple, its axiomatic treatment is complex. Early work of Hoare, Dijkstra, and Brinch [8] studied the resource control for concurrent programming. They relied on synchronization mechanisms and resource separation to control process interactions and minimize time-dependent errors.

Based on early work by Burstall, recently Reynolds [18] and independently O'Hearn and Ishtiaq [9] introduced a new technique (separation logic) to reason about resources. The main contribution of separation logic is a logical connector (separation conjunction) that facilities localizing operation effects on shared resources. The separation conjunction was used to achieve a sound form of local reasoning that guaranteed the invariance of resources not considered in preconditions.

In this paper, we present a novel operational framework to conventionally define the semantics of SIMD programs running on a hierarchical memory model. Our framework has two sets of inference rules. Using the first set, the semantics of executing a given program on a single machine, $m$, is captured on a universal memory state, say $(s, h)$, resulting on a new memory state $(s_m, h_m)$. The second set of inference rules combines the states $\{(s_m, h_m) \mid m \in M\}$ resulted from executing the program on different machines whose IDs are in $M$ into a single state, say $(s', h')$. Hence the state $(s', h')$ captures the semantics of executing the given program concurrently on the machines $M$.

The second contribution of this paper is an axiomatic semantics for SIMD. This semantics is a generalization of the separation's logical system (designed for sequential programs) to specify and prove partial correctness properties for SIMD programs. The main idea is to delineate the assertion $M \models \{P\}\ S\ \{Q\}$ to mean that

- the execution of $S$ on any machine $m \in M$ ($\subseteq M-$ set of all machines) at a state satisfying $P$ is guarantied not to abort.
- any state, at which the execution of $S$ on the machine $m$ is ended, will satisfy $Q$, and
- any state, at which the execution of $S$ on all the machine in $M$ is ended, will satisfy $Q$.

The assertions P and Q may depend upon the values of local variables on different machines as well as upon the program control locations. A scheme of axioms and inference rules system is presented, and a mathematical correctness justification for a simple program is introduced. Using the operational semantics presented in this the paper, the mathematical soundness of the logical system is shown.

$x \in$ lVar, $n \in \mathbb{Z}$, $i_{op} \in \mathbb{I}_{op}$, and $b_{op} \in \mathbb{B}_{op}$

$e \in$ AExpr $::= n \mid x \mid e_1 \; i_{op} \; e_2$.

$b \in$ BExpr $::=$ true $\mid$ false $\mid e_1 \; b_{op} \; e_2$.

$S \in$ Stmts $::=$ skip $\mid x := e \mid x := *e \mid *e_1 := e_2 \mid x := new(e) \mid x :=$ convert $(e, d)$
$\mid$ transmit $S$ from $n \mid x :=$ transmit $e$ from $n \mid S_1; S_2 \mid$ if $b$ then $S_t$ else $S_f$
$\mid$ while $b$ do $S_t$.

**Fig. 1.** A model for a SIMD Programming language

Obtaining a partial correctness proof like that in Figure 9.7 to a program like that in Figure 7 which is executed concurrently on 3 machines is a motivating example to the research.

Contributions of this paper are the following:

1. A novel operational framework to conventionally define the semantics of SIMD programs running on a hierarchical memory model.
2. A new axiomatic semantics to specify and prove partial correctness properties for SIMD programs.

The organization of this paper is as follows. In Section 2, the programming language and memory model are presented. Section 3 presenters the operational semantics of the language. In Section 4, we present the logical system including the assertion language, axioms, and inferences rules. Also Section 4 uses the semantics of Section 3 to provide a mathematical proof for the soundness of the proposed logical system. An example of using the logical system is presented in Section 4 as well. Related research is discussed in Section 5.

## 2    Programming Language and Memory Model

This section presents the programming language and memory model. The language of commands has the following syntax:

We define a class of local variable names (lVar), a class of integer-valued binary operations ($\mathbb{I}_{op}$), and a class of Boolean-valued binary operations ($\mathbb{B}_{op}$). Integer constants, program variables, and arithmetic operations are the components of arithmetic expressions. Boolean values and Boolean operations on arithmetic expressions are the components of Boolean expressions. The statements of the language incorporate the following.

- Empty statement, variable assignments, memory de-references and modifications, allocation statements.
- The statement $x := convert \; (e, d)$ which calculates the expression $e$ and assures that it evaluates to an address on a machine, say $m$. The machine $m$ has to be on a distance less than or equal to $d$ from the current machine.

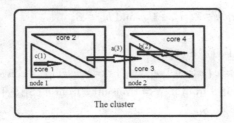

**Fig. 2.** Memory Hierarchy

- The statement *transmit S from n* which executes the statement *S* on the machine *n*.
- The statement *x := transmit e from n* which evaluates the expression *e* on the machine *n* and assigns the result to the local variable *x* of the current machine.
- Sequential compositions, conditionals, and loops statements.

This language is the *while* language enriched with pointer and concurrency constructions [10,12]. SIMD is the model of concurrency used in the language model. In SIMD, the same program is run on all machines.

The hierarchical memory model [12] is the most common model for concurrent computers. The main feature of this model is that each processor has its own local stores. In processors, hierarchies are represented by caches and local addresses. A very famous example is the cell game processor in which every SPE processor has its own local store. Then operations for memory moves make the local store of each SPE processor accessible from each other SPE processor. A grid memory can be partitioned into nodes that contain different cores and grouped into clusters like in Figure 2. We let *h* denote the height of the hierarchical memory. Hence, the interval [1,h] contains widths of all possible pointers.

Most PGAS [1] use memories that have two-levels of hierarchy. One of the two levels is local and is owned by a specific core. The other level is shared and available to all threads. A main idea in PGAS languages is to associate every pointer with the memory level that the pointer is allowed to reach. Figure 2 presents an example of a three-levels hierarchy (core, node, and cluster). In this figure, the pointer *c* is a core pointer and can only point at addresses on core 1. Examples of node and cluster pointers are *b* and *a*, respectively. To represent such type of pointers, every pointer is assigned a width (a number; arrow labels of Figure 2). Hardware research tends to increase levels of hierarchy. Hence it is quite important for programming languages to advantage from the hierarchy [1,12].

# 3   Operational Semantics

Figure 3 presents an operational semantics for executing statements of the programming langauge on a single machine. Configurations are states $(s, h)$ of local variables and addresses on all machines. Transitions are from one state to another or from a state to *abort* denoting run-time errors like de-referencing a value that is not an allocated memory cell.

We define the following composite semantics domains.

**Definition 1.**   *1. lAddrs denotes the set of local addresses located on each single machine.*

2. *The set of machine identifiers is $M = \{1, 2, \ldots, \delta\}$.*
3. *The set of global variables, denoted by gVar, is defined as $gVar = \{(x, m) \mid m \in M, x \in lVar\}$. Similarly $gAddrs = \{(a, m) \mid m \in M, a \in lAddrs\}$.*
4. *$v \in Values = \mathbb{Z} \cup gAddrs \cup \{true, false\}$.*
5. *$S = \{s \mid s : gVar \rightarrow Values\}$ and $H = \{h \mid h : gAddrs \rightarrow_{fin} Values\}$.*
6. *$(s, h) \in States = S \times H$.*
7. *For $(s, h) \in States$ and $m \in M$, $s \rceil m = s \rceil \{(x, m) \mid x \in lVar\}$, and $h \rceil m = h \rceil \{(a, m) \mid a \in lAddrs\}$.*

The proof of the following lemma is straightforward.

**Lemma 1.** *Suppose $(s, h) \in States$. Then*

- *$(h \rceil 1) \# (h \rceil 2) \# \ldots \# (h \rceil \delta)$.*
- *$(s, h) = (\cup_{m \in M}(s \rceil m), (h \rceil 1).(h \rceil 2). \ldots .(h \rceil \delta))$.*

*For maps $f$ and $g$, we let $f \# g$ denotes $dom(f) \cap dom(g) = \emptyset$ and $(f.g)$ denotes the union of the two maps.*

The following lemma is true because arithmetics are not allowed on addresses.

**Lemma 2.** *If $[\![e]\!]_m s \in gAddrs$, then $e \in lVar$.*

Judgments of the form $S : (s, h) \leadsto_m (s', h')$ are produced by rules of Figure 3. Such judgement denotes that running $S$ on the machine $m$ when the state of all the machines of $M$ is $(s, h)$ results in $(s', h')$. By convention $\delta[x \mapsto v] = \lambda y.$ if $y = x$ then $v$ else $\delta(y)$.

Some remarks on Figure 3 are in order. Arithmetics on addresses are not allowed according to rule (iexp$_2^s$). In line with rule $(:= *_2^s)$, the de-reference succeeds only if $e$ evaluates in the current state to an address on the machine $m$. On the one hand, the rule (trans$_1^s$) evaluates $S$ on the machine $n$ (the subscribe of $\leadsto$ in the precondition is $n$) and ends at the state at which the execution of $S$ ends. On the other hand, (trans$_2^s$) evaluates the expression $e$ on the machine $n$ and modifies the variable $x$ of the machine $m$. The rule (conv$_1^s$) evaluates $e$ on the machine $m$ and assigns the evaluated value to the variable $x$ of $m$ only if this value is an address on a machine on distance less than or equal to $n$. This rule assumes a function *hdist* that calculates the distance between machines. The remaining rules are self-explanatory.

$$[\![n]\!]_m s = n, \qquad [\![x]\!]_m s = s(x,m) \qquad \text{skip}: (s,h) \rightsquigarrow_m (s,h) \qquad \dfrac{[\![e_1]\!]_m s\ i_{op}\ [\![e_2]\!]_m s \notin \mathbb{Z}}{[\![e_1\ i_{op}\ e_2]\!]_m s = \text{abort}} \ (\text{iexp}_1^s)$$

$$\dfrac{[\![e_1]\!]_m s\ i_{op}\ [\![e_2]\!]_m s \in \mathbb{Z}}{[\![e_1\ i_{op}\ e_2]\!]_m s = [\![e_1]\!]_m s\ i_{op}\ [\![e_2]\!]_m s} \ (\text{iexp}_2^s) \qquad \dfrac{[\![e_1]\!]_m s\ b_{op}\ [\![e_2]\!]_m s \text{ is a Boolean value}}{[\![e_1\ b_{op}\ e_2]\!]_m s = [\![e_1]\!]_m s\ b_{op}\ [\![e_2]\!]_m s} \ (\text{Bexp}_1^s)$$

$$\dfrac{[\![e_1]\!]_m s\ b_{op}\ [\![e_2]\!]_m s \text{ is not a Boolean value}}{[\![e_1\ b_{op}\ e_2]\!]_m s = \text{abort}} \ (\text{Bexp}_2^s) \qquad \dfrac{[\![e]\!]_m s = \text{abort}}{x := e : (s,h) \rightsquigarrow_m \text{abort}} \ (:=_1^s)$$

$$\dfrac{[\![e]\!]_m s \neq \text{abort}}{\substack{x := e : (s,h) \rightsquigarrow_m \\ (s[(x,m) \mapsto [\![e]\!]_m s], h)}} \ (:=_2^s) \qquad \dfrac{[\![e]\!]_m s \notin \text{dom}(h{\upharpoonright}m)}{x := *e : (s,h) \rightsquigarrow_m \text{abort}} \ (:=*_1^s) \qquad \dfrac{[\![e]\!]_m s \in \text{dom}(h{\upharpoonright}m)}{\substack{x := *e : (s,h) \rightsquigarrow_m \\ (s[(x,m) \mapsto h([\![e]\!]_m s)], h)}} \ (:=*_2^s)$$

$$\dfrac{\substack{[\![e_1]\!]_m s \notin \text{dom}(h{\upharpoonright}m) \\ \text{Or } [\![e_2]\!]_m s = \text{abort}}}{*e_1 := e_2 : (s,h) \rightsquigarrow_m \text{abort}} \ (*:=_1^s) \qquad \dfrac{\substack{[\![e_1]\!]_m s \in \text{dom}(h{\upharpoonright}m) \\ [\![e_2]\!]_m s \neq \text{abort}}}{*e_1 := e_2 : (s,h) \rightsquigarrow_m (s, h[[\![e_1]\!]_m s \mapsto [\![e_2]\!]_m s])} \ (*:=_2^s)$$

$$\dfrac{S : (s,h) \rightsquigarrow_n st}{\text{transmit } S \text{ from } n : (s,h) \rightsquigarrow_m st} \ (\text{trans}_1^s) \qquad \dfrac{x := \text{transmit } e \text{ from } n : (s,h) \rightsquigarrow_m}{\begin{cases} (s[(x,m) \mapsto [\![e]\!]_n s], h), & [\![e]\!]_n s \neq \text{abort}; \\ \text{abort}, & \text{otherwise.} \end{cases}} \ (\text{trans}_2^s)$$

$$\dfrac{[\![e]\!]_m s \in dom(h{\upharpoonright}m') \quad \text{hdist}(m,m') \leq n}{\substack{x := \text{convert}(e,n) : (s,h) \rightsquigarrow_m \\ (s[(x,m) \mapsto [\![e]\!]_m s], h)}} \ (\text{conv}_1^s) \qquad \dfrac{[\![e]\!]_m s \notin dom(h{\upharpoonright}m') \vee \text{hdist}(m,m') > n}{x := \text{convert}(e,n) : (s,h) \rightsquigarrow_m \text{abort}} \ (\text{conv}_2^s)$$

$$\dfrac{[\![e]\!]_m s \neq \text{abort} \quad a \in \text{lAddrs} \quad a \text{ is fresh on } m}{x := new(e) : (x,h) \rightsquigarrow_m (s, h[(a,m) \mapsto [\![e]\!]_m s])} \ (\text{new}_1^s) \qquad \dfrac{[\![e]\!]_m s = \text{abort}}{x := new(e) : (x,h) \rightsquigarrow_m \text{abort}} \ (\text{new}_2^s)$$

$$\dfrac{\substack{S_1 : (s,h) \rightsquigarrow_m (s'',h'') \\ S_2 : (s'',h'') \rightsquigarrow_m st}}{S_1;S_2 : (s,h) \rightsquigarrow_m st} \ (\text{sq}_1^s) \qquad \dfrac{\substack{[\![b]\!]_m s = \text{abort}, \\ [\![b]\!]_m s = \text{true and } S_t : (s,h) \rightsquigarrow_m \text{abort, Or} \\ [\![b]\!]_m s = \text{false and } S_f : (s,h) \rightsquigarrow_m \text{abort}}}{\text{if } b \text{ then } S_t \text{ else } S_f : (s,h) \rightsquigarrow_m \text{abort}} \ (\text{if}_1^s)$$

$$\dfrac{S_1 : (s,h) \rightsquigarrow_m \text{abort}}{S_1;S_2 : (s,h) \rightsquigarrow_m \text{abort}} \ (\text{sq}_2^s) \qquad \dfrac{\substack{[\![b]\!]_m s = \text{true and } S_t : (s,h) \rightsquigarrow_m (s',h'), \text{ Or} \\ [\![b]\!]_m s = \text{false and } S_f : (s,h) \rightsquigarrow_m (s',h')}}{\text{if } b \text{ then } S_t \text{ else } S_f : (s,h) \rightsquigarrow_m (s',h')} \ (\text{if}_2^s)$$

$$\dfrac{\substack{[\![b]\!]_m s = \text{abort, Or} \\ [\![b]\!]_m s = \text{true and } S_t : (s,h) \rightsquigarrow_m \text{abort}}}{(\text{while } b \text{ do } S_t, \delta) \rightsquigarrow_m \text{abort}} \ (\text{whl}_1^s) \qquad \dfrac{[\![b]\!]_m s = \text{false}}{\text{while } b \text{ do } S_t : (s,h) \rightsquigarrow_m (s,h)} \ (\text{whl}_2^s)$$

$$\dfrac{\substack{[\![b]\!]_m s = \text{true} \\ S_t : (s,h) \rightsquigarrow_m (s'',h'') \\ \text{while } b \text{ do } S_t : (s'',h'') \rightsquigarrow_m st}}{\text{while } b \text{ do } S_t : (s,h) \rightsquigarrow_m st} \ (\text{whl}_3^s)$$

**Fig. 3.** Operational semantics for running a program on a single machine.

$$\frac{S \text{ is atomic} \qquad \forall m \in M.\; S : (s,h) \leadsto_m (s_m, h_m)}{M \dashv S : (s,h) \leftrightsquigarrow (s_M, h_M)} \;(\text{atom}^s_M)$$

$$\frac{M' \dashv S_1 : (s,h) \leftrightsquigarrow (s'', h'') \qquad M \dashv S_2 : (s'', h'') \leftrightsquigarrow (s', h') \qquad M \subseteq M'}{M \dashv S_1; S_2 : (s,h) \leftrightsquigarrow (s', h')} \;(\text{seq}^s_M)$$

$$\frac{\{n\} \dashv S : (s,h) \leftrightsquigarrow (s', h')}{\{n\} \cup M \dashv \text{transmit } S \text{ from } n : (s,h) \leftrightsquigarrow (s', h')} \;(\text{trans}^s_M)$$

$$\frac{\begin{array}{c} M_t \dashv S_t : (s,h) \leftrightsquigarrow (s_t, h_t) \\ M_f \dashv S_f : (s,h) \leftrightsquigarrow (s_f, h_f) \\ \forall m \in M_t.[\![b]\!]_m s = \text{true} \qquad \forall n \in M_f.[\![b]\!]_n s = \text{false} \end{array}}{\begin{array}{c} M_t \cup M_f \dashv \text{if } b \text{ then } S_t \text{ else } S_f : (s,h) \leftrightsquigarrow \\ \bigcup\{(s_t\rceil m, h_t\rceil m), (s_f\rceil n, h_f\rceil n), (s_f\rceil o, h_f\rceil o) \mid m \in M_t, n \in M_f, o \in M \setminus M_t \cup M_f\} \end{array}} \;(\text{if}^s_M)$$

$$\frac{\begin{array}{c} M_1 \dashv S_t : (s,h) \leftrightsquigarrow (s'', h'') \\ M_2 \dashv \text{while } b \text{ do } S_t : (s'', h'') \leftrightsquigarrow (s', h') \\ M_2 \subseteq M_1 \qquad \forall m \in M_1.[\![b]\!]_m s = \text{true} \qquad \forall n \in M_1 \setminus M_2.[\![b]\!]_n s = \text{false} \end{array}}{M_1 \dashv \text{while } b \text{ do } S_t : (s,h) \leftrightsquigarrow (s', h')} \;(\text{while}^{s_1}_M)$$

$$\frac{\forall m \in M.\; [\![b]\!]_m s = \text{false}}{M \dashv \text{while } b \text{ do } S_t : (s,h) \leftrightsquigarrow (s,h)} \;(\text{while}^{s_2}_M)$$

**Fig. 4.** Operational semantics for running a program concurrently on multiple machines

Figure 4 presents a generic set of inference rules to give semantics for the concurrent execution of statements on a set of machines. Judgments of the form $M \dashv S : (s,h) \leftrightsquigarrow (s', h')$ are produced by rules of Figure 4. Such judgement denotes that running $S$ concurrently on all the machines in $M$, provided that the state of all the machines of $\mathcal{M}$ is $(s,h)$, results in $(s', h')$. Definition 2 fixes the atomic statements of the programming langauge and introduces a way for mixing results of executing a statement on a set of machines.

**Definition 2.**
  – *Atomic statements of the language in Figure 1 are skip, $x := e$, $x := *e$, $*e_1 := e_2$, $x := new(e)$, $x := convert\,(e,d)$, $x :=$ transmit $e$ from $n$.*
  – *Suppose that $M \subseteq \mathcal{M}$, $(s,h)$ is a state, and $S$ is a statement. Also suppose that $\forall m \in M\;(S : (s,h) \leadsto_m (s_m, h_m))$. We define*

$$(s_M, h_M) = (\bigcup\{(s_m\rceil m, h_m\rceil m), (s\rceil l, h\rceil l) \mid (m \in M) \wedge (l \in \mathcal{M} \setminus M)\}).$$

Some remarks on Figure 4 are in order. Suppose that we are given a state $(s,h)$ and a Boolean condition $b$. Recall that $(s,h)$ represents the status of memories of all machines. Then for two different machines $m, n$, we may have $[\![b]\!]_m s = \text{true}$ and $[\![b]\!]_n s = \text{false}$. This means that $b$ may be true and false concurrently at the state $(s,h)$. Therefore the rule $(\text{if}^s_M)$ mixes $(s_t, h_t)$ representing machines $M_t$ on which the condition is *true* and $(s_f, h_f)$ representing machines $M_f$ on which the condition is *false*. The rule (while$_M^{s_2}$) captures the fact that if we start executing the *while* statement on the machines $M_1$, then eventually the execution ends (the condition becomes *false*) on some machines $M_1 \setminus M_2$ and continues on the others ($M_2$).

Lemma 3 supports the fact that the mix operation in ($s_M, h_M$) of Definition 2 is guaranteed to represent the concurrent execution of statements on different machines of $M$.

**Lemma 3.** *Suppose $M \dashv S : (s, h) \hookrightarrow (S', h')$. Suppose also that $loc(M)$ is the set of local variables and addresses on machines in $M$ and modvar($S$) is the set of variables and memory location modified by this execution. Then modvar($S$) $\subseteq loc(M)$.*

## 4   Assertion Language and Logical System

This section presents an extension to separation logic for concurrent programs running on hierarchical memories using the SIMD model. The assertion language of the logical system is presented first, then the axioms and inference rules of the logical system are presented. A formal mathematical proof for the soundness of the inference rules using the semantics of the previous section is also shown in this section. This section also presents a detailed example of the derivation of a partial correctness specification in the proposed logical system. Assertions of separation logic consist of Boolean expressions, first order quantification, classical connectives, and few assertions relevant to separation logic. We use the following modified version of these assertions.

- $emp_m$ means that the heap of machine $m$ is empty,
- $e_1 \mapsto_m e_2$, means that the heap of machine $m$ consists of a unique memory cell whose address is $e_1$ and content is $e_2$,
- $P * Q$, is named separating conjunction,
- $P \mathbin{-\!\!*} Q$, is named separating implication, and
- $\circledast_{i \in I} P_i$, is an repetitive version of separating conjunction.

---

$e, e_1, e_2 \in \text{AAExpr} ::= n \mid x^m \mid e_1 \; i_{op} \; e_2.$

$b \in \text{ABExpr} ::= \text{true} \mid \text{false} \mid \neg b \mid e_1 \; b_{op} \; e_2.$

$P, Q, R, J_1, J_2 \in \text{Asser} ::= b \mid P \wedge Q \mid P \vee Q \mid P \Rightarrow Q \mid \neg P \mid \forall x^m. \; P \mid \exists \; x^m. \; P \mid \text{ok}(e)$

$\qquad\qquad\qquad\qquad \mid emp_m \mid e_1 \mapsto_m e_2 \mid P * Q \mid P \mathbin{-\!\!*} Q \mid \circledast_{i \in I} P_i \mid \text{dist}(m, m') \leq n.$

where

$x \in \text{lVar}$, a finite set of variables, $n \in \mathbb{Z}$ (integers), $m \in \mathcal{M}$, $i_{op} \in \mathbb{I}_{op}$ (integer-valued binary operations), and $b_{op} \in \mathbb{B}_{op}$ (Boolean-valued binary operations).

---

**Fig. 5.** An assertion language

The precise definition of the assertion langauge is presented in Figure 5. Each assertion represents a set of states. A modeling relation $((s,h) \models P)$ models this representation. The meaning of this relation is that the assertion $P$ is satisfied in the state $(s,h)$. A precise definition of this modeling relation is given in Definition 3.

**Definition 3.**    $- [\![x^m]\!]s = [\![x]\!]_m s = s(x,m).$

$- [\![e_1 \; i_{op} \; e_2]\!]s = \begin{cases} [\![e_1]\!] \; i_{op} \; [\![e_2]\!]s, & \{[\![e_1]\!]s, \; [\![e_2]\!]s\} \subseteq \mathbb{Z}; \\ abort, & otherwise. \end{cases}$

$- (s,h) \models ok(e) \stackrel{def}{\Longleftrightarrow} [\![e]\!]s \neq abort.$

$- (s,h) \models emp_m \stackrel{def}{\Longleftrightarrow} dom(h_m) = \emptyset.$

$- (s,h) \models e_1 \mapsto_m e_2 \stackrel{def}{\Longleftrightarrow} h_m([\![e_1]\!]s) = [\![e_2]\!]s.$

$- (s,h) \models e \mapsto_m \_ \stackrel{def}{\Longleftrightarrow} [\![e_1]\!]s \in dom(h_m).$

$- (s,h) \models P_1 * P_2 \stackrel{def}{\Longleftrightarrow} \exists \; h',h''. \; h'\# h'', h = h'.h'', (s,h') \models P_1, and(s,h'') \models P_2.$

$- (s,h) \models P_1 \;-\!* \; P_2 \stackrel{def}{\Longleftrightarrow} \forall h'((h'\# h \; and \; (s,h') \models P_1) \Longrightarrow (s,h.h') \models P_2).$

The expression $h_1\# h_2$ denotes that $dom(h_1) \cap dom(h_2) = \emptyset$. The expression $h_1.h_2$ denotes the heaps union and is defined only if $h_1\# h_2$.

With respect to a given machine $m$, the following function calculates for each arithmetic expression of the programming language (Figure 1) an equivalent arithmetic expression in the assertion language (Figure 5).

**Definition 4.**

$$ e^m : AExpr \rightarrow AAExpr : x \mapsto x^m, n \mapsto n, \; and \; e_1 \; i_{op} \; e_2 \mapsto e_1^m \; i_{op} \; e_2^m. $$

The relationship between semantics of arithmetic expressions of the programming language and that of arithmetic expressions of the assertion language is clarified by Lemma 4 which is proved by structure induction on expressions.

**Lemma 4.** *Suppose* $e \in AExp$ *and* $m \in M$. *Then* $[\![e^m]\!]s = [\![e]\!]_m s$.

Judgments of the presented logical have the form $M \dashv \{P\} \; S \; \{Q\}$ where $P$ and $Q$ are the precondition and the postcondition, respectively. Informally, this judgement denotes that if S is executed concurrently on all the machines in $M$ from an initial state satisfying $P$, then $Q$ will be satisfied by the final state (if the statement terminates on all machines in $M$).

Figure 6 presents the axioms and inference rules of the logical system. Definition 5 presents the soundness concept of the specifications generated by the logical system.

**Definition 5.** *A judgment* $M \dashv \{P\} \; S \; \{Q\}$ *is sound if for every state* $(s,h)$ *satisfying* $(s,h) \models P$, *the following is true:*

*1.* $\forall m \in M. \; \neg(S : (s,h) \leadsto_m abort),$ *and*
*2. if* $S : (s,h) \leadsto_m (s_m, h_m),$ *then* $(s_m, h_m) \models Q.$
*3. if* $M \dashv S : (s,h) \hookrightarrow (s',h'),$ *then* $(s',h') \models Q.$

$$\frac{}{M \dashv \{Q\} \text{ skip } \{Q\}} \text{ (skip}^l) \qquad \frac{}{M \dashv \{\forall m \in M.\ ok(e^m) \land Q[e^m/x^m]\}\ x := e\ \{Q\}} \text{ (:=}^l)$$

$$\frac{}{M \dashv \{\forall m \in M.\ ok(e_2^m) \land \circledast_{m \in M} e_1^m \mapsto_m \_\}\ *e_1 := e_2\ \{\circledast_{m \in M} e_1^m \mapsto_m e_2^m\}} \text{ (}* :=^l)$$

$$\frac{\forall m.\ x^m \notin fv(e^m, e_1^m)}{M \dashv \{\circledast_{m \in M} e^m \mapsto_m e_1^m\}\ x := *e\ \{\circledast_{m \in M}(e^m \mapsto_m e_1^m \land x^m = e_1^m)\}} \text{ (:=}*^l)$$

$$\frac{x \notin fv(e)}{M \dashv \{\forall m \in M.\ ok(e^m) \land emp_m\}\ x := new(e)\ \{\circledast_{m \in M} x^m \mapsto_m e_m\}} \text{ (new}^l)$$

$$\frac{}{\begin{array}{c} M \dashv \{\forall m \in M.\ e^m = (a, m') \land dist(m, m') \le n\} \\ x := convert(e, n) \\ \{\forall m \in M.\ e^m = (a, m') \land x^m = e^m\} \end{array}} \text{ (conv}^l)$$

$$\frac{\{n\} \dashv \{P\}\ S\ \{Q\}}{M \cup \{n\} \dashv \{P\}\ \text{transmit } S \text{ from } n\ \{Q\}} \text{ (trans}_1^l)$$

$$\frac{}{M \dashv \{ok(e^n) \land \forall m \in M.\ Q[e^n/x^m]\}\ x := \text{transmit } e \text{ from } n\ \{Q\}} \text{ (trans}_2^l)$$

$$\frac{M' \dashv \{P\}\ S_1\ \{R\} \quad M \dashv \{R\}\ S_2\ \{Q\} \quad M \subseteq M'}{M \dashv \{P\}\ S_1; S_2\ \{Q\}} \text{ (Seq}^l) \qquad \frac{M \dashv \{P\}\ S\ \{Q\} \quad M \dashv \{P'\}\ S\ \{Q'\}}{M \dashv \{P \lor P'\}\ S\ \{Q \lor Q'\}} \text{ (Disj}^l)$$

$$\frac{\begin{array}{c} M_t \dashv \{\forall m \in M_t.\ P \land b^m\}\ S_t\ \{Q_t\} \\ M_f \dashv \{\forall m \in M_f.\ P \land \neg(b^m)\}\ S_f\ \{Q_f\} \\ fv(Q_t) \cap var(M_f) = \emptyset \land fv(Q_f) \cap var(M_t) = \emptyset \end{array}}{M_t \cup M_f \dashv \{P\}\ \text{if } b \text{ then } S_t \text{ else } S_f\ \{Q_t \lor Q_f\}} \text{ (if}^l)$$

$$\frac{M \dashv \{\forall m \in M.\ ok(b^m) \land P\}\ S_t\ \{\forall m \in M.\ ok(b^m) \land P\}}{M \dashv \{\forall m \in M.\ ok(b^m) \land P\}\ \text{while } b \text{ do } S_t\ \{\neg(b^m) \land P\}} \text{ (while}^l)$$

$$\frac{\begin{array}{c} M \dashv \{P\}\ S\ \{Q\} \\ P' \Longrightarrow P \quad Q \Longrightarrow Q' \end{array}}{M \dashv \{P'\}\ S\ \{Q'\}} \text{ (Conseq}^l) \qquad \frac{\begin{array}{c} M_1 \dashv \{P\}\ S\ \{Q\} \\ M_2 \dashv \{P'\}\ S\ \{Q'\} \end{array}}{M_1 \cup M_2 \dashv \{P \land P'\}\ S\ \{Q \land Q'\}} \text{ (Conj}^l)$$

$$\frac{\begin{array}{c} M_1 \dashv \{P\}\ S\ \{Q\} \\ M_2 \dashv \{R\}\ S\ \{R\} \\ M_1 \cap M_2 = \emptyset \end{array}}{M_1 \cup M_2 \dashv \{P * R\}\ S\ \{Q * R\}} \text{ (share}^l) \qquad \frac{\begin{array}{c} M \dashv \{P\}\ S\ \{Q\} \\ fv(R) \cap mod(S, M) = \emptyset \end{array}}{M \dashv \{P * R\}\ S\ \{Q * R\}} \text{ (frame}^l)$$

$$\frac{\begin{array}{c} \{m\} \dashv \{P\}\ S\ \{Q\} \\ \mathcal{M} \setminus \{m\} \dashv \{J_1\}\ S\ \{J_2\} \end{array}}{\mathcal{M} \dashv \{P \land J_1\}\ S\ \{Q \lor J_2\}} \text{ (glob}^l) \qquad \frac{\begin{array}{c} M \dashv \{P\}\ S\ \{Q\} \\ n \in M \quad x \notin fv(S) \end{array}}{M \dashv \{\exists x^n.\ P\}\ S\ \{\exists x^n.\ Q\}} \text{ (Ex}^l)$$

**Fig. 6.** The inference rules of the proposed logical system

Some of the rules in Figure 6 are specifically noteworthy. Rules $(* :=^l)$ and $(:= *^l)$ both obligate that all expressions of the statement contribute to the pre-condition: this guarantees that the memory addresses being read or modified are indeed allocated and that no expression evaluation aborts. The preconditions of these rules guarantee that no other machine is accessing the local location concurrently. The side-conditions guarantee the absence of data races on local variables of machines. The rule $(if^l)$ considers the possibility that the condition of the $if$ statement can be $true$ on one set of machines and $false$ on another set. Therefore this rule combines the specifications of the two sets. The is so if the postcondition of each specification puts restrictions on the local variables of the other set of machines. The rule $(trans_1^l)$ adds the machine $n$ to the set of machines $M$ to consider cases like that of the following example:

$$\{4\} \models \{P\}\text{transmit (transmit } x := 5; \text{ from 4) from 3}\{Q\},$$

whose statement eventually gets executed on the machine 4. The precondition of the rule $(trans_2^l)$ substitutes $e^n$ for $x^m$ as the expression is evaluated on the machine $n$ and the assignment is executed on the machine $m$. We let $fv$ denotes the set of free variables of an expression and $mod(S, M)$ denotes the set of locations that $S$ modifies on the machines in $M$. The rule $(share^l)$ acknowledges composing assertions for two disjoint sets of machines in parallel provided that their preconditions characterize disjoint parts of the whole heap. While guaranteeing that $R$ is still satisfied at the postcondition, the rule $(frame^l)$ is useful for avoiding part of the heap, the frame $R$, that is not accessed by the statement.

The following theorem proves the soundness of the proposed logical system.

**Theorem 1.** *For a statement $S$, every specification $M \dashv \{P\}\ S\ \{Q\}$ obtained by the logical system above is sound (Definition 5).*

*Proof.* The proof is by structure induction on inference rules of the logical system as follows:

- The case of the rule $(conv^l)$: in this case
  - $S = x := \text{convert}(e, n)$,
  - $P$ is $\forall m.\ e^m = (a, m') \wedge \text{dist}(m, m') \le n$, and
  - $Q$ is $\forall m.\ e^m = (a, m') \wedge x^m = e^m$.
  Fix an $m \in M$ and suppose that for state $(s, h)$, $(s, h) \models P$. Hence $[\![e^m]\!]s = [\![e]\!]_m s = (a, m')$ and $\text{hdist}(m, m') \le n$ which implies $[\![e]\!]_m s \in dom(h_{m'})$. Therefore $x := \text{convert } (e, n)$ does not abort at $(s, h)$ on the machine $m$, i.e. $x := \text{convert } (e, n) : (s, h) \not\leadsto_m \text{abort}$ . Now suppose that $x := \text{convert } (e, n) : (s, h) \leadsto_m (s_m, h_m)$. Then $(s_m, h_m) = (s[(x, m) \mapsto [\![e]\!]_m s], h)$. By Lemma 2, $e$ is a variable. If $e = x$, then $[\![e^m]\!]s_m = (a, m')$. If $e \ne x$, then $[\![e^m]\!]s_m]\!]m = [\![e^m]\!]s]\!]m = (a, m')$. This shows that $(s_m]\!]m, h_m]\!]m) \models (e^m = (a, m') \wedge x^m = e^m)$. This implies $(s_m, h_m \models Q)$ and $(\cup\{(s_m]\!]m, h_m]\!]m) \mid m \in M\} \models (\forall m.\ e^m = (a, m') \wedge x^m = e^m) = Q$ which implies $(s_M, h_M) = (s', h') \models Q$.
- The case of the rule $(trans_1^l)$: in this case $S = \text{transmit } S'$ from $n$ for some statement $S'$. Fix an $m \in M$ and suppose that for state $(s, h)$, $(s, h) \models P$.

We have $\{n\} \dashv \{P\}\ S'\ \{Q\}$. Therefore by induction hypothesis $S' : (s,h)\ \not\leadsto_n$ abort. Hence by $(\text{trans}_1^s)$, $\forall m \in M \cup \{n\}$ $(S = \text{transmit }S'\text{ from }n : (s,h)\ \not\leadsto_m$ abort). This proves the first requirement of soundness. The second and third requirements (Definition 5) are as follows. Suppose $M \dashv \text{transmit }S'\text{ from }n :$ $(s,h) \hookrightarrow (s',h')$. Then by $(\text{trans}_M^s)$, $\{n\} \dashv S' : (s,h) \hookrightarrow (s',h')$. By induction hypothesis and since $\{n\} \dashv \{P\}\ S'\ \{Q\}$, the required is satisfied.

- The case of the rule $(\text{trans}_2^s)$: in this case
  - $S = x := \text{transmit }e\text{ from }n$, and
  - $P = \text{ok}(e^n) \wedge \forall m \in M.\ Q[e^n/x^m]$.

Fix an $m \in M$ and suppose that for state $(s,h)$, $(s,h) \models P$. This implies $(s,h) \models \text{ok}(e^n)$ which means that $\llbracket e^n \rrbracket s \neq$ abort. Hence by Lemma 4, $\llbracket e \rrbracket_n s \neq$ abort. By $(\text{trans}_2^s)$, $\forall m \in M$ $(S = x := \text{transmit }e\text{ from }n :$ $(s,h)\ \not\leadsto_m$ abort). This shows the first part of soundness whose second and third requirements (Definition 5) are proved as follows. Suppose $x := \text{transmit }e\text{ from }n : (s,h) \leadsto_m (s_m,h_m)$. Then by $(\text{trans}_2^s)$, $(s_m,h_m) = (s[(x,m) \mapsto \llbracket e \rrbracket_n s],h)$. We have $(s,h) \models Q[e^n/x^m]$ which implies $(s_m,h_m) = (s[(x,m) \mapsto \llbracket e \rrbracket_n s],h) \models Q$. Suppose $M \dashv x := \text{transmit }e\text{ from }n : (s,h) \hookrightarrow$ $(s',h')$. Then by $(\text{atom}_M^s)$, $x := \text{transmit }e\text{ from }n : (s,h) \leadsto_m (s_m,h_m)$. Then by $(\text{trans}_2^s)$, $(s_m,h_m) = (s[(x,m) \mapsto \llbracket e \rrbracket_n s],h)$. We have $(s,h) \models Q[e^n/x^m]$ which implies $(s_m,h_m \models Q)$ and $(s[(x,m) \mapsto \llbracket e \rrbracket_n s],h) \models Q$. This implies $(s',h') = (s_{m \in M \cup \{n\}}, h_{m \in M \cup \{n\}}) \models Q$ because $S$ modifies only variables on the machine $m$ and $(s[(x,m) \mapsto \llbracket e \rrbracket_n s],h) \models Q$ is true for every $m \in M$.

- The case of the rule $(:= *^l)$: in this case
  - $S = x := *e$,
  - $P = \circledast_{m \in M} e^m \mapsto_m e_1^m$, and
  - $Q = \circledast_{m \in M}(e^m \mapsto_m e_1^m \wedge x^m = e_1^m)$.

Fix an $m \in M$ and suppose that for state $(s,h)$, $(s,h) \models P$. This implies $(s,h) \models e^m \mapsto_m e_1^m$ which means that $\llbracket e^m \rrbracket s \in \text{gAddrs}$ and $h(\llbracket e^m \rrbracket s) = \llbracket e_1^m \rrbracket s$ implying $h(\llbracket e \rrbracket_m s) = \llbracket e_1 \rrbracket_m s$. Hence by Lemma 4, $h(\llbracket e \rrbracket_m s) \in \text{dom}(h)$. By $(:= *_2^s)$, $\forall m \in M$ $(S = x := *e : (s,h)\ \not\leadsto_m$ abort). This shows the first part of soundness whose second and third requirements (Definition 5) are proved as follows. Suppose $M \dashv x := *e : (s,h) \hookrightarrow (s',h')$. Then by $(\text{atom}_M^s)$, $x := *e :$ $(s,h) \leadsto_m (s_m,h_m)$. Then by $(:= *_2^s)$, $(s_m,h_m) = (s[(x,m) \mapsto h(\llbracket e \rrbracket_m s)],h) =$ $(s[(x,m) \mapsto \llbracket e_1 \rrbracket_m s],h)$. Hence $(s_m,h_m) \models (x^m = e_1^m)$. Since $x^m \notin \text{fv}(e^m, e_1^m)$, it is true that $(s_m,h_m) \models (e^m \mapsto_m e_1^m)$. Therefore $(s_m,h_m) \models (e^m \mapsto_m e_1^m \wedge x^m = e_1^m)$ which clearly implies $(s_m,h_m \models Q)$ and $(s_m]m,h_m]m) \models (e^m \mapsto_m e_1^m \wedge x^m = e_1^m)$. Since $m$ is arbitrary and the heaps $\{h_m \mid m \in M\}$ are separated, we conclude $(s_M,h_M) = (s',h') \models \circledast_{m \in M}(e^m \mapsto_m e_1^m \wedge x^m = e_1^m) = Q$.

- The case of the rule $(:= *^l)$: in this case
  - $S = *e_1 := e_2$,
  - $P = \forall m \in M.\ \text{ok}(e_2^m) \wedge \circledast_{m \in M} e_1^m \mapsto_m \_$, and
  - $Q = \circledast_{m \in M} e_1^m \mapsto_m e_2^m$.

Fix an $m \in M$ and suppose that for state $(s,h)$, $(s,h) \models P$. This implies $(s,h) \models \text{ok}(e_2^m)$ which means that $\llbracket e_2^m \rrbracket s = \llbracket e_2 \rrbracket_m s \neq$ abort. Also in this case $(s,h) \models (e_1^m \mapsto_m \_)$ which means that $\llbracket e_1^m \rrbracket s \in \text{dom}(h)$. Hence by Lemma 4, $h(\llbracket e_1 \rrbracket_m s) \in \text{dom}(h)$. By $(* := _2^s)$, $\forall m \in M$ $(S = *e_1 := e_2 : (s,h)\ \not\leadsto_m$ abort).

This shows the first part of soundness whose second and third requirements (Definition 5) are proved as follows. Suppose $M \dashv x := *e : (s,h) \leftrightarrow (s',h')$. Then by $(\text{atom}_M^s)$, $x := *e : (s,h) \leadsto_m (s_m, h_m)$. Then by $(:= *_2^s)$, $(s_m, h_m) = (s, h[[\![e_1]\!]_m s \mapsto [\![e_2]\!]_m s])$. Hence $(s_m, h_m) \models (e_1^m \mapsto_m e_2^m)$ which clearly implies $(s_m, h_m \models Q)$ and $(s_m]m, h_m]m) \models (e_1^m \mapsto_m e_2^m)$. Since $m$ is arbitrary and the heaps $\{h_m \mid m \in M\}$ are separated, we conclude $(s_M, h_M) = (s', h') \models \circledast_{m \in M}(e_1^m \mapsto_m e_2^m) = Q$.

- The case of the rule (Seq$^l$): in this case
  - $S = S_1; S_2$
  - $M' \dashv S_1 : (s,h) \leftrightarrow (s'', h'')$ & $M' \dashv \{P\} S_1 \{R\}$,
  - $M \dashv S_2 : (s'', h'') \leftrightarrow (s', h')$ & $M \dashv \{R\} S_2 \{Q\}$, and
  - $M \subseteq M'$.

Fix an $m \in M$ and suppose that for state $(s,h)$, $(s,h) \models P$. This implies $m \in M'$ and by induction hypothesis $\neg(S_1 : (s,h) \leadsto_m$ abort). If $S_1 : (s,h) \leadsto_m (s_m, h_m)$, then by induction hypothesis $(s_m, h_m) \models Q$. Then again by induction hypothesis $\neg(S_2 : (s_m, h_m) \leadsto_m$ abort) since $M \dashv \{R\} S_2 \{Q\}$. Therefore by $(\text{seq}_1^s)$, $\neg(S_1; S_2 : (s,h) \leadsto_m$ abort). Now suppose $M \dashv S_1; S_2 : (s,h) \leftrightarrow (s', h')$. Therefore by induction hypothesis we have $(s'', h'') \models R$ which implies $(s', h') \models Q$.

- The case of the rule (if$^l$): in this case
  - $S = \text{if } b \text{ then } S_t \text{ else } S_f \wedge M = M_t \cup M_f$,
  - $M_t \dashv S_1 : (s,h) \leftrightarrow (s_t, h_f)$ & $M_t \dashv \{P\} S_1 \{Q_t\}$,
  - $M_f \dashv S_2 : (s,h) \leftrightarrow (s_f, h_f)$ & $M_f \dashv \{P\} S_2 \{Q_f\}$,
  - $fv(Q_t) \cap var(M_f) = \emptyset \wedge fv(Q_f) \cap var(M_t) = \emptyset$, and
  - $(s', h') = \bigcup\{(s_t]m, h_t]m), (s_f]n, h_f]n), (s_f]o, h_f]o) \mid m \in M_t, n \in M_f, o \in M \setminus M_t \cup M_f\}$.

Fix an $m \in M_t$ and suppose that for state $(s,h)$, $(s,h) \models b^m \wedge P$. Then by induction hypothesis $\neg(S_t : (s,h) \leadsto_m$ abort) which implies by (if$_2^s$) that $\neg(\text{if } b \text{ then } S_t \text{ else } S_f : (s,h) \leadsto_m$ abort). If if $b$ then $S_t$ else $S_f : (s,h) \leadsto_m (s_m, h_m)$, then if $b$ then $S_t$ else $S_f : (s,h) \leadsto_m (s_m, h_m)$ which by induction hypothesis implics $(s_m, h_m) \models Q_t$. Hence $(s_m, h_m) \models Q_t \vee Q_f$. Now suppose $M \dashv$ if $b$ then $S_t$ else $S_f : (s,h) \leftrightarrow (s', h')$. Therefore by induction hypothesis we have $(s_t, h_f) \models Q_t$. Since $fv(Q_t) \cap var(M_f) = \emptyset$, we get $(s', h') \models Q_t$ which implies $(s', h') \models Q_t$. Similar proof works for the case $m \in M_f$. This completes the proof.

## Example

An example program executed on three machines is presented in Figures 7. Figure 8 illustrates the content of three machines before and after executing the program provided that the machine heaps are empty before running the program. Figure 9 presents a detailed derivation for a logical specification using the rules of Figure 6. In this derivation we let $M = \{1, 2, 3\}$.

$y := \text{new } (x);$
$z := \text{transmit } y \text{ from } 2;$
$w := \text{convert } (z, 2);$
$x := *w$

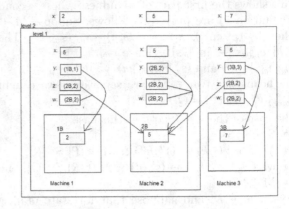

**Fig. 7.** An example program

**Fig. 8.** Memory description after executing the program of Figure 7

1. By (new$^l$),
$M \dashv \{\forall m. \text{ok}(x^m) \wedge \text{emp}_m\}$
$y := \text{new } (x)$
$\{\circledast_{i \in \{1,2,3\}} y^i \mapsto_i x^i\},$
$\implies M \dashv \{(x^1 = 2 \wedge x^2 = 5 \wedge x^3 = 7) *$
$\forall m. \text{ok}(x^m) \wedge \text{emp}_m\}$
$y := \text{new } (x)$
$\{(x^1 = 2 \wedge x^2 = 5 \wedge x^3 = 7) * \circledast_{i \in \{1,2,3\}} y^i \mapsto_i x^i\},$
by (frame$^l$)

2. $M \dashv \{y^2 \mapsto_2 x^2\}$
$z := \text{transmit } y \text{ from } 2$
$\{\circledast_{i \in \{1,2,3\}} z^i \mapsto_i x^2\}$
$\implies M \dashv \{(x^1 = 2 \wedge x^2 = 5 \wedge x^3 = 7) * y^2 \mapsto_2 x^2\}$
$z := \text{transmit } y \text{ from } 2$
$\{(x^1 = 2 \wedge x^2 = 5 \wedge x^3 = 7) \circledast_{i \in \{1,2,3\}} z^i \mapsto_i x^2\}$

3. By 1 & 2,
$M \dashv \{(x^1 = 2 \wedge x^2 = 5 \wedge x^3 = 7) *$
$\forall m. \text{ok}(x^m) \wedge \text{emp}_m\}$
$y := \text{new } (x);$
$z := \text{transmit } y \text{ from } 2$
$\{(x^1 = 2 \wedge x^2 = 5 \wedge x^3 = 7) \circledast_{i \in \{1,2,3\}} z^i \mapsto_i x^2\}$

4. By (conv$^l$), $M \dashv \{\forall m. z^m = (a_m, m') \wedge dist(m, m') \leq n\}$
$w := \text{convert } (z, n)$
$\{\forall m. z^m = (a_m, m') \wedge w^m = z^m\}$
$\implies M \dashv \{(x^1 = 2 \wedge x^2 = 5 \wedge x^3 = 7) \wedge \forall m. (z^m = (a_m, m') \wedge dist(m, m') \leq n)\}$
$w := \text{convert } (z, n)$
$\{(x^1 = 2 \wedge x^2 = 5 \wedge x^3 = 7) \wedge \forall m. (z^m = (a_m, m') \wedge w^m = z^m)\}$
$\implies M \dashv \{(x^1 = 2 \wedge x^2 = 5 \wedge x^3 = 7) \wedge \forall m. (z^m = (a_m, m') \wedge dist(m, m') \leq n) \circledast_{i \in \{1,2,3\}} z^i \mapsto_i x^2\}$
$w := \text{convert } (z, n)$
$\{(x^1 = 2 \wedge x^2 = 5 \wedge x^3 = 7) \wedge \forall m. (z^m = (a_m, m') \wedge w^m = z^m) \circledast_{i \in \{1,2,3\}} z^i \mapsto_i x^2\}$

5. By 3 & 4,
$M \dashv \{(x^1 = 2 \wedge x^2 = 5 \wedge x^3 = 7) *$
$\forall m. \text{ok}(x^m) \wedge \text{emp}_m\}$
$y := \text{new } (x);$
$z := \text{transmit } y \text{ from } 2;$
$w := \text{convert } (z, n)$
$\{(x^1 = 2 \wedge x^2 = 5 \wedge x^3 = 7) \wedge$
$\forall m. (z^m = (a_m, m') \wedge w^m = z^m) \circledast_{i \in \{1,2,3\}}$
$z^i \mapsto_i x^2\}$

6. $\{\forall m. w^m \mapsto_m 5\}$
$x := *w$
$\{\forall m. (w^m \mapsto_m 5 \wedge x^m = 5)\}$
$\implies M \dashv \{(\forall m. w^m \mapsto_m 5) \wedge \circledast_{i \in \{1,2,3\}} z^i \mapsto_i 5\}$
$x := *w$
$\{(\forall m. (w^m \mapsto_m 5 \wedge x^m = 5)) \wedge \circledast_{i \in \{1,2,3\}} z^i \mapsto_i 5\}$

7. By 5 & 6,
$M \dashv \{(x^1 = 2 \wedge x^2 = 5 \wedge x^3 = 7) *$
$\forall m. \text{ok}(x^m) \wedge \text{emp}_m\}$
$y := \text{new } (x);$
$z := \text{transmit } y \text{ from } 2;$
$w := \text{convert } (z, n)$
$x := *w$
$\{(\forall m. (w^m \mapsto_m 5 \wedge x^m = 5)) \wedge \circledast_{i \in \{1,2,3\}} z^i \mapsto_i 5\}$

**Fig. 9.** A specification derivation for the program in Figure 7

# 5  Discussion

Concurrent separation logic (CSL) was proposed by Peter O'Hearn [14] to verify parallel programs that have pointer constructs. CSL is an application of local-reasoning feature of separation logic [9,18]. In CSL, specifications of separation logic grab resources ownership [15]. To show structures at their side conditions, in [16] the rules of concurrent separation as well as sequential logic were reformulated utilizing the concepts of syntactic control of interference [13]. Using the concept of fractional permissions [2], the work in [16] embellishes the classical syntactic control of interference to develop a cogent variant. A modernized version of separation logic [18] was then formulated using this system in [16]. The roots of this work goes back to the work in [17] which presents a set of rules called "syntactic control of interference"(SCI). In [20], a direct technique (in the form of a standard operational semantics) is presented to show soundness of CSL. Among advances of this work are the inclusion of the framing concept of separation logic, the removal of requirement of precise resource invariants, and the adaption ability to concurrent separation logic extensions [4]. However none of the techniques mentioned above provides a smooth way to reason about SIMD programs running on hierarchical memories. The current paper can be realized as an attempt to fill this gap similarly to [7].

A simple version of the langauge that we use in the current paper can be found in [10] where [10] achieves some important static analyses for parallel programs on parallel machines with hierarchical memoirs. Using a two-level hierarchy of memory, [11] studies sharing features and locality information of pointers in the form of a constraint-based analysis. Due to their importance, distributed programs and separation logic have been the focus of much research activates [5–7]. One important problem is data racing bugs [3] which results from parallel access of cores of a distributed program to a memory that is physically distributed. The work in [3] presents DRARS, an algorithm to avoid and replay the data race in distributed programs. The main idea of DRARS is to assist debugging distributed programs. Catching and testing the parallel and classical relationships in distributed programs is an important issue.

# References

1. Soubhagya Sankar Barpanda and Durga Prasad Mohapatra: Dynamic slicing of distributed object-oriented programs. IET Software 5(5), 425–433 (2011)
2. Bornat, R., Calcagno, C., O'Hearn, P.W., Parkinson, M.J.: Permission accounting in separation logic. In: Palsberg, J., Abadi, M. (eds.) POPL, pp. 259–270. ACM (2005)
3. Chiu, Y.-C., Shieh, C.-K., Huang, T.-C., Liang, T.-Y., Chu, K.-C.: Data race avoidance and replay scheme for developing and debugging parallel programs on distributed shared memory systems. Parallel Computing 37(1), 11–25 (2011)
4. Dinsdale-Young, T., Dodds, M., Gardner, P., Parkinson, M.J., Vafeiadis, V.: Concurrent abstract predicates. In: D'Hondt, T. (ed.) ECOOP 2010. LNCS, vol. 6183, pp. 504–528. Springer, Heidelberg (2010)

5. El-Zawawy, M.A.: Dead code elimination based pointer analysis for multithreaded programs. Journal of the Egyptian Mathematical Society **20**(1), 28–37 (2012)
6. El-Zawawy, M.A.: Frequent statement and de-reference elimination for distributed programs. In: Murgante, B., Misra, S., Carlini, M., Torre, C.M., Nguyen, H.-Q., Taniar, D., Apduhan, B.O., Gervasi, O. (eds.) ICCSA 2013, Part III. LNCS, vol. 7973, pp. 82–97. Springer, Heidelberg (2013)
7. El-Zawawy, M.A.: Testing automation of context-oriented programs using separation logic. Applied Comp. Int. Soft Computing **2014** (2014)
8. Hoare, C.A.R.: Monitors: An operating system structuring concept. Commun. ACM **17**(10), 549–557 (1974)
9. Ishtiaq, S.S., O'Hearn, P.W.: Bi as an assertion language for mutable data structures. In: Hankin, C., Schmidt, D. (eds.) POPL, pp. 14–26. ACM (2001)
10. Kamil, A., Yelick, K.A.: Hierarchical pointer analysis for distributed programs. In: Riis Nielson, H., Filé, G. (eds.) SAS 2007. LNCS, vol. 4634, pp. 281–297. Springer, Heidelberg (2007)
11. Liblit, B., Aiken, A., Yelick, K.A.: Type systems for distributed data sharing. In: Cousot, R. (ed.) SAS 2003. LNCS, vol. 2694, pp. 273–294. Springer, Heidelberg (2003)
12. Lindberg, P., Leingang, J., Lysaker, D., Khan, S.U., Li, J.: Comparison and analysis of eight scheduling heuristics for the optimization of energy consumption and makespan in large-scale distributed systems. The Journal of Supercomputing **59**(1), 323–360 (2012)
13. O'Hearn, P.W.: Linear logic and interference control. In: Pitt, D.H., Curien, P.-L., Abramsky, S., Pitts, A.M., Poigné, A., Rydeheard, D.E. (eds.) Category Theory and Computer Science. LNCS, vol. 530, pp. 74–93. Springer, Heidelberg (1991)
14. O'Hearn, P.W.: Resources, concurrency, and local reasoning. Theor. Comput. Sci. **375**(1–3), 271–307 (2007)
15. Owicki, S.S., Gries, D.: Verifying properties of parallel programs: An axiomatic approach. Commun. ACM **19**(5), 279–285 (1976)
16. Reddy, U.S., Reynolds, J.C.: Syntactic control of interference for separation logic. In: Field, J., Hicks, M. (eds) POPL, pp. 323–336. ACM (2012)
17. Reynolds, J.C.: Syntactic control of interference. In: Aho, A.V., Zilles, S.N., Szymanski, T.G. (eds.) POPL, pp. 39–46. ACM Press (1978)
18. Reynolds, J.C.: Separation logic: a logic for shared mutable data structures. In: LICS, pp. 55–74. IEEE Computer Society (2002)
19. Udaya Shankar, A.: Distributed Programming: Theory and Practice. Springer (2013)
20. Vafeiadis, V.: Concurrent separation logic and operational semantics. Electr. Notes Theor. Comput. Sci. **276**, 335–351 (2011)
21. Yelick, K.A., Bonachea, D., Chen, W.-Y., Colella, P., Datta, K., Duell, J., Graham, S.L., Hargrove, P., Hilfinger, P.N., Husbands, P., Iancu, C., Kamil, A., Nishtala, R., Su, J., Welcome, M.L., Wen, T.: Productivity and performance using partitioned global address space languages. In: Maza, M.M., Watt, S.M. (eds.) PASCO, pp. 24–32. ACM (2007)

# Analysis of Web Accessibility in Social Networking Services Through Blind Users' Perspective and an Accessible Prototype

Janaína Rolan Loureiro(✉), Maria Istela Cagnin, and Débora Maria Barroso Paiva

College of Computing, Federal University of Mato Grosso do Sul, Av. Costa e Silva, s/n, CP 549, Campo Grande, MS 79070-900, Brazil

{janrloureiro,istela,dmbpaiva}@gmail.com.br,
{istela,dmbpaiva}@gmail.com

**Abstract.** It is not just in architectural contexts that accessibility concern is present: accessibility in web portals is also a right guaranteed by law to people with any kind of disability in many countries, and Brazil takes part in this group. Although extensive researches focused on web accessibility were developed recently, not all areas were contemplated with these efforts. Social networking services are a great example of sites that ask for more attention in this field. They are important tools for integration of disabled people, but still lack web accessibility features. This paper presents an evaluation of the web accessibility of three major social networking services – Facebook, LinkedIn and Twitter – by the perspective of blind users, applying the WCAG 2.0 success criteria. Analysis demonstrated that there are many issues to be addressed in order to improve web accessibility for visually impaired in this domain. Besides that, a prototype of an accessible social networking service was proposed based on an instance of Elgg framework, where the experiences about web accessibility gathered with this study were applied.

**Keywords:** Web accessibility · Evaluation · WCAG 2.0 · Social networking services · Blind users

## 1 Introduction

According to the Demographic Census 2000 [7] and 2010 [8], the number of Brazilians presenting some kind of disability increased from 24,5 to 46 million in ten years, what corresponds to almost 25% of the current Brazilian population. However, the most concerning fact is that the data collected in last Census also showed that inequalities faced by disabled people reflect on their monthly incomes, employment and schooling rates, all of them being lower compared to people without any disability.

These evidences indicate that, for some reason, disabled people are having fewer opportunities than the rest of the population in Brazil. Aware of this reality, the Brazilian Government, as had already been done by other countries, published in 2004 decree-law nº 5,296 [2] in order to assure equal rights for people with disabilities. It comprehends accessibility not only in the architectural and transportation scopes, but also in the access to information, where the Web accessibility fits in. Even with this, W3C's business development officer Karen Myers pointed out that only 2% of

© Springer International Publishing Switzerland 2015
O. Gervasi et al. (Eds.): ICCSA 2015, Part V, LNCS 9159, pp. 117–131, 2015.
DOI: 10.1007/978-3-319-21413-9_9

webpages can be considered as accessible for disabled [14]. Therefore, it is clear that a lot of effort is still needed in this area.

Meantime, a comScore's report [4] showed that in 2012 Brazilian Internet users spent great part of their online time in social media sites, increasing 167% comparing to 2011. Putting into numbers, from the 27 hours they spent online monthly, 9.3 were used to access social networking services. Considering the significant growth of this type of website and that it presents an interesting alternative for the social inclusion of disabled people helping them to overcome socialization difficulties, the present study suggested and conducted a web accessibility evaluation in this domain, in order to better understand what barriers disabled people still have to face daily while attempting to browse the Internet.

The problem was addressed regarding visually impaired users, because this is the most common disability in Brazil, affecting 18.8% of the population [8]. Besides that, a study by Ruth-Janneck's [16] indicated that web accessibility is very important for visually impaired people, since this is the kind of disability that imposes more barriers for the access on social networking services.

In order to investigate the web accessibility for the visually impaired on social networking services, it was chosen to address the problem by the perspective of the end user. According to Melo *et al.* [13], an evaluation based on end users allows the researchers to observe the interaction strategies built by different types of users while performing common tasks, in distinct contexts of use and applying assistive technologies, to identify the difficulties they face. Thus, this study could be seen as an extension of Loureiro's *et al.* [12], where the web accessibility in social networking services was also assessed, but using evaluations done by automated tools and experts.

Making use of Loureiro's *et al.* observations and experience obtained with blind users' evaluation, it was possible to configure an instance of a social networking over Elgg framework, and adapt it to meet conformance level A of WCAG 2.0 in order to become a promising prototype of an accessible social networking.

Section 2 summarizes related works and important concepts for this paper, Section 3 sums up Loureiro's *et al.* evaluations [12] and their results, while Section 4 describes the planning and execution of the evaluation with end users, as well as reports results obtained. Section 5 presents the accessibility improvements performed in a prototype social network, Section 6 brings a discussion about what was previously presented, with future works explained and acknowledgments given.

## 2    Literature Background

### 2.1    WCAG 2.0

One of the design principles supported by World Wide Web Consortium (W3C), an international community in charge of developing Web standards to lead the Web to its full potential [20], is called Web for All. In this project's context, appears the Web Accessibility Initiative (WAI) that has as primary goal to make the Internet available to all people, no matter if they present or not any kind of physical or mental disability [22]. Therefore, a set of guidelines for web accessibility was developed, known as WCAG (Web Content Accessibility Guidelines). Currently in its 2.0 version, four principles are the main core of the WCAG (Perceivable, Operable, Understandable and Robust), around which 12 guidelines are distributed for making web content more accessible [21].

Each guideline has success criteria (CS) associated to it, and a level of web accessibility is attributed to a success criterion, through level A (the lowest) to AAA (the highest). According to how many and which success criteria are satisfied by a service, a conformance level of web accessibility can be assigned [10] to it. To reach the conformance level A, all success criteria from level A should be met; level AA requires that all success criteria from levels A and AA are met; and all success criteria from levels A, AA and AAA have to be met in order to be classified as conformance level AAA.

## 2.2   Sample of Social Networking Services

Three social networking services from the four previously analyzed by Loureiro *et al.* in [12] were chosen to compose the current study sample: Facebook [6], LinkedIn [11] and Twitter [18]. They represent the most popular social media among Brazilian users, as reported by [1]. ResearchGate [15] was not included in this evaluation because it does not present a version of its pages in Portuguese, Brazilian native language, what could become a barrier to navigation for the users and influence the evaluation of web accessibility issues.

## 2.3   Elgg Platform

The Elgg platform [5] is an environment, Open Source licensed under the terms of the GNU General Public License v2 MIT, that provides a robust framework for building social networking services. The Elgg platform allows high level of adaptability and the ability to follow evolutions, since it is software that only has a minimal fixed core structure, which can be customized through the addition of plug-ins [9]. Moreover, it offers a variety of options in terms of ease for social networking, with intrinsic characteristics of web blogging, file storage, personal pages [3]. The Elgg was chosen because of its customization flexibility, what allowed to structure a social network with features normally found on websites of this domain, adapting the interface related plug-ins in order to reach level A of WCAG 2.0 conformity.

## 2.4   Related Works

The decision to address the visually impaired was based in the study of Ruth-Janneck [15], in which a group of users with different disabilities (visual, hearing, mental and motor) was interviewed. The study showed that users with total visual impairment are the ones that most consider web accessibility as important, crediting to it much of its social inclusion by enabling their access to the Internet. Thus, this paper lists which success criteria from the WCAG 2.0 have more influence in web accessibility to the visually impaired.

Loureiro *et al.* [12] initiated an evaluation process of web accessibility in social networking services, submitting the sample to lexical analysis, performed by automated validators, and semantic analysis, executed by experts. In this study, the success criteria that affect the accessibility in the sample, preventing it to reach the level A of WCAG 2.0 conformance, are highlighted.

Sik-Lanyi in [17] also conducted a study evaluating four social networking services (Facebook, iWiW, MySpace and YouTube), submitting their homepages to an automated validator developed by the author, based on WCAG 2.0 guidelines.

The evaluations indicated which problems could compromise these sites accessibility, emphasizing that this indicated lack of commitment regarding web accessibility from the development teams.

## 3     Evaluation by Automated Tools and Experts

A web accessibility evaluation should consider more than one perspective during the assessment to fulfill its goals and highlight the real obstacles disabled people face while trying to browse the Internet, helping developers and researchers discuss about what can be done to improve the experience for everybody. Therefore, Loureiro *et al.* [12] conducted lexical evaluations, by applying automatic validation tools, and semantic evaluations, performed by specialists, about web accessibility in a sample of social networking services. In order to make the assessment more objective, a subset of the success criteria considered as most relevant to the visually impaired was selected, based on the study of Ruth-Janneck [16]. The results showed that none of the websites in the sample reached the level A of WCAG 2.0 conformance, and a list of the success criteria, among the 20 of this level, which were considered as not met, thus presenting barriers to the sample accessibility, can be found in Table 1.

**Table 1.** Detected Problems In Each Success Criteria According To Automated Tools And Experts Evaluations

| WCAG 2.0 SC | Detected Problems |
| --- | --- |
| 1.1.1: Non-text Content | Images processed just as a links, without meaningful alternative text. Images are not directly related to their description. |
| 1.3.1: Info and Relationships also presented non-visually | Required fields are not stated. Error messages without feedback to the user, or highlighting field in red, without additional information for the screen reader. Association between text and elements are not clear. |
| 1.4.1: Color is not the only means of conveying information | If the password is incorrect, error feedback requests to correct highlighted field, preventing reproduction by screen reader. |
| 2.1.1: All content is operable through a keyboard | Information in right sidebar is not accessible via keyboard because the site autoreloads posts. There are some buttons and modal windows that are not accessible via keyboard. |
| 2.4.1: Bypass Blocks | It is not possible to skip blocks of information. |
| 2.4.4: Link Purpose (In Context) | Links identifications are not enough to indicate their purpose. Links without alternative text. |
| 3.3.1: Error identification | Error feedback without information that can help users understand what went wrong. |
| 3.3.2: Labels or Instructions | Not all forms have their purpose indicated. Description is not provided in text entry fields. |
| 4.1.1: Code Parsing | Page is not well formed, errors were detected by the HTML validator from W3C. |
| 4.1.2: Name, Role, Value | Use of not standard HTML tags. |

# 4    Evaluation by End Users

In this context, the evaluation with end users reported here comes as an upgrade to the web accessibility in social networking services, observing how the previously listed accessibility issues influence disabled users' experience in practice, taking notes of their complaints and difficulties faced during the execution of tasks, and also collecting quantitative data about their performance.

## 4.1    User's Profile

The Ethics Committee for Research Involving Human Subjects of the Platform Brazil (Plataforma Brasil) CEP/CONEP has given its approval to this project execution in December 2013, under the report number 497406 and CAAE 25054113.0.0000.0021. All participants of this study signed an informed consent form, voluntarily agreeing to take the proposed assessments and allowing the researchers to use the generated data.

An assorted sample of users was considered in this study to prevent from bias evaluations, composed by people from different genders, ages, schooling and income levels. The nine volunteers that took part in this study are assisted by the Institute of Mato Grosso do Sul for Blinds Florivaldo Vargas (ISMAC), located at Campo Grande, MS, Brazil. They were on average 31 years old old, with mean deviation of 7 years, and blind for at least six years, with 33% of the volunteers being women. All participants stated that they are able to surf well the Internet, that are used to screen readers and that had already made use of a social network before. Their families incomes are about R$ 1400.00 (equivalent to two minimum wage in Brazil, or about $500.00) and more than 66% of the participants have completed high school.

## 4.2    Evaluation Planning

Each end user was asked to perform six simple tasks in one of the three social networking services, in order to submit each site to three different evaluators. Those tasks were planned to contemplate the most relevant WCAG 2.0 success criteria for blinds, as stated by Ruth-Janneck in [16] and also applied in Loureiro et al. evaluations [12]. Only success criteria from conformance level A were considered, once that the previous evaluations showed that the sample websites do not meet this lowest classification.

The evaluations executed by automated tools and experts in [12] considered a subset of 20 success criteria from level A of the WCAG 2.0, but in the current evaluation with end users 4.1.1 and 4.1.2 criteria were not included, once that both of them require technical background on HTML for a correct assessment. Success criteria 1.2.1, 1.2.2, 1.2.3 and 1.4.2 were not included in the LinkedIn evaluation either, since they refer to audio and/or video content, and the website does not support this kind of content.

## 4.3    Proposed Tasks

The six proposed tasks for each sample's sites are presented here. Although it was necessary to formulate slightly different activities for each website, since they offer distinct resources and functionalities, the aim of each task remained equivalent in all cases. A test account was created and populated with diverse content in each of three sites, and the sign in information (user and password) was provided to the volunteers.

### Facebook

FT1) Log in on Facebook;

FT2) In the "What's on your mind?" field, post a message;

FT3) In the "Search for people, places and things" field, search for the singer Xuxa and visit her profile;

FT4) At Xuxa's profile, search for her shared photos, and then select the first one. Report if it is possible to understand it;

FT5) Return to homepage, find your friends' posts on the news' feed and choose one to like;

FT6) Visit your profile's page, search for your last shared video and try to play it. Report if it is possible to understand it.

### LinkedIn

LT1) Log in on LinkedIn;

LT2) In the "Share an update..." field, post a message;

LT3) In the "Search for people, jobs, companies, and more..." field, search for Bill Gates and visit his profile;

LT4) At Bill Gates' profile, find his experience information;

LT5) Return to homepage, find your friends' updates on the news' feed and choose one to share;

LT6) On homepage, search for a friend's update that contains a photo. Report if it is possible to understand it.

### Twitter

TT1) Log in on Twitter;

TT2) In the "Compose a new Tweet" field, post a message;

TT3) In the "Search" field, search for the singer Xuxa and visit her profile;

TT4) At Xuxa's profile, search for her shared photos, and then select the first one. Report if it is possible to understand it;

TT5) Return to homepage, search for the "Tweets" field and retweet a tweet from someone you follow.

TT6) Visit your profile's page, search for your last shared video and try to play it. Report if it is possible to understand it.

### 4.4    Evaluation Conduction and Results

The evaluation with end users was conducted in February and March of 2014, at IS-MAC's facilities, where the volunteers normally attend to browse the Internet, aided by a screen reader. One at a time, the volunteers received the instructions for performing the evaluation. The researcher, who took notes of their performance, assisted them; help was offered to proceed or quit the task when they showed signs of mood change, since it was very important to ensure their well-being during the evaluation.

### Facebook

The three participants that evaluated Facebook showed no difficulties to log in (FT1). Finding a specific element was a hard assignment (FT2 to FT6), in which they had to go through the page elements more than once to understand they had reached the goal. Users were not able to identify the destination of some links (FT3 to FT6), nor the content of an image (FT4) or video (FT6) due to lack of alternatives texts for this kind of content. Besides that, videos were reported as image content for users, hindering the execution of FT6.

They complained that the Enter (Return) key did not directly post a status message, and that no feedback was provided after action, such as when a search was completed, or image/video content is been displayed. Table 2 indicates the success criteria not met according to difficulties faced by at least one of the blind end users during the navigation in Facebook.

**Table 2.** Problems faced by users in Facebook

| SC | Observations |
|---|---|
| 1.1.1 | None of the users was able to understand what the photos were about. |
| 1.2.1 | An alternative text for prerecorded media was not provided. |
| 1.2.2 | Captions for prerecorded media were not always provided, and when provided, the screen reader was not able to reproduce it. |
| 1.2.3 | An alternative description for prerecorded media was not provided. |
| 1.3.1 | Users could not identify required fields. |
| 1.3.2 | In some pages, advertisements were presented for users before the main content. |
| 1.3.3 | Users did not receive a non-visual feedback when searches were completed nor posts were published. |
| 2.4.1 | Users were not able to bypass blocks of repetitive content, unless they make use of a shortcut to list all links in the page. |
| 2.4.4 | The purposes of some links were not clear for the users. Some links did not even have a description, being reported for users only as a 'link'. |
| 3.3.2 | Users did not fully understand what each field required as input. |

**LinkedIn**

The feedback provided by LinkedIn for entering a wrong password (LT1), was displayed in a different color to highlight the information, in a way that was not noticed by blind users. Some links and buttons with similar descriptions caused misunderstanding among users (LT2 and LT5), as well as images with not meaningful alternative text (LT6).

When asked to search for a user profile (LT3), the volunteers did not notice that the results were being presented while they were typing. As no feedback was given to the users, they had to go through the whole page to understand it and choose a profile link to visit. At a profile page, they were not able to distinguish text fields of different information (LT4).

Again, users complained that it was not possible to post a message only by pressing enter key (LT2). Table 3 makes the association between problems faced by users and WCAG 2.0 success criteria that should be met to solve them.

**Table 3.** Problems faced by users in LinkedIn

| SC | Observations |
| --- | --- |
| 1.1.1 | Images were reported as links to the screen reader, and when visited, an alternative text was not always provided. |
| 1.3.1 | Users could not identify required fields. |
| 1.3.2 | In some pages, advertisements were presented before the main content for the users. |
| 1.3.3 | Users could not distinguish when a text field with personal information in a profile had started and ended. |
| 1.4.1 | Log in error feedback tells the user to correct the marked fields. |
| 2.1.1 | The right sidebar could not be accessed via keyboard. |
| 2.4.1 | User wanted to avoid going through all menu items, but was not able to. |
| 2.4.4 | Not every link's purpose was presented for the users. |
| 3.3.1 | Error feedback was not perceived by the user when he entered the wrong password while trying to log in. |
| 3.3.2 | Instructions were not provided for input fields, and users were not sure about what information they were required to enter. |

**Twitter**

Logging in Twitter as an issue, because users were not sure where to enter their username and password, since the purpose of the input fields were not described (TT1). For the same reason, finding where to write a message to be published took some time as well (TT2). Users which evaluated Twitter complained that Enter/Return key do not submit a form (TT2).

The search engine starts showing the found results as the user is typing, but the user didn't notice these partial results, since they didn't receive any feedback about it. The result page for the search was considered confusing, since it didn't show only other users, but also tweets that matches the context (TT3).

Besides difficulties to find photos that another user shared, blind users could not understand what is been displayed because of the lack of alternative text and the fact that images are opened in a modal window, not accessible via keyboard (TT4). The same was observed regarding video contents (TT6). The retweet function was also considered hard by the users, which after some tries, gave up the task without concluding it (TT5). Table 4 relates the success criteria not met by Twitter, responsible for the troubles faced by the users in cach task.

**Table 4.** Problems faced by users in Twitter

| SC | Observations |
| --- | --- |
| 1.1.1 | Text alternative was not provided for images. |
| 1.2.1 | An alternative text for prerecorded media was not provided. |
| 1.2.2 | Captions for prerecorded media were not provided. |
| 1.2.3 | An alternative description for prerecorded media was not provided. |
| 1.3.1 | Required fields were not stated. |
| 1.3.2 | Complementary information was presented before the main content of the page. |

**Table 4.** (*Continued.*)

| SC | Observations |
|---|---|
| 1.4.1 | When users tried to post a message with more than 140 characters, they were not notified that it was not allowed, because the error was indicated highlighting the extra characters in another color. |
| 2.1.1 | Photos, videos and retweet function opened in an overlayed window, not accessible via keyboard. |
| 2.4.1 | Users could only bypass blocks of content if they made use of the shortcut that lists the links contained in the page. |
| 2.4.2 | In users subpages, such as photos and videos, the title was maintained as it was the main user profile page. |
| 2.4.4 | Link purposes of images were not meaningful, and some links did not even have a description of its purpose. |
| 3.3.1 | Log in errors were reported in a general way. |
| 3.3.2 | Users had some trouble finding specific input fields, because their purpose were not clear to the user. |

## 5    Prototype of an Accessible Social Networking Service

A site was instantiated upon the Elgg platform in order to contemplate the basic features of a social networking service, such as profile pages, friends, comments, information and photo sharing, thereby creating a functional prototype of a social networking. It was built in PHP5, with MySQL as database and Apache Server, generating HMTL5 pages. To evaluate the prototype accessibility, it was submitted to automated tools, showing that it did not meet the requirements for conformance level A of the WCAG 2.0, as happened with the other social networking services in our sample. Thus, the goal was to improve the prototype accessibility, in order to meet the success criteria of WCAG 2.0 conformance level A.

Under these circumstances, the changes made aimed to contemplate the issues rise during the lexical evaluation of the prototype, and also the observations made by Loureiro's *et al.* evaluations [12], not forgetting to meet the remarks compiled during the evaluations with visually impaired users. Towards this direction, success criteria considered not met by this assessment were confronted with those of [12], indicating in Table 5 which SC compromise web accessibility of the sample as a whole, and because of that these success criteria were considered as priority during the prototype development.

Looking forward to meet the success criteria in Table 5, the solutions discussed in Table 6 were applied in the social networking prototype, by altering the login and theme plug-ins, responsible for site interface. After the implementation of the modifications, the prototype was submitted again to a lexical evaluation, this time meeting the requirements for the conformance level A of the WCAG 2.0.

**Table 5.** Success criteria not met by the sample of social networking services

| SC | Automated Tools | Experts | End Users |
|---|---|---|---|
| 1.1.1 | X | X | X |
| 1.3.1 | X | X | X |
| 1.3.2 | | | X |
| 1.4.1 | | X | |
| 2.1.1 | | X | X |
| 2.4.1 | X | X | X |
| 2.4.4 | | X | X |
| 3.3.1 | | X | X |
| 3.3.2 | | X | X |
| 4.1.1 | X | X | |
| 4.1.2 | X | | |
| 1.1.1 | X | X | X |
| 1.3.1 | X | X | X |

**Table 6.** Solutions applied to met each success criterion mentioned in Table 5

| SC | Solutions |
|---|---|
| 1.1.1 | It was included an attribute *alt* with a meaningful description of the image for every *img* tag. The purpose of each *input* tag was described by filling an attribute *title* or associating a tag *label* through the attribute *for*. |
| 1.3.1 | The purpose of each *input* tag was described by filling an attribute *title* or associating a tag *label* through the attribute *for*. Attribute *required* was included in *input* tags whose fillings were required. Page content and layout were separated with CSS. Each page has at least a level 1 heading *h1*, hierarchy of headings were respected (*h1>h2>h3>h4>h5>h6*), which helped to define the content structure. |
| 1.3.2 | Related contents were organized in such a way that their information was complementary. |
| 1.4.1 | A textual alternative conveying the same information was provided when color was applied to highlight page elements. |
| 2.1.1 | Autoloading pages were avoided, and modal pages receive the keyboard focus when opened. |
| 2.4.1 | Each page starts with a link to its main content, and links to skip content blocks were provided. |
| 2.4.4 | Attribute title with link description was defined for each tag *a*. |
| 3.3.1 | Errors were reported to the user with clear messages to help them understand what went wrong, and these feedbacks receive keyboard focus when detected. |
| 3.3.2 | The purpose of each *input* tag was described by filling an attribute *title* or associating a tag *label* through the attribute *for*. |
| 4.1.1 | HTML ambiguities were eliminated, adapting the pages to the current standards established by W3C [19]. Elements of a page received a unique *id* attribute. |

**Table 7.** (*Continued.*)

| SC | Solutions |
|---|---|
| 4.1.2 | If used, new interface components developed would be properly applied in association with assistive technologies. |
| 1.1.1 | It was included an attribute *alt* with a meaningful description of the image for every *img* tag. The purpose of each *input* tag was described by filling an attribute *title* or associating a tag *label* through the attribute *for*. |
| 1.3.1 | The purpose of each *input* tag was described by filling an attribute *title* or associating a tag *label* through the attribute *for*. Attribute *required* was included in *input* tags whose fillings were required. Page content and layout were separated with CSS. Each page has at least a level 1 heading *h1*, hierarchy of headings were respected (*h1>h2>h3>h4>h5>h6*), which helped to define the content structure. |

The solutions proposed try to achieve conformance level A of WCAG 2.0 by providing more complete and meaningful information to the users while navigating through the social networking prototype, especially for blind users that employ screen readers as assistive technology. Among the changes made, two of the most important will be reported in this paper. The first one was on the login page, in the input field where it is expected that the user enter his/her username or e-mail to access the site. As shown in Fig. 1, when this field receives keyboard focus in the original login page, the screen reader reports it just as "edit text blank", with no further information about what the user should enter.

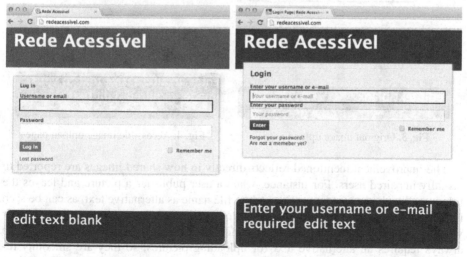

**Fig. 1.** Original login page          **Fig. 2.** Accessible login page

In the accessible version of the login page, displayed in Fig. 2, the message conveyed by the screen reader has become more significant: "Enter your username or e-mail required edit text", allowing that visually impaired ones could interact better with the social networking interface. This was achieved by associating the label tag with its respective input field, through association of attributes for and id. Also, the message was rewritten to better describe the intended purpose. These improvements contribute to contemplate success criteria 1.1.1, 1.3.1, 3.3.2 and 4.1.2. The same problem was observed in almost every input field of the original instantiation, and corrected as described.

Another very significant accessibility enhancement developed, which is worthy being reported here, is related to success criterion 1.1.1. In the original version of the prototype, when a user wanted to share an image, he/she had the option to enter a title for the picture, which would be treated also as alternative text by screen readers (Fig. 3). However the user could just leave this field blank if he/she wants, publishing it anyway. This was corrected in the accessible version of the prototype, by setting this field as required. Thus, the user has to enter an alternative text for the image he/she wants to share with friends in the social networking (Fig. 4).

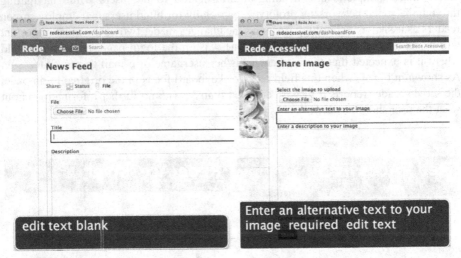

**Fig. 3.** Original image upload page          **Fig. 4.** Accessible image upload page

The improvement mentioned reflects directly in how shared images are reported to visually impaired users. For instance, when a user publishes a picture and leaves the title input blank, the system uses a random file name as alternative text, as can be seen in Fig. 5. Needless to say, with that behavior the image become incomprehensible for visually impaired users. On the other hand, the accessible version of the prototype always requires an alternative text for uploading pictures, so they are smoothly reported by screen readers, as shown in Fig. 6.

Fig. 5. Original image view page          Fig. 6. Accessible image view page

## 6    Final Remarks

### 6.1    Discussion

Knowing the importance of social inclusion that social networking services can provide to the visually impaired, it is important to ensure that the domain meets satisfactorily the needs of this population quota. This study intended to assess the status of Web accessibility within a sample of social networking services, taking into account lexical and syntactic evaluations, in addition to end users' reviews.

It was observed that the social networking services lack attention from developers and administrators regarding accessibility, since they do not meet requirements even for conformance level A of the WCAG 2.0, what make navigation on these websites tiring and not objective for the visually impaired.

Therefore, compiling what was noticed during this end user evaluation with the previously results of [12], it was possible to list the success criteria that represent potential problems to web accessibility of visually impaired within the domain of social networking services: 1.1.1, 1.3.1, 1.3.2, 1.4.1, 2.1.1, 2.4.1, 2.4.4, 3.3.1, 3.3.2, 4.1.1 and 4.1.2.

The experience gained from the assessments was gathered and allowed the implementation of a functional and accessible prototype of a social networking site based on the Elgg platform, in which the improvements needed to achieve this goal were identified and reported. This proved that, with the support of the development team, building an accessible website for blind people is a feasible task.

### 6.2    Future Works

As future work, it is intended to add other features to the social networking prototype presented such as video support, friends' birthday reminder, support to events' organization, pages and groups creation around a common interest and files sharing, always focusing on accessibility, looking forward to achieve progressively higher conformance levels of WCAG 2.0 (AA and AAA), thus generating a complete and accessible social networking service.

Furthermore, it would be an interesting idea to apply other evaluations (experts and end users) to the social networking prototype in order to validate its accessibility before publishing it on the Internet.

Back to the social networking domain, it would be interesting to select another sample of social networking services and replicate evaluations under the three perspectives assessed in order to get more data about web accessibility in this field.

**Acknowledgment.** Our thanks to the ISMAC staff for allowing us to conduct the evaluation in their installation, and to the visually impaired attended by them that agreed to take part in the study.

# References

1. Alexa. Alexa top-sites in Brazil (2014). http://www.alexa.com/topsites/countries/BR
2. Brasil Government. Decree-law n° 5,296 from December 2nd (2004)
3. Chau, J. : A developer's challenges on an e-portfolio journey. In: ICT: Providing choices for learners and learning - Proceedings ascilite Singapore 2007, pp. 145–148 (2007). comScore, Inc. Brazil digital future in focus (2013)
4. Elgg (2014). http://elgg.org/
5. Facebook (2014). https://www.facebook.com/
6. IBGE, Instituro Brasileiro de Geografia e Estatística. Census 2000 –advanced tab – preliminary results of sample (2000)
7. IBGE, Instituro Brasileiro de Geografia e Estatística. Census 2010 – general characteristics of the population, religion and people with disabilities (2010)
8. Kraker, J., Cörvers, R., Valkering, P., Hermans, M. e Rikers, J.: Learning for sustainable regional development: towards learning networks 2.0 Journal of Cleaner Production, vol. 49 (2013), pp. 114–122 (2013)
9. de Lara, S.M.A.: Support mechanisms for usability and accessibility in older adults interaction on the web. PhD thesis, Institute of Math Sciences and Computing, University of São Paulo, São Carlos, Brazil. In Portuguese (2012)
10. LinkedIn (2014). https://www.linkedin.com/
11. Loureiro, J.R., Cagnin, M.I., Paiva, D.M.: Web accessibility in social networking services. In: Murgante, B., Misra, S., Rocha, A.M.A., Torre, C., Rocha, J.G., Falcão, M.I., Taniar, D., Apduhan, B.O., Gervasi, O. (eds.) ICCSA 2014, Part V. LNCS, vol. 8583, pp. 586–601. Springer, Heidelberg (2014)
12. Melo, A.M., Baranauskas, M.C.C., Bonilha, F.F.G. : Evaluation of web accessibility with the user participation - a case study. In: VI Symposium on Human Factors in Computing Systems - Facilitating and Transforming Everyday, Curitiba, October 2004. UFPR, CEIHC—SBC, pp. 165–168 (2004)
13. Myers, K. : Lecture: Trends in web standards for accessibility and mobile payments. CIAB-Febraban (2012)
14. ResearchGate (2014). http://www.researchgate.net/

15. Ruth-Janneck, D.: Experienced barriers in web applications and their comparison to the wcag guidelines. In: Holzinger, A., Simonic, K.-M. (eds.) USAB 2011. LNCS, vol. 7058, pp. 283–300. Springer, Heidelberg (2011)
16. Sik-Lányi, C. : Accessibility testing of social websites. In: Furht, B. (ed.) Handbook of Social Network Technologies and Applications, pp. 409–425. Springer (2010)
17. Twitter (2014). https://twitter.com/
18. W3C, World Wide Web Consortium. Markup validation service (2012). http://validator.w3.org/
19. W3C, World Wide Web Consortium. W3C Mission (2012). http://www.w3.org/Consortium/mission
20. W3C, World Wide Web Consortium. Web content accessibility guidelines (WCAG) 2.0 (2008). http://www.w3.org/WAI/intro/wcag
21. WAI, Web Accessibility Initiative (2008). http://www.w3.org/WAI/

# Systematic Mapping Studies in Modularity in IT Courses

Pablo Anderson de L. Lima, Gustavo da C.C. Franco Fraga,
Eudisley G. dos Anjos, and Danielle Rousy D. da Silva(✉)

Centro de Informática, Universidade Federal da Paraíba, João Pessoa, Paraíba, Brazil
{pablo.luna.lima,eudisley,danielle.rousy}@gmail.com,
gutexp@hotmail.com

**Abstract.** Modularity is one of the most important quality attributes during system development. Its concepts are commonly used in disciplines of information technology courses, mainly in subjects as software project, software architecture, and others. However, it is notable among certain groups of students that this issue is not fully absorbed in a practical way. Although some researchers and practitioners have approach themes like this, there is still a lack of research about how modularity can be approached in IT courses. This paper presents a systematic mapping study about how the modularity is addressed in education. The main objective is to understand what are the main areas in this field and find more interesting points of research to improve the practice of modularity during IT disciplines.

**Keywords:** Modularity · IT learning · Software engineering

## 1 Introduction

Modularity, also used to define flexibility and understanding of any system, is a very important concept for any software development project [22]. This can be clearly seen into architectures developed by companies. Every complex system we can find is usually made of small subsystems working into an integrated way.

Knowledge about the use of modularity can help in various situations such as software maintainability, code changes. However, the misuse of this knowledge generates bad codes, undermining the system understanding and operation [12].

A study has shown that students know the importance of modularity in a project and understand the theoretical issues of this concept [5]. The biggest problem begins when they need to apply this concept in practical cases. It was concluded that there is a difference between the understanding of the concept of modularity, and its realization[5]. This demonstrates the need to combine the practice of this concept, to improve the student's learning.

Despite its importance the studies address important issues about modularity, such as: why modularity is so little emphasized in IT education; why the students, even aware of its importance, find difficulties to produce a less complex system; what are the most common errors related to the lack of modularity presented by the students; where should we improve the education of modularity; and so forth.

© Springer International Publishing Switzerland 2015
O. Gervasi et al. (Eds.): ICCSA 2015, Part V, LNCS 9159, pp. 132–146, 2015.
DOI: 10.1007/978-3-319-21413-9_10

These questions have raised issues approached in many research works, addressing different ways to teach modularity in the classroom for IT training professionals. The objective is to detect certain problems of the students, the cause of these problems, and how they could be solved.

This paper approaches how Modularity is taught and learned in IT courses in order to detect the needs of general studies on this area. For this, the authors performed a systematic study about modularity and education, identifying the main topics in current researches.

## 2    Theoretical Foundation

In this chapter we will discuss the main concepts to be used as basis for understanding this research. The topics include: systematic mapping, software architecture, modularity and education in information technology.

### 2.1    Systematic Mapping

To build the knowledge surrounding this paper, a search for valid resources, related to first studied cases, was needed. This search would align with the research objectives, guiding the work to a reliable starting point for the relation between theory and the obtained results.

The systematic mapping was a methodology uncovered recently by Software Engineering scholars [23]. The systematic mapping is set in five steps according to Petersen and other authors:

1. **Definition of research issues.** Questions evaluation can be the base for a successful work.
2. **Document search for first studies.** Identified by search from a planned string to find the results.
3. **Review of essential results applying inclusion and exclusion criteria.** Important to determinate the final number of valid resources.
4. **Definition of keywords related to the topic.** Important to sort the results by areas of interest.
5. **Systematic mapping.** The final step to reveal all the frequency of studies by areas of interest.

The systematic study is an ideal tool to begin researches about a certain area, reinforcing the importance of viewing results [23].

It is necessary to extract results of the main concepts included in the mapping, so as to help the search on the subject studied, which has close proximity to the concepts of Software Engineering and Computer Science. This mapping is beneficial to define the sequence of procedures to raise points and counterpoints to about the desired area of knowledge [4].

## 2.2    Software Architecture

We need to approach on the issue of architecture, which emerges as a crucial part of design process. Over the years, the system came from a condition where programmers only obeyed technical requirements [2].

Through this area, a group of properties may be designated to be implemented into a system. We can say the architecture is a result from a set of influences taking place on the software implementation.

Several people or authorities may provide certain weight on the software implementation, and these are end users, developers, maintenance staff, vendors and others.

Opening a special consideration in this regard, when the people are in process of formation in IT course (as in a university), there is less influence of stakeholders. It could be asked, how much it affects the construction of values involving the Software Architecture. After all, the modularity, with its proposal of a system division into pieces, helps people to comprehend the entire system and the relations between them components and functions.

## 2.3    Modularity Definitions

Modularity is the mechanism which can be used to increase system flexibility, understanding and allows reduction in software development time [22]. The modules generate components that are able to provide the use of services from other system components. Each module may be composed by other simpler parts.

The module is a unit where structural elements can be strongly connected to each other and somehow loosely connected with other units. There are degrees of connection between elements and units, and degrees of modularity.

The modularity definition is difficult to make in terms of functions, which are stationary. We can say that the vision of modularity is essential in some acts as production and use of products in the industrial segments [1].

On the issue of production, modularity became useful on the processes simplification. More complex products had their production divided into "cells" or modules. And also, it was something very common since the early automobile industry, for example.

Finally, the use of modular products is the use of elements initially independent, to make a union of these elements which will form a new product consistent with tastes and goals of the users. This case is often used in manufacturing, where independent materials are purchased and used to be finally united and delivered to an end user.

## 2.4    Difficulties in Modularity Understanding and Use

There are some problems with insufficient use of modularity coming from certain programs [12]. It happens even in Software Engineering companies, where there is learning performance mode and system design of the entity [11]. Then we can ask why so many problems arise, directly or indirectly related to modularity.

The fact is: even decades ago, there was a concern about the modularity within older languages. There was a comparison between parameters of an automatic program marker (called Automark) and human markers within the McMaster University in Ontario, Canada. The automatic marking was intended to emphasize economy, modularity, simplicity and structure of the codes. And the difference between automatic and human markers was always great because humans maybe consider superfluous factors [13].

What has changed since then? Why the concern for modularity among the institutions is lost? And why students only made functional programs, without taking into account code structure and the software architecture concepts learned earlier?

It is assumed that what has happened is the development factor of nowadays, where computers are in constant development, and a much more available memory than the first computers of history. Nevertheless, it should be necessary to stimulate students somehow to they detect their own problems and resolve them, in an effective way.

The situation becomes even more serious when we observe in institutions where students must have some basic knowledge about architecture and software projects [5]. Despite the fact that modularity is one of the most important principles, when students conclude their studies, they cannot evolve in that area, unless they learn about it at work [11].

There is not a vision of the simple or complex software to maintaining for the students, because a return on the implementation does not exist [5]. This causes a common habit among students that is to make the program "works", and nothing else, since issues such as separation of duties and code encapsulation are rarely evaluated by teachers. This creates a big problem in learning to the students, as it is quite common the presence of low quality codes, where modules have a level of dependence among themselves much higher than the indicated. There is the risk of students with wrong concept of modularity, spread the deficiency among computing industries, even with the help of training companies [5].

So, it makes necessary to think how these problems occur and, then, avoid them. Besides that, even the better students can have this issue, acquired by a bad programming habit that does not changed through the course. Students are barely able to perform this auto-detection, since a review process does not exist, in most cases. So, it is impossible for the students become more conscious about these bad habits.

Some problems may arise to the students during the course, like bad understanding of architectures, wrong software project concepts, indifference about design, and others. The problems come from mistakes about modularity in low-level, and even during the roles division, when a project is initiated. [17].

Education gives the basic principles, but even after teaching them, there are gaps that new professionals in IT must explore, after the course [11]. The major problem is when the programmer will be able to focus on your problem, and should be done as soon as possible, during the formation of the student.

## 2.5   Solutions Within IT Learning

The rescue of knowledge, in order to improve it, can be done as a reminder of the principles of Architecture and Software Engineering, even in other disciplines. If students have some experience in design, implementation and test of small programs; a few number of them actually joined a system implementation that requires skill, teamwork and discipline, in order to build high quality software [17].

The students' understanding was the main objective, as they had project delivery time problems in the past. The most important is that modularity is a central concept for the project. Not only in code, but also in everything that involves the division of tasks and problems [17].

The concept of modularity used to begin a course that put together the disciplines of Architecture Hardware, Compilers and Software Engineering at four American universities (Hebrew University, Haifa University and Harvard Harzliya IDC) [27]. Throughout the course, students feel lost about the computer science concepts, as the development of systems. There were many troubles, among them, to make abstractions, interfaces, and other things. Their objective would be uniting the hardware, software compilers and engineering issues, so, with that, the learning would be more profitable.

Modularity within the work can be observed in the way the project was divided in three main stages. Each project can be broken into pieces that can be tested separately. The multiple division of the system gives various points of view as user, architect and programmer. At that time, we can see the level of concern that exists in relation to the modularity.

The implementation-level systems must "reflect" the architecture of a company. This happens in most of companies, and also happens at the academic base. So, the final project of the course may reflect directly the structure on which the system was made, the communication among teammates (division of labor), and the knowledge sharing [7].

Anyway, it makes necessary to keep in mind that modularity is an important topic to be discussed as far as possible, during the course. There is also different ways to teach it. The correlations between design products and organizations, processes for learning management and knowledge, and competitive strategy, they all exist. The concepts for loosely coupled systems, where the concept of modularity ends being heavily involved, were used in this context [25].

In addition to the modularity itself, the concept of hierarchy was used, a principle organizational of complex systems assembled by subsystems, correlated with the concept of semi-separable systems where the subsystems would be almost independent, relating to each other by a sharing of data or variables [25].

The proposed teaching functions components and design by the authors can be applied both for training companies and for schools, adjusting the teachers to their needs.

According to the authors, four forms of education can be made: Incremental (limited variations of functions and components in existing architecture) Modular (new variations with existing architecture), Architectural (variations of limited functions and components with new architectures) and Extreme (new varieties with new architectures). Everything depends on the current focus of the course [25].

The modularity problem in courses can also be solved by curricular issues. Sometimes, the students fail to enter the reality of their profession, because there is a problem with their formation. So, a new curriculum style was proposed to be taught [18].

With insufficient consensus about a specialized curriculum in software engineering, there is a gap in certain areas of knowledge, and some concepts that could be improved. New curricula have been proposed to suggested updates in sub-areas of knowledge. In some areas, the use of modularity can be clearly seen, which allows the base of work allocation, and best management projects practice [18].

# 3    Methodology

The methodology used for this work was important to define deadlines and review points. In every level of the research, different questions and activities were made.

The first activity was the study of concepts from related areas. It was necessary to understand what subjects are the most important in our research. The review of papers was required to understand the different ways that the main concepts are used.

The second activity was to define the subject to be studied, creating a brainstorm to discuss the main ideas. The daily coexistence with IT education, the ability to find some material and the existing local problems with modularity, made the conclusions clear, guiding the research to the definition of study parameters.

The decision to use Systematic Mapping was essential to obtain the necessary results to this research. It is important to mention the mapping of all current works in this field has enabled a broader view of studies about modularity in education [3].

The strings used in the systematic mapping were defined through research questions. For all digital libraries the string was a variation of: ("modularity") AND ("learning" OR "education"). The digital libraries were chosen by the quality of papers such as: ACM Digital Library, IEEE Xplore, Springer Link and Scopus.

After the first mapping interaction, reading title and abstracts, twenty-eight results were selected. These documents were considered potentially relevant for this research. In the Table 1, it is possible to see the initial results and the final amount of selected works.

**Table 1.** Results of the first systematic mapping filter

| Search Engine | Results Achieved | Number of Relevant Results (after reading titles and text summaries) |
|---|---|---|
| ACM Digital Library | 287 results | 19 results |
| IEEE Xplore | 32 results | 2 results |
| Scopus | 700 results | 3 results (There was one repetition in IEEE Xplore) |
| Springer | 1753 results | 4 results |
| TOTAL: | 2772 results (without repetitions) | 28 results |

The results were taken and the papers could be divided in effective and discarded ones. During the reading of the papers, some questions were evaluated to understand better the purpose of each research. The following shows the questions selected to be evaluated during the reading of the papers.

- Is the document related to modularity?
- Is the document related to education in IT courses?
- The purpose of document
- The Identification of problem or hypothesis raised by the research
- The Solution of the problem raised
- The Stage of the work (theoretical application, practical application, case study, questionnaire or experiment)
- The Approach used to teach modularity
- Type of course and public target addressed by the document
- Teaching technique (if had some)
- The problems that the authors had

After these questions, nineteen results were considered and nine were discarded. Whit this, was possible to bring responses about the general research of modularity in IT courses. The criteria of inclusion were:

- The result is related to modularity or IT education.
- The result offers some contribution to the research, answering some of the questions made.
  And the criteria of exclusion were:
- The approach of modularity and IT education was superficial into the document.
- The document brought a specific subject that escapes the main area of interest.

## 4    Results and Discussion

As seen in methodology, from the documents search in systematic mapping, the answers of questions made in methodology were searched and found. It was expected to find information relevant to the base of the work. Nineteen papers were defined as research properly aligned to work.

From all papers, seventeen talk about modularity, to a greater or lesser degree (about 89,47%). And three documents cited some sort of concept related to this factor (about 15,78%). The concepts found from these studies were very relevant to help in the study on the modularity concept in the systematic study.

It is necessary to keep in mind that modularity is a concept very widespread in various areas of knowledge today, even outside the computing. Regarding the development of technologies, the presence of this concept is even higher, so the majority of documents consider that the concept is something that is already known by scholars who engage in the area.

Table 2. List of relevant documents in System Mapping

| Document Title | Author(s) | Year |
|---|---|---|
| A 2007 Model Curriculum For a Liberal Arts Degree in Computer Science | RICHARDS, B. | 2007 |
| A Network Analysis of Student Groups in Threaded Discussions | KANG, J.; KIM, J; SHAW, E. | 2010 |
| A Programming Competition for High School Students Emphasizing Process | SHERRELL, L.; MCCAULEY, L. | 2004 |
| A Synthesis Course in Hardware, Architecture, Compilers and Software Engineering | SCHOCKEN, S.; NISAN, N.; ARMONI, M. | 2009 |
| Can Students Reengineer? | LEACH, R; BURGE, L; KEELING, H. | 2008 |
| Comparison of Manual and Automated Marking of Student Programs | FLEMING, W.; REDISH, K.; SMYTH, W. | 1988 |
| Computer Science Curricula 2013 | DRAFT. S; | 2013 |
| Computing Curricula 2001 Computer Science | ENGEL, G; ROBERTS, E. | 2001 |
| Exploring Experienced Professionals' Reflections on Computing Education | EXTER, M; TURNAGE, N. | 2012 |
| IS'97: Model Curriculum and Guidelines for Undergraduate Degree Programs in Information Systems | DAVIS, G.; GORGONE, J.; COUGER, J. | 1997 |
| Metrics-Based Evaluation of Learning Object Reusability | SANZ-RODRIGUEZ, J.; DODERO, J.; SANCHEZ-ALONSO, S. | 2010 |
| Modularity, Flexibility, and Knowledge Management in Product and organization Design | SANCHEZ, R.; MAHONEY, J. | 1998 |
| MSIS 2000: Model Curriculum and Guidelines for Graduate Degree Programs in Information Systems | GORGONE, J; GRAY, P. | 2000 |
| MSIS 2006: Model Curriculum and Guidelines for Graduate Degree Programs in Information Systems | GORGONE, J; GRAY, P. | 2006 |
| Student Competitions and Bots in an Introductory Programming Course | LADD, B.; HARCOURT, E. | 2005 |
| Teaching Software Engineering From a Maintenance-Centric View | GOKHALE, S; MCCARTNEY, R;SMITH, T. | 2013 |
| Teaching Software Engineering in a Compiler Project Course | GRISWOLD, W. | 2002 |
| The Education of a Software Engineer | JAZAYERI, M. | 2004 |
| The Risks and Benefits of Teaching Purely Functional Programming in First Year | CHAKRAVERTY, M. M. T.; KELLER, G. | 2004 |

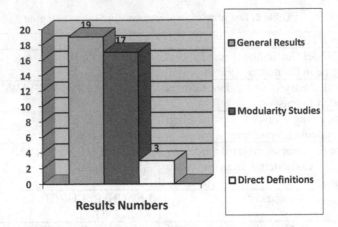

**Fig. 1.** Graphic of direct or indirect modularity citations

By reading all studies, it was possible to build a solid base to work in order to report about the techniques and teaching approaches used in the classroom. Furthermore, it is also possible to define the problems that were generated from the reports of studies and experiments of modularity teaching in the classroom.

The aim was always to answer the questions raised at the beginning of systematic mapping, and take advantage of answers based in problems and solutions of information technology courses.

The more information as possible was extracted, considering any relevant information contained in articles so that the information found become part of a constituted knowledge.

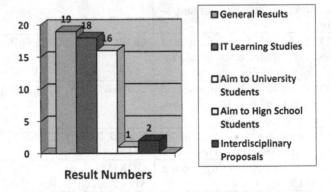

**Fig. 2.** Graphic of IT Learning related results

Eighteen of nineteen results found (about 94,73%) treated in some way about IT Courses. Most of them focused in teaching of subjects such as software engineering, but two papers proposed an interdisciplinary activity, to profit the learning.

One of those papers proposed the building of a virtual machine, through the reading of logic gates, coming to the interpretation of languages, to finalize with the emulation of a game, involving hardware, compilers and engineering software issues [27].

The other paper suggested the implementation of a compiler project, which would translate directly the syntax of a language to another. There is a separation roles between developer and tester, and the need for documentation development. [17].

The base of sixteen from nineteen results (about 84,21%) had as the main field, the academic area. However, it is worth mentioning a paper, aimed at students of an American High School. The article depicted the importance of knowledge about software design concepts for those who would enter to a course focused on IT [28].

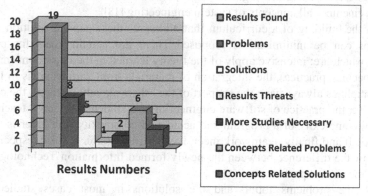

**Fig. 3.** Graphic of results related to problems and solutions found

Among all the results of the systematic study, eight of them define problems, being six of them, one way or the other related to teaching approach of certain computing concepts, including the modularity, in a theoretical or practical level (approximately 42,10% of articles found in general).

An article involves advantages and disadvantages in education functional programming in the first university levels [6]. Another article focuses on the issue of students learning in a discussion group [19]. The major focus was in educational problems of concepts, which undermine learning an important issue as the modularity.

The profile of well-defined mapping results, it is possible to make the panorama that was portrayed of the situation through them. The design of the situation in this area makes possible to construct this relationship.

These more specific problems arise from the very beginning. One of the examples is the distinction between human marker and automatic marker of programs. The primary concepts of a program at this time were the economy, simplicity, modularity and code structure. Thus, it was concluded at this time that the human markers were giving too much emphasis on cosmetic issues of a program. [27].

Problems in disciplines related to compiler projects are common. Students often do not finish the compiler projects at the right time, maybe because fail communication between group members, mistakes with compilers concepts, and others [17].

In another paper, a reengineering of modules with old codes from NASA paper was proposed to students. This research was restricted in types of modularity engagement. Students should include this information in an assignment table of modules. So, the correlation between the types of modular coupling and interaction levels, had been very low, which means the majority students in this research have a worrying general lag, then, participants were not able to make a reengineering analysis accurately. And theoretical understanding of principles such as modularity was absent in practice [21].

The problem is the care given to certain aspects of a professional in software engineering. In a way, the modularity can be one of those aspects, depending on the structure of certain courses. A greater approach of certain subjects was proposed to solve this problem, however, it could be impossible to be treated in just one or two subjects related to software engineering. If it were done, it would be very difficult for the student to remember all concepts of system engineering [18].

With the building of a curriculum that addresses these gaps, a solution where the problems can be minimized is proposed. There are certain goals for curriculum reform, which are: intensive apply of the theory learned in the classroom (making this theory become practice), the integration of materials from various areas in IT (so that new disciplines always have the support of the others already learned), the correct use of tools for the practice of software engineering and students' reviews, teaching about essential complexity of a program and accidental complexity (the latter is which can be avoided) and finally, teamwork incentive. The discussion is valid, especially when you know the difference between the newly formed Information Technology individual and the market professional [11].

Thus, the problems found and the solutions in most cases studied are explained. Between three articles that do not showed immediate solutions, one of them addressed the possibility of fail in the research, given the number of students with low performance reengineering [21], for the possibility of wrong approach to teaching structure, the absence of motivation by the students, in addition to teachers' own teaching method.

Another study was proposed, to make the automatic marker more credible to separate cosmetic factors to fundamental programming [13]. We can say that in these cases, the identification is made from the problem, so it will be possible to study about the identified factors.

Another paper search for more answers to set more specific problems of the gap between academic community and the professionals inserted into the IT area. The analysis will be done at that point and will occur a relation to stage of the results [11].

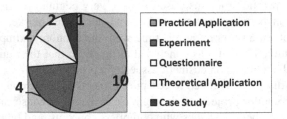

**Fig. 4.** Graphic of Stage of Studies

The graph in Figure 1 presents an overview by the progress of work on the results in general. Studies were also included not specified, that did not enter into this paper.

So we can see ten documents among the results with practical application (about 52,63% of results). Four documents are experiments (21,05% of results), two documents are theoretical (about 10,52%), two documents are questionnaires (exactly 10,52% of results) and the other valid resource is a case study (about 5,26% of results).

As for the theoretical approach of teaching, four papers presented models for teaching in the classroom (approximately 21,05% of results), namely: The Concept of Meaningful Learning, or Teaching by Performing Tasks [27], a Study by Primary Factors on Schedule [13], Problem Solving by Divide and Conquer [17] and The Use of Functional Language for Explanation of Initial Computing Concepts [6].

In addition, nine documents demonstrated techniques, or teaching direct activities (47,36% of the results). With the union of approaches and teaching techniques, important conclusions about the student learning can be noted. By observing techniques, there was a match between two or more studies using the same techniques.

There is a list of the high-lighted acts to minimize the learning problems: practical interdisciplinary projects, continuous experiments with students, use of free software tools, reinforcement the sense of responsibility, reinforcement of communication and teamwork, personal teaching and priority of actions into the programming activities.

There are some articles with difficulties and menaces to study, three of them (15,78%). Maybe, with some new approaches and techniques, next studies will be easier. There are also some curricula models from United States and Canada between the results (about 31,57%), and they will be useful for future researches.

Then, that makes possible to create the structure of the systematic mapping with the found results. The stage of the studies was chosen to sort the results, because of the easier division by groups.

**Table 2.** Results of frequency, by comparison between stage of studies and research objects of the documents

| Stage of Studies VS. Research Objects | Theoretical Application | Practical Application | Case Study | Experiment | Question-naire | Total |
|---|---|---|---|---|---|---|
| Modularity Studies | 2 | 8 | 1 | 4 | 2 | 17 |
| IT Learning Studies | 2 | 9 | 1 | 4 | 2 | 18 |
| Found Problems | 1 | 2 | 0 | 3 | 2 | 8 |
| Found Solutions | 1 | 2 | 0 | 1 | 1 | 5 |
| Techniques | 1 | 3 | 1 | 3 | 1 | 9 |
| Approaches | 1 | 2 | 0 | 1 | 1 | 4 |
| Study Difficulties | 0 | 1 | 0 | 1 | 1 | 3 |
| Modularity Definitions | 2 | 1 | 0 | 0 | 0 | 3 |
| Aim to Academic Community | 1 | 9 | 1 | 3 | 2 | 16 |
| Curricula Models | 0 | 6 | 0 | 0 | 0 | 6 |

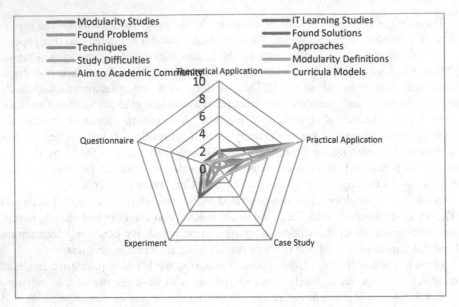

**Fig. 5.** Radar chart of results frequency, by comparison between stage of studies and research objects of the studies

## 5     Conclusion

The results obtained in this paper reveal different points of view about modularity and how to learn this concept. The systematic mapping brought up major details about the importance of modularity in IT area and how the current works analyze them. It was possible to determine different factors, which direct influence strongly IT learning. Also, there are too many possibilities to be explored.

### 5.1     Thoughts About the Results

Among many ways to be explored, it makes the necessity to keep in mind how the professor shares the knowledge about important concepts like modularity in a classroom. It is possible to know about some education techniques and approaches, but not all of them.

The environment where the students are inserted has also great importance, as the focus of a specific IT course. So, we need to look for a most accurate technique for each specific situation. To find that, the professor must evaluate the students sporadically, seeing if the students are absorbing the subject of the classes or not.

The possibility of verifying modules from simple programs or complete systems made by students exists. Also, the structural side of an IT course is important. With some curricula models found by systematic study, it makes possible to begin more researches about the general trends related to the modularity into IT courses and their learning.

## 5.2    Possibility of Future Researches

So we can say that the results into the mapping led to another points of interest. These points can be:

- Research about techniques and approaches into IT Courses, and their efficiency among the students;
- Students' knowledge about modular software projects;
- Students' experience into recognizing modularity errors;
- Research about disciplines from IT courses and their structure.

With some of these possibilities, we need to reflect about some points of view on the modularity studies. From the first to the third point, the research can be made in a more specific way, from IT courses occurring in a region of interest. The research objects can be registered from classes of some professors, and also programming activities made by the students.

The fourth point can be made in wider perspective. Disciplines from IT courses institutions around the world can be searched and, then, analyzed by a modularity perspective to evaluate how this concept is focused, from various courses and different places.

## References

1. Baldwin, C.Y., Clark, K.B.: Design Rules: The Power of Modularity, vol. 1. Massachusetts Institute of Technology (2000)
2. Bass, L., Clements, P., Kazman, R.: Software Architecture in Practice, 2nd edn. Addison Wesley (2013)
3. Brereton, P., et al.: Using mapping studies in software engineering. In: Proceedings of PPIG (2008)
4. Budgen, D., Brereton, P.: Performing systematic literature in software engineering. In: Proceeding of the 28th International Conference on Software Engineering ICSE 2006. ACM Press, New York (2006)
5. Cai, Y., et al.: Introducing Tool-Supported Architecture Review into Software Design Education (2013)
6. Chakravarty, M.M.T., Keller, G.: The Risks and Benefits of Teaching Purely Functional Programming in First Year. Journal of Functional Programming 14, 113–123 (2004)
7. Colfer, L., Baldwin, C.Y.: The Mirroring Hypothesis: Theory, Evidence and Exceptions. Harvard Business School (2010)
8. Davis, G.B., et al.: IS 1997: Model Curriculum and Guidelines for Undergraduate Degree Programs in Information Systems (1997)
9. Draft, S.: Computer Science Curricula 2013. ACM, IEEE Computer Society (2013)
10. Engel, G., Roberts, E.: Computing Curricula 2001 Computer Science. ACM, IEEE Computer Society (2001)
11. Exter, M., Turnage, N.: Exploring Experienced Professionals' Reflections on Computing Education. ACM Transactions on Computer Education 12, 1–23 (2012)
12. Fernandes, J.M.M., Carneiro, G.F.: Estratégias e Perfis de Programadores Iniciantes na Idenrificação de Anomalias de Modularidade de Software. UNIFACS, Bahia (2012)

13. Fleming, W.H., Redish, K.A., Smyth, W.F.: Comparison of manual and automated marking of student programs. Information and Software Technology **30**, 547–552 (1988)
14. Gokhale, S., Mccartney, R., Smith, T.: Teaching Software Engineering form a Maintenance-Centric View. Consortium for Computing Sciences in Colleges (2013)
15. Gorgone, J.T., Gray, P.: MSIS 2000: Model Curriculum and Guidelines for Graduate Degree Programs in Information Systems (2000)
16. Gorgone, J.T., et al.: MSIS 2006: Model Curriculum and Guidelines for Graduate Degree Programs in Information Systems (2006)
17. Griswold, W.G.: Teaching Software Engineering in a Compiler Project Course. Journal on Educational Resources in Computing, 1–18 (2002)
18. Jazayeri, M.: The education of a software engineer. In: Proceedings of the 19th International Conference on Automated Software Engineering (ASE 2004) (2004)
19. Kang, J.-H., Kim, J., Shaw, E.: A network analysis of student groups in threaded discussions. In: Aleven, V., Kay, J., Mostow, J. (eds.) ITS 2010, Part II. LNCS, vol. 6095, pp. 359–361. Springer, Heidelberg (2010)
20. Ladd, B., Harcourt, E.: Student Competitions and Bots in an Introductory Programming Course. Journal of Computing Sciences in Colleges, 274–284 (2005)
21. Leach, R.J., Burge, L.L., Keeling, H.N.: Can Students Reengineer? (2008)
22. Parnas, D.L.: On the Criteria to be Used in Decomposing Systems into Modules. Comunications of ACM **15** (1972)
23. Petersen, K., et al.: Systematic Mapping Studies in Software Engineering (2008)
24. Richards, B.: A 2007 Model Curriculum for a Liberal Arts Degree in Computer Science. ACM Journal on Educational Resources in Computing **7** (2007)
25. Sanchez, R., Mahoney, J.T.: Modularity, Flexibility, and Knowledge Management in Product and Organization Design. Strategic Management Journal **17** (1996)
26. Sanz-Rodrigues, J., Deodoro, J.M., Sanchez-Alonso, S.: Metrics-based evaluation of learning object reusability. Software Quality Journal, 121–140 (2010)
27. Schocken, S., Nisan, N.; Armoni, M.: A synthesis course in hardware architecture, compilers and software engineering. ACM SIGCSE Bulletin, 443–447 (2009)
28. Sherrell, L., Mccauley, L.: A programming competition for high school students emphasizing process. In: Proceedings at the 2nd Annual Mid-South College Computing Conference, pp. 173–182 (2004)
29. Soares, S.: O que é modularidade? December 18, 2014. http://www.cin.ufpe.br/~scbs/ceut/fundamentosES/EXTRA_Modularidade.pdf

# Mobile Application Verification:
# A Systematic Mapping Study

Mehmet Sahinoglu[1,3(✉)], Koray Incki[2], and Mehmet S. Aktas[3]

[1] TUBITAK BILGEM Center for Software Test and Quality Assessment, Kocaeli, Turkey
[2] Department of Computer Engineering, Adana Science and Technology University,
Seyhan/Adana, Turkey
kincki@adanabtu.edu.tr
[3] Department of Computer Engineering, Yıldız Technical University, Istanbul, Turkey
mehmet.sahinoglu@tubitak.gov.tr, mehmet@ce.yildiz.edu.tr

**Abstract.** The proliferation of mobile devices and applications has seen an un-precedented rise in recent years. Application domains of mobile systems range from personal assistants to point-of-care health informatics systems. Software development for such diverse application domains requires stringent and well-defined development process. Software testing is a type of verification that is required to achieve more reliable system. Even though, Software Engineering literature contains many research studies that address challenging issues in mobile application development, we could not have identified a comprehensive literature review study on this subject. In this paper, we present a systematic mapping of the Software Verification in the field of mobile applications. We provide definitive metrics and publications about mobile application testing, which we believe will allow fellow researchers to identify gaps and research opportunities in this field.

**Keywords:** Verification · Software testing · Mobile application · Systematic mapping · Literature review

## 1 Introduction

Software Testing is the most frequently utilized software verification technique, which aims to ensure bug-free and reliable software products on the market. That is why, it is one of the most important topics in software development and quality assurance. Craig and Jaskiel describe software testing as "A concurrent life cycle process of engineering, using and maintaining test ware in order to measure and improve the quality of the software being tested." [24]. The purpose of testing is to improve software quality by finding bugs before these bugs cause serious effects [25].

In today's world, all sorts of mobile devices including mobile phones and personal digital assistants (PDAs) changed the way we use technology and live our daily lives. Such a world-wide network of uniquely addressable, interconnected objects (Cisco predicts 50 billion connected objects by 2020) [27] are heading into a new era of ubiquity, where the "users" of the Internet will be counted in billions and where humans may become the minority as generators and receivers of traffic.

© Springer International Publishing Switzerland 2015
O. Gervasi et al. (Eds.): ICCSA 2015, Part V, LNCS 9159, pp. 147–163, 2015.
DOI: 10.1007/978-3-319-21413-9_11

Mobile Application Testing is defined as "Mobile application testing is a process by which application software developed for hand held mobile devices is tested for its functionality, usability and consistency."[30] Thus, mobile application testing is equally critical as the verification of general purpose computer programs; moreover, it is a trending topic as shown by Google trends in Figure-1 [21]. Figure 1 shows us number of searches on Google for a specific term in a period of time. One can see in Figure-1, the number of searches for "Mobile Application Testing" is increasing over time.

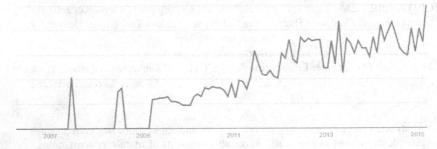

**Fig. 1.** Search Volume Index for "Mobile Application Testing" [21]

When we scan literature for secondary studies, we found a survey study for model-based GUI testing[22] and another survey study for usability testing[23]. As opposed to the importance and popularity of the topic (Figure-1), we could not find any literature review study that covers mobile application (software) testing in general.

In this paper, we provide the results of our review on the subject matter in order to identify gaps, trending opportunities and facilitate future research directions by presenting them in the form of a systematic mapping. We provide definitive metrics and publications about mobile application testing, which we believe will allow fellow researchers to identify gaps and research opportunities in this field.

The rest of the paper is structured as follows: Section 2 Systematic Mapping Process where we defined our steps while doing this literature review study. Section 3 answers the research questions with synthesized data that extracted from primary studies. Section 4 describes the threats to the validity for the mapping study, and Section 5 presents a conclusion and future work of the study.

## 2    Systematic Mapping Process

A primary study is defined as a firsthand empirical work in an area, where secondary study is defined as "A study that reviews all the primary studies relating to a specific research question with the aim of integrating/synthesising evidence related to a specific research question." [29]

We have applied systematic mapping study method to our secondary study research[3][29] on topic of Mobile Application Verification: A Systematic Mapping Study.

## 2.1    Definition of Research Scope

We asked specific questions to define body of knowledge about related literature. Research questions and their main motivations to research are shown below.

**Table 1.** Research Questions

| Research Questions | Main Motivation |
|---|---|
| RQ1: What are the most frequently used test types for mobile applications? (Compatibility, Concurrent, Conformance, Performance, Security, Usability) | Identify trends and opportunities for mobile application testing. |
| RQ2: Which research issues in mobile application testing are addressed and how many papers cover the different research issues? (Test Execution Automation, Test Case Generation, Test Environment Management, Testing on Cloud, Model Based Testing) | Identify which research areas are widely used in mobile application testing papers. |
| RQ3: At what test level have researchers' studies most frequently? (Unit, Component, Integration, System, Acceptance) | Provide information for most frequently studied test level |
| RQ4: What is the paper-publication frequency? | Provide a trend analyze graph |
| RQ5: Which journals include papers on mobile application testing? | Identify journals which researchers may publish their studies |

## 2.2    Search for Primary Studies

As a first step of a literature review study, we searched digital databases to get primary studies [3]. We selected keywords: mobile, application and test, then used their synonyms to construct search strings (Table 2).

**Table 2.** Search Strings

| |
|---|
| (Mobile) |
| **AND** |
| (Application OR Software) |
| **AND** |
| (Test OR Testing OR Validation OR Verification) |

After deciding search strings we applied same search strings to 4 digital library and collected the results without any time constraint (Table 3). In order to monitor popularity increase of the topic, we did not limit search by year. In turn, we get the results those are published until June 2014. Thus; we did not include studies that published after June 2014. Downloading, managing and working on collection of papers are complex tasks that cannot be done with manual techniques using spreadsheet applications. Hence, we used several reference management tools such as Zotero [4] and JabRef [5]. We used Zotero to download references automatically from digital databases, and used JabRef to manage all downloaded studies, to merge them, to apply filter and to apply exclusion criteria.

**Table 3.** Search Result in Digital Libraries

| No | Name | Result |
|----|------|--------|
| 1 | IEEE | 8.157 |
| 2 | ACM | 1.908 |
| 3 | Science Direct | 1.218 |
| 4 | Springer Link | 1.516 |
| | **Total** | **12799** |

We collected search results and merged them with eliminating duplicate studies that were downloaded from different digital databases. We applied three filters (Figure 2) to get primary studies. The filters that we applied to search results (12799 papers) described in Table-4. At the end of the filtering process, we got 123 primary studies (see Appendix A).

**Table 4.** Filters Applied to Search Results

| Filters | Description |
|---------|-------------|
| Filter-1 | We excluded studies from irrelevant journals and conferences. |
| Filter-2 | We read the study title and abstract and to eliminate clearly irrelevant papers. |
| Filter-3 | We read abstracts deeply and some cases full text to apply our inclusion and exclusion criteria. |

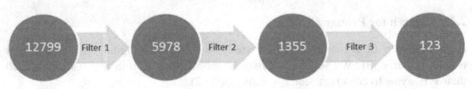

**Fig. 2.** This figure illustrates the changing number of target papers as the filters, described in Table-4, applied

## 2.3    Inclusion and Exclusion Criteria

In systematic mapping studies, we need to select some papers in purpose of answering the research questions. We use inclusion and exclusion criteria to have an objective study selection process. We define these criteria at the beginning of the selection process, to avoid selection bias.[1][28]

We applied inclusion and exclusion criteria to filter two results on purpose of select studies as our primary studies.

**Inclusion Criteria:** We included studies that contribute mobile application (software) testing in a particular way such as a theory, a solution, a practice, a strategy or an approach. In addition to regular research papers, we also took into account short papers. When a study has several versions, we accepted the new one, except the case that if there was any extension to the original study. In that case, we included both versions.

**Exclusion Criteria:** We excluded studies that do not clearly linked to mobile application testing, and do not clearly investigate the topic. Book chapters, secondary studies, thesis, and experimental studies, those do not propose a solution or an approach, were also excluded. Studies on testing mobile devices, hardware of mobile devices or hardware used by mobile application were also excluded, because we limit our mapping study only with mobile software testing.

## 2.4    Classification Schema

We grouped 123 studies, which we had after filtering process, with a view of software testing as test levels [19] and test types [20]. In order to classify researchers' contribution areas [20] we also categorized the research issues. We classified studies according to Table 5.

**Table 5.** Classification Framework

| Category | Subcategory | Definition |
|---|---|---|
| Test Levels | System Testing | "Testing conducted on a complete, integrated system to evaluate the system's compliance with its specified requirements." [12] |
| | Acceptance Testing | "Formal testing conducted to enable a user, customer, or other authorized entity to determine whether to accept a system or component." [13] |
| | Unit Testing | "Testing of individual hardware or software units or groups of related units." [12] |
| | Component Testing | "Testing of individual hardware or software components or groups of related components." [12] |
| | Integration Testing | "Testing in which software components, hardware components, or both are combined and tested to evaluate the interaction between them." [12] |
| Test Types | Compatibility | "The ability of two or more systems or components to perform their required functions while sharing the same hardware or software environment." [12] |
| | Concurrency Testing | "Testing to determine how the occurrence of two or more activities within the same interval of time, achieved either by interleaving the activities or by simultaneous execution, is handled by the component or system." [12] |
| | Conformance Testing | "Conformance testing is testing to see if an implementation meets the requirements of a standard or specification." [15] |
| | Performance Testing | "Testing conducted to evaluate the compliance of a system or component with specified performance requirements." [12] |
| | Security Testing | "Testing to determine the security of the software product." [14] |
| | Usability Testing | "Testing to determine the extent to which the software product is ATA understood, easy to learn, easy to operate and attractive to the users under specified conditions." [12] |

**Table 5.** (*Continued.*)

| | | |
|---|---|---|
| Research Issues | Test Execution Automation | "The use of software, e.g. capture/playback tools, to control the execution of tests, the comparison of actual results to expected results, the setting up of test preconditions, and other test control and reporting functions." [14] |
| | Test Case Generation | "A computational method for identifying test cases from data, logical relationships or other software requirements information." [17] |
| | Test Environment: | "An environment containing hardware, instrumentation, simulators, software tools, and other support elements needed to conduct a test." [12] |
| | Cloud Testing | "Cloud Testing uses cloud infrastructure for software testing. Cloud computing offers use of virtualized hardware, effectively unlimited storage, and software services that can aid in reducing the execution time of large test suites in a cost-effective manner." [18] |
| | Model-based testing | "Testing based on a model of the component or system under test, e.g., reliability growth models, usage models such as operational profiles or behavioral models such as decision table or state transition diagram." [14] |

## 2.5    Data Extraction and Mapping of the Literature

After selecting primary research papers we extracted data from studies, on purpose of analyze them. We used a data extraction form shown in Table 6. During extraction process, if a paper did not clearly define research issue or testing type, we left this section blank, to avoid making any assumptions.

Our data extraction methodology is as follows: Before starting data extraction process for 123 studies, we randomly chose a small dataset. Then, We extracted data according to Table-6. A different author also did the extraction process for same data set. We did a cross-check. When we found a conflict, we went back to this study to make an inspection, and identified the reason caused to conflict. After we satisfied all these points worked on the dataset, we applied data extraction process to 123 papers to get the results.

**Table 6.** Form for Data Extraction

| No | Data Extraction Columns |
|---|---|
| 1. | Study ID (example: S1) |
| 2. | Title: Paper's title (example: Big Data Processing Research: A Systematic Review) |
| 3. | Author: Author's name |
| 4. | Year: Publication year (example: 2014) |
| 5. | Conference or journal name: Any conference or journal name (ex: IEEE Big Data 2014) |
| 6. | Test Levels: System testing or unit testing |
| 7. | Test Type: Performance, conformance |
| 8. | Research Issues: Test automation, Test environment management |
| 9. | Test Environment: Real Device, Target System |
| 10. | Contribution Type: Process / method / model / framework / metric |
| 11. | Research Method: Theory / Survey / Experiment / Own Experience / Review |
| 12. | Study Context: Student or simple projects / Professional Projects / None |

# 3    Results

This section discusses outputs of the data extraction process that are described in Section 2.3. Each research question was answered and evaluated separately but, in order to improve understanding, we grouped the results according to the Table.5.

## 3.1    RQ1: What are the Most Frequently Used Test Types for Mobile Applications? (Compatibility, Concurrent, Conformance, etc.)

The distribution of publications according to testing type/technique is shown in Figure-3. The most frequently researched testing techniques are functional testing and usability testing.

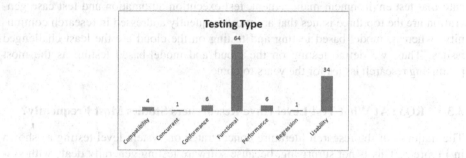

**Fig. 3.** Frequency of mobile application testing type in published papers

Mobile applications tend to prevail as the dominant computing platform for the upcoming decades; because, proliferation of internet with the Internet of Things concept shall entail employing much more mobile applications in very diverse domains. Thus, performance issues will be of great interest in resource constrained embedded devices.

RQ1 seeks to identify research trends and opportunities in testing techniques. As one can tell from Figure-3, functional and usability testing types have already been studied in many research papers. We think that performance testing may provide opportunities in terms of research in mobile application testing research field.

## 3.2    RQ2: Which Research Issues in Mobile Application Testing are Addressed and How Many Papers Cover the Different Research Issues?

We have described several research issues that relate to research in mobile testing. The literature review we conducted yielded the results shown in Figure-4. After careful inspection and applying inclusion-exclusion criteria, a total of ninety five (95) publications are classified into six (6) different research issues.

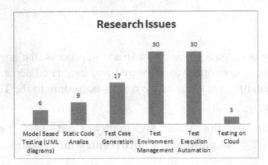

**Fig. 4.** The distribution of publications according to research issues

RQ2 seeks to identify the opportunities for open research issues. The results indicate that test environment management, test execution automation and test case generation are the top three issues that are most frequently addressed in research community; whereas, model-based testing and testing on the cloud are the least challenged issues. Thus, we define testing on the cloud and model-based testing as the most promising research issues for the years to come.

### 3.3    RQ3 : At What Test Level Have Researchers Studies Most Frequently?

The majority of the research literature concentrated on system level testing as shown in Figure-5. This is not surprising, because software testing generally deals with system level challenges. This infers that any new research to be conducted should concentrate on system level testing problems.

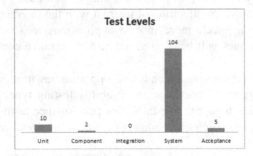

**Fig. 5.** The distribution of publications according to Mobile Application Test Levels

### 3.4    RQ4. What Is the Papers Publication Frequency?

We have identified that the literature on mobile application domain picked up an upward momentum with the proliferation of mobile devices in everyday life (Figure-6). The results shown in Figure-6 indicate that the total number of publications per individual year. We think that this is revealing a correlation with the maturity of the subject domain.

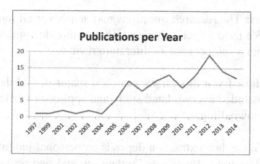

**Fig. 6.** Publication frequency per year

## 3.5    RQ5: Which Journals Include Papers on Mobile Application Testing?

**Table 7.** Journal Names

| Journal No | Journal Names |
| --- | --- |
| 1 | Computer |
| 2 | Information and Software Technology |
| 3 | IT Professional |
| 4 | Procedia Computer Science |
| 5 | Procedia Technology |
| 6 | Software Engineering, IEEE Transactions on |
| 7 | Software, IEEE |
| 8 | Advances in Computers |
| 9 | Software Engineering Notes |
| 10 | Universal Access in the Information Society |
| 11 | Computing |
| 12 | Systems and Service-Oriented Engineering |
| 13 | Journal of Interaction Science |
| 14 | Software Testing, Verification & Reliability |
| 15 | Mobile Networks and Applications |
| 16 | Personal and Ubiquitous Computing |
| 17 | Multimedia Tools and Applications |
| 18 | Interactions |
| 19 | Software Quality Journal |
| 20 | The Journal of Supercomputing |
| 21 | Tsinghua Science & Technology |

## 4    Threats to Validity

In general, we may consider wrong decision-makings in the followings as threats to validity for the mapping study: research questions, publication selection, and data extraction. We describe each possible treat below and give our methodology to eliminate them.

**Research Questions:** The research questions may not cover all mobile software testing area in detail. We tried to get only a snapshot of mobile application testing literature. This study is not a subject specific literature review.

**Publication Selection:** Even though we searched major digital databases we cannot guarantee that we included every related study and applied our criteria.[2] To mitigate that threat we did snowballing.

**Data Extraction:** Since data extraction depends on personal judgment, there is possibility of incorrect decision. We selected a data set, and two researchers did the data extraction. Then we compared with others answers. When there is a disagreement, we both checked, talked and decide to a settlement.

## 5     Conclusion and Future Work

Our motivation was to summarize existing studies in mobile application testing area and conduct a gap analysis. The main contribution of this study is to provide a mapping of the current-state of art on mobile application testing.

By analyzing the results shown in Figure-3, Figure-4 and Figure 5, we conclude that software testing research is open to new contributions in general. In particular, research on performance testing of mobile applications may provide more research opportunities as we observe lack of research studies in that area. The results also indicate immerging research needs on *mobile application testing on the cloud*, which deal with *test execution automation* or *test environment management* for *system level functional testing.* Analyzing the results, we also observe that previous research gives high importance to system level testing problems.

Work remains in extending this study to conduct a systematic literature review on mobile application testing. As future work, we will also investigate how to devise a framework for mobile system testing framework on cloud computing platforms.

## References

1. Novais, R.L., Torres, A., Mendes, T.S., Mendonça, M., Zazworka, N.: Software evolution visualization: A systematic mapping study. Inf. Softw. Technol. **55**, 11 (2013)
2. Jamshidi, P., Ahmad, A., Pahl, C.: Cloud Migration Research: A Systematic Review. IEEE Transactions on Cloud Computing **1**(2), 142–157 (2013)
3. Petersen, K., Feldt, R., Mujtaba, S., Mattsson, M.: Systematic mapping studies in software engineering. In: Proceedings of the 12th International Conference on Evaluation and Assessment in Software Engineering (EASE 2008) (2008)
4. Web Site for Zatero Reference Management Tool (access date: February 15, 2015). https://www.zotero.org/
5. Web Site for JabRef Reference Management Tool (access date: February 15, 2015). http://jabref.sourceforge.net/
6. Web Site for Journal (access date: February 15, 2015). http://www.journals.elsevier.com/acta-astronautica/

7. Web Site for Journal (access date: February 15, 2015). http://www.journals.elsevier.com/analytica-chimica-acta/

8. Web Site for Journal (access date: February 15, 2015). http://www.journals.elsevier.com/biosensors-and-bioelectronics/

9. Web Site for the Journal (access date: February 15, 2015). http://www.sciencedirect.com/science/journal/07317085

10. Web Site for Journal (access date: February 15, 2015). http://www.vlsisymposium.org/

11. Web Site for Journal (access date: February 15, 2015). http://www.ieeevtc.org/

12. IEEE Standard Computer Dictionary: A Compilation of IEEE Standard Computer Glossaries. IEEE Std 610, pp. 1–217, January 18, 1991

13. IEEE Standard for System and Software Verification and Validation. IEEE Std 1012–2012 (Revision of IEEE Std 1012-2004), pp. 1–223, May 25, 2012

14. ISTQB® Glossary of Testing Terms Version: 2.4 (access date: February 15, 2015). http://www.istqb.org/downloads/viewcategory/20.html

15. Overview of Conformance Testing (access date: February 15, 2015). http://www.nist.gov/itl/ssd/is/overview.cfm

16. ISO/IEC, ISO/IEC 9126-1 Software engineering- Product quality- Part 1: Quality model (2001)

17. American Psychological Association (APA): The Free On-line Dictionary of Computing. (access date: February 15, 2015). Dictionary.com website: http://dictionary.reference.com/browse/algorithmictestcasegeneration (retrieved February 22, 2015)

18. Tilley, S.; Parveen, T.: Migrating software testing to the cloud. In: 2010 IEEE International Conference on Software Maintenance (ICSM), p. 1, September 12–18, 2010. doi: 10.1109/ICSM.2010.5610422

19. Craig, R.D., Jaskiel, S.P.: Systematic Software Testing. Artech House Inc., Norwood (2002)

20. İnçki, K., Ari, I., Sozer, H.: A Survey of Software Testing in the Cloud. In: 2012 IEEE Sixth International Conference on Software Security and Reliability Companion (SERE-C), pp. 18–23, June 20–22, 2012

21. Google Trend Result for "Mobile Application Testing" (access data: February 15, 2015). http://www.google.com/trends/explore?hl=en-US&q=mobile+application+testing&cmpt=date&tz&tz&content=1

22. Janicki, M., Katara, M., Pääkkönen, T.: Obstacles and opportunities in deploying model-based GUI testing of mobile software: a survey. Softw. Test. Verif. Reliab. **22**, 5 (2012)

23. Zhang, D., Adipat, B.: Challenges, Methodologies, and Issues in the Usability Testing of Mobile Applications. International Journal of Human-Computer Interaction, Taylor & Francis **18**(3), 293–308 (2005)

24. Craig, R.D., Jaskiel, S.P.: Systematic Software Testing. Artech House Publishers, Boston (2002)

25. Kaner, C., Falk, J., Nguyen, H.Q.: Testing Computer Software, 2 edn. Dreamtech Press (2000)

26. Budgen, D., Brereton, P.: Performing systematic literature reviews in software engineering. In: Proceedings of the 28th International Conference on Software Engineering (ICSE 2006) (2006)

27. Evans, D.: The Internet of Things How the Next Evolution of the Internet Is Changing Everything (2011) (access date: February 15, 2015). Web Site for white paper: http://www.cisco.com/web/about/ac79/docs/innov/IoT_IBSG_0411FINAL.pdf

28. Budgen, D., Brereton, P.: Performing systematic literature reviews in software engineering. In: Proceedings of the 28th International Conference on Software Engineering (ICSE 2006) (2006)
29. Kitchenham, B., Charters, S.: Guidelines for performing Systematic Literature Reviews in Software Engineering (2007)
30. Web Site for smartbear (access date: February 15, 2015). http://smartbear.com/all-resources/articles/what-is-mobile-testing/

## Appendix A.  Selected Primary Studies

| | References |
|---|---|
| S01 | Amalfitano, D.; Fasolino, A.R.; Tramontana, P., "A GUI Crawling-Based Technique for Android Mobile Application Testing," Software Testing, Verification and Validation Workshops (ICSTW), 2011 IEEE Fourth International Conference on , vol., no., pp.252,261, 21-25 March 2011 |
| S02 | Amalfitano, D.; Fasolino, A.R.; Tramontana, P.; Amatucci, N., "Considering Context Events in Event-Based Testing of Mobile Applications," Software Testing, Verification and Validation Workshops (ICSTW), 2013 IEEE Sixth International Conference on , vol., no., pp.126,133, 18-22 March 2013 |
| S03 | Amalfitano, D.; Fasolino, A.R.; Tramontana, P.; De Carmine, S.; Memon, A.M., "Using GUI ripping for automated testing of Android applications," Automated Software Engineering (ASE), 2012 Proceedings of the 27th IEEE/ACM International Conference on , vol., no., pp.258,261, 3-7 Sept. 2012 |
| S04 | Amalfitano, D.; Fasolino, A.; Tramontana, P.; Ta, B.; Memon, A., "MobiGUITAR -- A Tool for Automated Model-Based Testing of Mobile Apps," Software, IEEE , vol.PP, no.99, pp.1,1 |
| S05 | Stephan Arlt, Cristiano Bertolini, Simon Pahl, Martin Schäf, Chapter 6 - Trends in Model-based GUI Testing, In: Ali Hurson and Atif Memon, Editor(s), Advances in Computers, Elsevier, 2012, Volume 86, |
| S06 | Rebecca Baker, Xiaoning Sun, and Bob Hendrich. 2011. Testing touch: emulators vs. devices. In Proceedings of the 4th international conference on Internationalization, design and global development (IDGD'11), P. L. Patrick Rau (Ed.). Springer-Verlag, Berlin, Heidelberg, 135-142. |
| S07 | Florence Balagtas-Fernandez and Heinrich Hussmann. 2009. A Methodology and Framework to Simplify Usability Analysis of Mobile Applications. In Proceedings of the 2009 IEEE/ACM International Conference on Automated Software Engineering (ASE '09). IEEE Computer Society, Washington, DC, USA, 520-524. |
| S08 | Srikanth Baride and Kamlesh Dutta. 2011. A cloud based software testing paradigm for mobile applications. SIGSOFT Softw. Eng. Notes 36, 3 (May 2011 |
| S09 | Ana Correia de Barros, Roxanne Leitão, Jorge Ribeiro, Design and Evaluation of a Mobile User Interface for Older Adults: Navigation, Interaction and Visual Design Recommendations, Procedia Computer Science, Volume 27, 2014, Pages 369-378, |
| S10 | Selin Benli, Anthony Habash, Andy Herrmann, Tyler Loftis, and Devon Simmonds. 2012. A Comparative Evaluation of Unit Testing Techniques on a Mobile Platform. In Proceedings of the 2012 Ninth International Conference on Information Technology - New Generations (ITNG '12). |
| S11 | Birgitta Bergvall-Kåreborn and Staffan Larsson. 2008. A case study of real-world testing. In Proceedings of the 7th International Conference on Mobile and Ubiquitous Multimedia (MUM '08). |
| S12 | Berkenbrock, C.D.M.; da Silva, A.P.C.; Hirata, C.M., "Designing and evaluating interfaces for mobile groupware systems," Computer Supported Cooperative Work in Design, 2009. CSCWD 2009. 13th International Conference on , vol., no., pp.368,373, 22-24 April 2009 |
| S13 | Marco Billi, Laura Burzagli, Tiziana Catarci, Giuseppe Santucci, Enrico Bertini, Francesco Gabbanini, and Enrico Palchetti. 2010. A unified methodology for the evaluation of accessibility and usability of mobile applications. Univers. Access Inf. Soc. 9, 4 (November 2010), |
| S14 | Robert V. Binder and James E. Hanlon. 2005. The advanced mobile application testing environment. In Proceedings of the 1st international workshop on Advances in model-based testing (A-MOST '05) |

S15  Jiang Bo, Long Xiang, and Gao Xiaopeng. 2007. MobileTest: A Tool Supporting Automatic Black Box Test for Software on Smart Mobile Devices. In Proceedings of the Second International Workshop on Automation of Software Test (AST '07)

S16  Willem-Paul Brinkman, Reinder Haakma, and Don G. Bouwhuis. 2004. Empirical usability testing in a component-based environment: improving test efficiency with component-specific usability measures. In Proceedings of the 2004 international conference on Engineering Human Computer Interaction and Interactive Systems (EHCI-DSVIS'04)

S17  Canfora, G.; Mercaldo, F.; Visaggio, C.A.; D'Angelo, M.; Furno, A.; Manganelli, C., "A Case Study of Automating User Experience-Oriented Performance Testing on Smartphones," Software Testing, Verification and Validation (ICST), 2013 IEEE Sixth International Conference on , vol., no., pp.66,69, 18-22 March 2013

S18  Vivien Chinnapongse, Insup Lee, Oleg Sokolsky, Shaohui Wang, and Paul L. Jones. 2009. Model-Based Testing of GUI-Driven Applications. In Proceedings of the 7th IFIP WG 10.2 International Workshop on Software Technologies for Embedded and Ubiquitous Systems (SEUS '09)

S19  Corral, L.; Sillitti, A.; & Succi, G.; (2014) Software assurance practices for mobile applications. A Survey of the State of the Art. Computing. Springer.

S20  Cira Cuadrat Seix, Montserrat Sendín Veloso, and Juan José Rodríguez Soler. 2012. Towards the validation of a method for quantitative mobile usability testing based on desktop eyetracking. In Proceedings of the 13th International Conference on Interacción Persona-Ordenador (INTERACCION '12)

S21  Dantas, V.L.L.; Marinho, F.G.; da Costa, A.L.; Andrade, R.M.C., "Testing requirements for mobile applications," Computer and Information Sciences, 2009. ISCIS 2009. 24th International Symposium on , vol., no., pp.555,560, 14-16 Sept. 2009

S22  M. E. Delamaro, A. M. R. Vincenzi, and J. C. Maldonado. 2006. A strategy to perform coverage testing of mobile applications. In Proceedings of the 2006 international workshop on Automation of software test

S23  Dhanapal, K.B.; Deepak, K.S.; Sharma, S.; Joglekar, S.P.; Narang, A.; Vashistha, A.; Salunkhe, P.; Rai, H.G.N.; Somasundara, A.A.; Paul, S., "An Innovative System for Remote and Automated Testing of Mobile Phone Applications," SRII Global Conference (SRII), 2012 Annual , vol., no., pp.44,54, 24-27 July 2012

S24  Edmondson, J.; Gokhale, A.; Sandeep Neema, "Automating testing of service-oriented mobile applications with distributed knowledge and reasoning," Service-Oriented Computing and Applications (SOCA), 2011 IEEE International Conference on , vol., no., pp.1,4, 12-14 Dec. 2011

S25  Ermalai, I.; Onita, M.; Vasiu, R., "Testing the viability of podcasting in a particular eLearning system," Electronics and Telecommunications (ISETC), 2010 9th International Symposium on , vol., no., pp.411,414,

S26  Esipchuk, I.A.; Vavilov, D.O., "PTF-based Test Automation for JAVA Applications on Mobile Phones," Consumer Electronics, 2006. ISCE '06. 2006 IEEE Tenth International Symposium on , vol., no., pp.1,3

S27  Eugster, P.; Garbinato, B.; Holzer, A., "Pervaho: A Development & Test Platform for Mobile Ad hoc Applications," Mobile and Ubiquitous Systems: Networking & Services, 2006 Third Annual International Conference on , vol., no., pp.1,5, July 2006

S28  Fetaji, M.; Dika, Z.; Fetaji, B., "Usability testing and evaluation of a mobile software solution: A case study," Information Technology Interfaces, 2008. ITI 2008. 30th International Conference on , vol., no., pp.501,506, 23-26 June 2008

S29  André L. L. de Figueiredo, Wilkerson L. Andrade, and Patrícia D. L. Machado. 2006. Generating interaction test cases for mobile phone systems from use case specifications. SIGSOFT Softw. Eng. Notes 31, 6

S30  Derek Flood, Rachel Harrison, and Claudia Iacob. 2012. Lessons learned from evaluating the usability of mobile spreadsheet applications. 4th international conference on Human-Centered Software Engineering)

S31  Philip W. L. Fong. 2004. Proof Linking: A Modular Verification Architecture for Mobile Code Systems. Ph.D. Dissertation. Simon Fraser University, Burnaby, BC, Canada

S32  Franke, D.; Kowalewski, S.; Weise, C.; Prakobkosol, N., "Testing Conformance of Life Cycle Dependent Properties of Mobile Applications," Software Testing, Verification and Validation (ICST), 2012 IEEE Fifth International Conference on , vol., no., pp.241,250, 17-21 April 2012

S33  Franke, D.; Kowalewski, S.; Weise, C.; Prakobkosol, N., "Testing Conformance of Life Cycle Dependent Properties of Mobile Applications," Software Testing, Verification and Validation (ICST), 2012 IEEE Fifth International Conference on , vol., no., pp.241,250, 17-21 April 2012

S34  Fritz, G.; Paletta, L., "Semantic analysis of human visual attention in mobile eye tracking applications," Image Processing (ICIP), 2010 17th IEEE International Conference on , vol., no., pp.4565,4568, 26-29

S35    Stefano Gandini, Danilo Ravotto, Walter Ruzzarin, Ernesto Sanchez, Giovanni Squillero, and Alberto Tonda. 2009. Automatic detection of software defects: an industrial experience. In Proceedings of the 11th Annual conference on Genetic and evolutionary computation (GECCO '09)

S36    Hendrik Gani, Caspar Ryan, and Pablo Rossi. 2006. Runtime metrics collection for middleware supported adaptation of mobile applications. In Proceedings of the 5th workshop on Adaptive and reflective middleware (ARM '06)

S37    Gao, J.; Xiaoying Bai; Wei-Tek Tsai; Uehara, T., "Mobile Application Testing: A Tutorial," Computer , vol.47, no.2, pp.46,55, Feb. 2014

S38    Garcia Laborda, J.; Magal-Royo, T.; Gimenez Lopez, J.L., "Common problems of mobile applications for foreign language testing," Interactive Collaborative Learning (ICL), 2011 14th International Conference on , vol., no., pp.95,97, 21-23 Sept. 2011

S39    Gatsou, C.; Politis, A.; Zevgolis, D., "Exploring inexperienced user performance of a mobile tablet application through usability testing.," Computer Science and Information Systems (FedCSIS), 2013 Federated Conference on , vol., no., pp.557,564, 8-11 Sept. 2013

S40    Ghiron, S.L.; Sposato, S.; Medaglia, C.M.; Moroni, A., "NFC Ticketing: A Prototype and Usability Test of an NFC-Based Virtual Ticketing Application," Near Field Communication, 2009. NFC '09. First International Workshop on , vol., no., pp.45,50, 24-24 Feb. 2009

S41    Vlado Glavinic, Sandi Ljubic, and Mihael Kukec. 2011. Supporting universal usability of mobile software: touchscreen usability meta-test. In Proceedings of the 6th international conference on Universal access in human-computer interaction: context diversity - Volume Part III (UAHCI'11)

S42    Wolfgang Gottesheim, Stefan Mitsch, Rene Prokop, and Johannes Schonbock. 2007. Evaluation of a Mobile Multimodal Application Design - Major Usability Criteria and Usability Test Results. In Proceedings of the International Conference on the Management of Mobile Business (ICMB '07)

S43    Varun Gupta, D. S. Chauhan, and Kamlesh Dutta. 2012. Regression Testing-Based Requirement Prioritization of Mobile Applications. Int. J. Syst. Serv.-Oriented Eng. 3, 4 (October 2012)

S44    Walter Hargassner, Thomas Hofer, Claus Klammer, Josef Pichler, and Gernot Reisinger. 2008. A Script-Based Testbed for Mobile Software Frameworks. In Proceedings of the 2008 International Conference on Software Testing, Verification, and Validation (ICST '08)

S45    Harrison, R., Flood, D. and Duce, D., Usability of mobile applications: literature review and rationale for a new usability model, Journal of Interaction Science, Springer-Verlag, 2013, Vol. 1(1), pp. 1-16-

S46    Henry Ho, Simon Fong, and Zhuang Yan. 2008. User Acceptance Testing of Mobile Payment in Various Scenarios. In Proceedings of the 2008 IEEE International Conference on e-Business Engineering

S47    Cuixiong Hu and Iulian Neamtiu. 2011. A GUI bug finding framework for Android applications. In Proceedings of the 2011 ACM Symposium on Applied Computing (SAC '11)

S48    Huang, J.-F.; Gong, Y.-Z., "Remote mobile test system: a mobile phone cloud for application testing," Cloud Computing Technology and Science (CloudCom), 2012 IEEE 4th International Conference on , vol., no., pp.1,4, 3-6 Dec. 2012

S49    Shah Rukh Humayoun and Yael Dubinsky. 2014. MobiGolog: formal task modelling for testing user gestures interaction in mobile applications. In Proceedings of the 1st International Conference on Mobile Software Engineering and Systems (MOBILESoft 2014)

S50    Sun-Myung Hwang; Hyeon-Cheol Chae, "Design & Implementation of Mobile GUI Testing Tool," Convergence and Hybrid Information Technology, 2008. ICHIT '08. International Conference on , vol., no., pp.704,707, 28-30 Aug. 2008

S51    Jaaskelainen, A.; Kervinen, A.; Katara, M., "Creating a Test Model Library for GUI Testing of Smartphone Applications (Short Paper)," Quality Software, 2008. QSIC '08. The Eighth International Conference on , vol., no., pp.276,282, 12-13 Aug. 2008

S52    Marek Janicki, Mika Katara, and Tuula Pääkkönen. 2012. Obstacles and opportunities in deploying model-based GUI testing of mobile software: a survey. Softw. Test. Verif. Reliab. 22, 5 (August 2012)

S53    Casper S. Jensen, Mukul R. Prasad, and Anders Møller. 2013. Automated testing with targeted event sequence generation. In Proceedings of the 2013 International Symposium on Software Testing and Analysis

S54    Jouko Kaasila, Denzil Ferreira, Vassilis Kostakos, and Timo Ojala. 2012. Testdroid: automated remote UI testing on Android. In Proceedings of the 11th International Conference on Mobile and Ubiquitous Multimedia (MUM '12)

Luiz Kawakami, André Knabben, Douglas Rechia, Denise Bastos, Otavio Pereira, Ricardo Pereira e Silva, and Luiz C. V. dos Santos.

S55    2007. An object-oriented framework for improving software reuse on automated testing of mobile phones. In Proceedings of the 19th IFIP TC6/WG6.1 international conference, and 7th international conference on Testing of Software and Communicating Systems

Heejin Kim, Byoungju Choi, and Seokjin Yoon. 2009. Performance testing based on test-driven development for mobile

S56    applications. In Proceedings of the 3rd International Conference on Ubiquitous Information Management and Communication (ICUIMC '09)

S57    Thomas W. Knych and Ashwin Baliga. 2014. Android application development and testability. In Proceedings of the 1st International Conference on Mobile Software Engineering and Systems

K. Kuutti, K. Battarbee, S. Säde, T. Mattelmäki, T. Keinonen, T. Teirikko, and A. Tornberg. 2001. Virtual Prototypes in

S58    Usability Testing. In Proceedings of the 34th Annual Hawaii International Conference on System Sciences ( HICSS-34)-Volume 5 - Volume 5 (HICSS '01)

S59    Oh-Hyun Kwon and Sun-Myung Hwang. 2008. Mobile GUI Testing Tool based on Image Flow. In Proceedings of the Seventh IEEE/ACIS International Conference on Computer and Information Science (icis 2008) (ICIS '08)

S60    H. D Lambright. 1997. Automated Verification of Mobile Code. Technical Report. University of Arizona, Tucson, AZ, USA.

S61    Patrice Laurençot and Sébastien Salva. 2004. Testing mobile and distributed systems: method and experimentation. In Proceedings of the 8th international conference on Principles of Distributed Systems

Jae-Ho Lee; Yeung-Ho Kim; Sun-Ja Kim, "Design and Implementation of a Linux Phone Emulator Supporting Automated

S62    Application Testing," Convergence and Hybrid Information Technology, 2008. ICCIT '08. Third International Conference on , vol.2, no., pp.256,259, 11-13 Nov. 2008

S63    Li, Q., Wang, T., Wang, J. and Li, Y., Case study of usability testing methodology on mobile learning course, Advanced Intelligence and Awareness Internet (AIAI 2011), 2011 International Conference on, 2011, pp. 408-412

S64    Lima, L., Iyoda, J., Sampaio, A. and Aranha, E., Test case prioritization based on data reuse an experimental study, Empirical Software Engineering and Measurement, 2009. ESEM 2009. 3rd International Symposium on, 2009, pp. 279-29

S65    Lingling, W. and Ruitao, L., The Research of Orthogonal Experiment Applied in Mobile Phone's Software Test Case Generation, Information Technology and Applications (IFITA), 2010 International Forum on, 2010, Vol. 2, pp. 345-348

S66    Liu, Z., Gao, X. and Long, X., Adaptive random testing of mobile application, Computer Engineering and Technology (ICCET), 2010 2nd International Conference on, 2010, Vol. 2, pp.

S67    Fang Liu, Z., Liu, B. and peng Gao, X., SOA based mobile application software test framework, Reliability, Maintainability and Safety, 2009. ICRMS 2009. 8th International Conference on, 2009, pp. 765-769

S68    Lopes, R. and Cortes, O., An Ubiquitous Testing System for m-Learning Environments, Systems and Networks Communications, 2007. ICSNC 2007. Second International Conference on, 2007, pp. 32-32

S69    Ma, X., Yan, B., Chen, G., Zhang, C., Huang, K., Drury, J. and Wang, L., Design and Implementation of a Toolkit for Usability Testing of Mobile Apps, 2013, Vol. 18(1), pp. 81-97

S70    Maciel, F. R., PALMA: Usability Testing of an Application for Adult Literacy in Brazil, DUXU'13, Springer-Verlag, 2013, pp. 229-237

S71    Maly, I., Mikovec, Z. and Vystrcil, J., Interactive analytical tool for usability analysis of mobile indoor navigation application, Human System Interactions (HSI), 2010 3rd Conference on, 2010, pp. 259-266

S72    Mansar, S. L., Jariwala, S., Shahzad, M., Anggraini, A., Behih, N. and AlZeyara, A., A Usability Testing Experiment For A Localized Weight Loss Mobile Application, Procedia Technology, 2012, Vol. 5(0), pp. 839-848

S73    Manzoor, U., Irfan, J. and Nefti, S., Autonomous agents for Testing and Verification of Softwares after Deployment over Network, Internet Security (WorldCIS), 2011 World Congress on, 2011, pp. 36-41

S74    Mazlan, M. A., Stress Test on J2ME Compatible Mobile Device, Innovations in Information Technology, 2006, 2006, pp. 1-5

S75    Memon, A. and Cohen, M., Automated testing of GUI applications: Models, tools, and controlling flakiness, Software Engineering (ICSE), 2013 35th International Conference on, 2013, pp. 1479-1480

S76    van der Merwe, H., van der Merwe, B. and Visser, W., Verifying Android Applications Using Java PathFinder, 2012, Vol. 37(6), pp. 1-5

S77    Mtibaa, A., Harras, K. and Fahim, A., Towards Computational Offloading in Mobile Device Clouds, Cloud Computing Technology and Science (CloudCom), 2013 IEEE 5th International Conference on, 2013, Vol. 1, pp. 331-338

S78    Nagowah, L. and Sowamber, G., A novel approach of automation testing on mobile devices, Computer & Information Science (ICCIS), 2012 International Conference on, 2012, Vol. 2, pp. 924-930

S79    do Nascimento, L. H. O. and Machado, P. D. L., An Experimental Evaluation of Approaches to Feature Testing in the Mobile Phone Applications Domain, DOSTA '07, ACM, 2007, pp. 27-33

S80    Nguyen, M. D., Waeselynck, H. and Riveire, N., Testing Mobile Computing Applications: Toward a Scenario Language and Tools, WODA '08, ACM, 008, pp. 29-35

S81    Were Oyomno, Pekka Jäppinen, Esa Kerttula, Kari Heikkinen, Usability study of ME2.0, 2013, Vol. 17(2), pp. 305-319

S82    Park, B., Song, S., Kim, J., Park, W. and Jang, H., User Customization Methods Based on Mental Models: Modular UI Optimized for Customizing in Handheld Device, HCI'07, Springer-Verlag, 2007, pp. 445-451

S83    Payet, Ā. and Spoto, F., Static analysis of Android programs, Information and Software Technology, 2012, Vol. 54(11), pp. 1192-1201

S84    Pesonen, J., Extending Software Integration Testing Using Aspects in Symbian OS, Testing: Academic and Industrial Conference - Practice And Research Techniques, 2006. TAIC PART 2006. Proceedings, 2006, pp. 147-151

S85    Pichler, J. and Ramler, R., How to Test the Intangible Properties of Graphical User Interfaces?, Software Testing, Verification, and Validation, 2008 1st International Conference on, 2008, pp. 494-497

S86    Piotr Chynał, Jerzy M. Szymański, Janusz Sobecki, Using Eyetracking in a Mobile Applications Usability Testing, ACIIDS'12, Springer-Verlag, 2012, pp. 178-186

S87    Wilson Prata, Claudia Renata Mont' Alvão, Manuela Quaresma, Usability Testing of Mobile Applications Store: Purchase, Search and Reviews, DUXU'13, Springer-Verlag, 2013, pp. 714-722

S88    Qiu, Y.-F., Chui, Y.-P. and Helander, M., Usability Analysis of Mobile Phone Camera Software Systems, Cybernetics and Intelligent Systems, 2006 IEEE Conference on, 2006, pp. 1-6

S89    Ridene, Y. and Barbier, F., A Model-driven Approach for Automating Mobile Applications Testing, ECSA '11, ACM, 2011, pp. 9:1-9:7

S90    Ridene, Y., Belloir, N., Barbier, F. and Couture, N., A DSML for Mobile Phone Applications Testing, DSM '10, ACM, 2010, pp. 3:1-3:6

S91    Roy Choudhary, S., Cross-platform Testing and Maintenance of Web and Mobile Applications, ICSE Companion 2014, ACM, 2014, pp. 642-645

S92    Ryan, C. and Gonsalves, A., The Effect of Context and Application Type on Mobile Usability: An Empirical Study, ACSC '05, Australian Computer, Society, Inc., 2005, pp. 115-124

S93    Ryan, C. and Rossi, P., Software, performance and resource utilisation metrics for context-aware mobile applications, Software Metrics, 2005. 11th IEEE International Symposium, 2005

S94    Marco Sá and Luís Carriço., An Evaluation Framework for Mobile User Interfaces, INTERACT '09, Springer-Verlag, 2009, pp. 708-721

S95    Satoh, I., Software testing for mobile and ubiquitous computing, Autonomous Decentralized Systems, 2003. ISADS 2003. The Sixth International Symposium on, 2003, pp. 185-192

S96    Satoh, I., A testing framework for mobile computing software, Software Engineering, IEEE Transactions on, 2003, Vol. 29(12), pp. 1112-1121

S97    Satoh, I., Flying Emulator: Rapid Building and Testing of Networked Applications for Mobile Computers, MA '01, Springer-Verlag, 2002, pp. 103-118

S98    Wolfgang Schönfeld and Jörg Pommnitz, A Testbed for Mobile Multimedia Applications, 1999, Vol. 9(1), pp. 29-42

S99    Schultz, D., 10 Usability Tips Tricks for Testing Mobile Applications, 2006, Vol. 13(6), pp. 14-15

S100    Seffah, A., Donyaee, M., Kline, R. B. and Padda, H. K., Usability measurement and metrics: A consolidated model, 2006, Vol. 14(2), pp. 159-178

S101    Shabtai, A., Fledel, Y. and Elovici, Y., Automated Static Code Analysis for Classifying Android Applications Using Machine Learning, Computational Intelligence and Security (CIS), 2010 International Conference on, 2010, pp. 329-333

S102    Shahriar, H., North, S. and Mawangi, E., Testing of Memory Leak in Android Applications, High-Assurance Systems Engineering (HASE), 2014 IEEE 15th International Symposium on, 2014, pp. 176-183

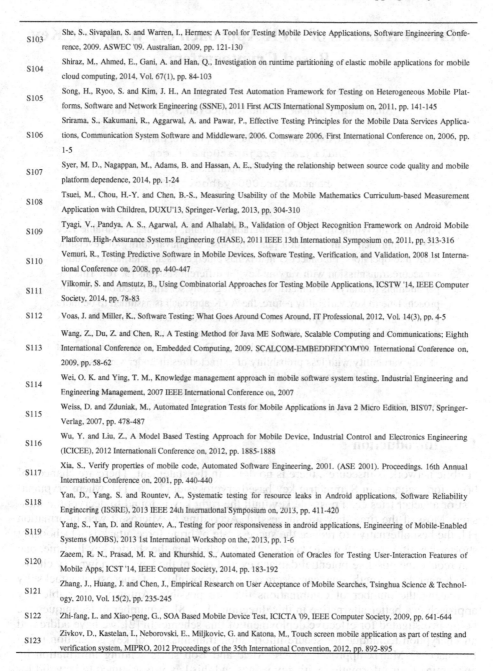

S103    She, S., Sivapalan, S. and Warren, I., Hermes: A Tool for Testing Mobile Device Applications, Software Engineering Conference, 2009. ASWEC '09. Australian, 2009, pp. 121-130

S104    Shiraz, M., Ahmed, E., Gani, A. and Han, Q., Investigation on runtime partitioning of elastic mobile applications for mobile cloud computing, 2014, Vol. 67(1), pp. 84-103

S105    Song, H., Ryoo, S. and Kim, J. H., An Integrated Test Automation Framework for Testing on Heterogeneous Mobile Platforms, Software and Network Engineering (SSNE), 2011 First ACIS International Symposium on, 2011, pp. 141-145

S106    Srirama, S., Kakumani, R., Aggarwal, A. and Pawar, P., Effective Testing Principles for the Mobile Data Services Applications, Communication System Software and Middleware, 2006. Comsware 2006. First International Conference on, 2006, pp. 1-5

S107    Syer, M. D., Nagappan, M., Adams, B. and Hassan, A. E., Studying the relationship between source code quality and mobile platform dependence, 2014, pp. 1-24

S108    Tsuei, M., Chou, H.-Y. and Chen, B.-S., Measuring Usability of the Mobile Mathematics Curriculum-based Measurement Application with Children, DUXU'13, Springer-Verlag, 2013, pp. 304-310

S109    Tyagi, V., Pandya, A. S., Agarwal, A. and Alhalabi, B., Validation of Object Recognition Framework on Android Mobile Platform, High-Assurance Systems Engineering (HASE), 2011 IEEE 13th International Symposium on, 2011, pp. 313-316

S110    Vemuri, R., Testing Predictive Software in Mobile Devices, Software Testing, Verification, and Validation, 2008 1st International Conference on, 2008, pp. 440-447

S111    Vilkomir, S. and Amstutz, B., Using Combinatorial Approaches for Testing Mobile Applications, ICSTW '14, IEEE Computer Society, 2014, pp. 78-83

S112    Voas, J. and Miller, K., Software Testing: What Goes Around Comes Around, IT Professional, 2012, Vol. 14(3), pp. 4-5

S113    Wang, Z., Du, Z. and Chen, R., A Testing Method for Java ME Software, Scalable Computing and Communications; Eighth International Conference on, Embedded Computing, 2009. SCALCOM-EMBEDDEDCOM'09. International Conference on, 2009, pp. 58-62

S114    Wei, O. K. and Ying, T. M., Knowledge management approach in mobile software system testing, Industrial Engineering and Engineering Management, 2007 IEEE International Conference on, 2007

S115    Weiss, D. and Zduniak, M., Automated Integration Tests for Mobile Applications in Java 2 Micro Edition, BIS'07, Springer-Verlag, 2007, pp. 478-487

S116    Wu, Y. and Liu, Z., A Model Based Testing Approach for Mobile Device, Industrial Control and Electronics Engineering (ICICEE), 2012 Internationali Conference on, 2012, pp. 1885-1888

S117    Xia, S., Verify properties of mobile code, Automated Software Engineering, 2001. (ASE 2001). Proceedings. 16th Annual International Conference on, 2001, pp. 440-440

S118    Yan, D., Yang, S. and Rountev, A., Systematic testing for resource leaks in Android applications, Software Reliability Engineering (ISSRE), 2013 IEEE 24th International Symposium on, 2013, pp. 411-420

S119    Yang, S., Yan, D. and Rountev, A., Testing for poor responsiveness in android applications, Engineering of Mobile-Enabled Systems (MOBS), 2013 1st International Workshop on the, 2013, pp. 1-6

S120    Zaeem, R. N., Prasad, M. R. and Khurshid, S., Automated Generation of Oracles for Testing User-Interaction Features of Mobile Apps, ICST '14, IEEE Computer Society, 2014, pp. 183-192

S121    Zhang, J., Huang, J. and Chen, J., Empirical Research on User Acceptance of Mobile Searches, Tsinghua Science & Technology, 2010, Vol. 15(2), pp. 235-245

S122    Zhi-fang, L. and Xiao-peng, G., SOA Based Mobile Device Test, ICICTA '09, IEEE Computer Society, 2009, pp. 641-644

S123    Zivkov, D., Kastelan, I., Neborovski, E., Miljkovic, G. and Katona, M., Touch screen mobile application as part of testing and verification system, MIPRO, 2012 Proceedings of the 35th International Convention, 2012, pp. 892-895

# Markov Analysis of AVK Approach of Symmetric Key Based Cryptosystem

Shaligram Prajapat[1(✉)] and Ramjeevan Singh Thakur[2]

[1] MANIT, Bhopal and DAVV, Indore, India
Shaligram.prajapat@gmail.com
[2] MANIT, Bhopal, India
ramthakur2000@yahoo.com

**Abstract.** In Symmetric Key Cryptography domain, Automatic Variable Key (AVK) approach is in inception phase because of unavailability of reversible XOR like operators. Fibonacci-Q matrix has emerged as an alternative solution for secure transmission with varying key for different sessions [3, 10]. This paper attempts to analyze symmetric key cryptography scheme based on AVK approach. Due to key variability nature, the AVK approach is assumed to be more powerful, efficient and optimal but its analysis from hackers' point of view is demonstrated in this paper. This paper also assumes various situations under which mining of future keys can be achieved. The paper also discusses concept of Key variability with less probability of extracted result under various scenario with the different degree of difficulty in key mining.

**Keywords:** Symmetric key cryptography · AVK · Key mining

## 1 Introduction

"Public network is insecure", there is no doubt in the statement. There are chances of brute force attacks in Symmetric key based cryptosystem [1, 5, 11], where cryptanalyst or attacker tries each possible key until the right key is found to decrypt the message. Most of the time they are successful .According to Microsoft web information [1], the best alternative to reduce the success rate of brute force attack is: (1)choosing shorter key lifetimes (2)Longer key lengths. By choosing the shorter key lifetime, one can reduce the possible potential damage even if one of the keys is known .By choosing longer key length one can decrease the probability of successful attacks by increasing the number of combinations that are possible. Automatic Variable key approach is a better alternative in this direction [4, 7, 8]. A number of techniques are being investigated for effective deployment of this scheme. In [2], we have addressed an encryption technique for secure information transmission of key generation using Fibonacci-Q Matrix approach, where Alice and Bob starts sending information by encrypting their information with key values and this key was assumed to be valid for a given session. For the next slot, key is differed.

After a quick review of necessary terminology in section 2, Subsequent section of the paper shows behavior of key generation. This approach has been analyzed for future behavior of Fibonacci-Q based symmetric cryptosystem under different possible situations.

O. Gervasi et al. (Eds.): ICCSA 2015, Part V, LNCS 9159, pp. 164–176, 2015.
DOI: 10.1007/978-3-319-21413-9_12

## 2     Basic Terminologies

Symmetric Cryptosystem: Any symmetric encryption algorithm [5,11] is an invertible transformation/mappings of the message space (M) into the cipher space (C) using finite length key k from key space (K).

$$E_k: M \rightarrow C, \text{ such that:} \tag{1}$$

$$E_k(m) = c \quad \text{where } k \in K, m \in M, c \in C \tag{2}$$

An inverse decryption algorithm would be $D_k = E^{-1}{}_k$:

$$D_k: C \rightarrow M, \text{ such that} \tag{3}$$

$$D_k(c) = D_k[E_k(m)] = m \tag{4}$$

According to [1],

The keys should be unique i.e., $E_{k1}(m) \neq E_{k2}(m)$ where $k1 \neq k2$. \tag{5}

### 2.1     Markov Process, Markov Chain

A random process $\{X(t)\}$ will be a Markov process [17], if it satisfies criteria in equation (6):

$$P[X(t_n)=q_n/X(t_{n-1})=q_{n-1}, X(t_{n-2})=q_{n-2},\ldots\ldots\ldots, X(t_2)=q_2, X(t_1)=q_1] \text{ for all } t_1<t_2<\ldots<t_n \tag{6}$$

In Markov process, future behavior of a process depends on the present value but not on the past. If in the process equation (6) is satisfied for all n then the constants

$$\{ q_n, q_{n-1}, q_{n-2},\ldots\ldots\ldots, q_2, q_1\} \tag{7}$$

are the states of the Markov chain. Pictorially markov model can be expressed by a directed graph G= (V, E) with vertices representing states V= {q1, q2... qn} and edges or arcs E= {< qi, qj > where qi, qj $\in$ V} shows transition from state i to state j. Each edge is associated with a probability pij of transition from qi to qj. Also it is worthwhile to note here that, the probability of future transitions depends only on current state, not on the earlier states.

### 2.2     One Step Transition Probability (TP)

The conditional transitional probability $P[X(t_n) = q_j/X(t_{n-1})=q_i]$ is one step transition probability[17] from the state $q_I$ to $q_j$ at the nth step and it is denoted by $P_{ij}$ ( n-1, n ).

If one step transition probability does not depend on the step i.e. Pij (n-1, n) =P ij (m-1, m). Then the Markov chain is said to be homogeneous.

## 2.3      Transition Probability Matrix

In homogeneous, the one-step transition probability[17] is denoted by $p_{ij}$. The matrix $P=(p_{ij})$ satisfies two conditions:

1. $p_{ij \geq 0}$
2. $\sum P_{ij}=1$ for all i. i.e. the sum of the elements of any row of the t.p.m. is 1.

# 3      Review of Literature

The concept of Fibonacci Q-matrices was applied in AVK [2] based a symmetric algorithm in [5]. Non -AVK based approach was described nicely by Stakhov et al [3,10]. This approach assumes an initial message in the form of square matrix M of size (p+1) x (p+1) where p = 0, 1, 2, 3...

Now choose the Fibonacci Q $_p$-matrix, Q $_p^n$, of size (p+1) x (p+1) as an encryption (key) matrix and its inverse matrix, $Q_p^{-n}$, of the same size as decryption (key) matrix. Therefore, the encryption and decryption are defined by parameters n and p.

## 3.1      Encryption Process

The working of above symmetric key encryption algorithm based on classical Q-matrix is beautifully illustrated in [8, 16].

**Step 1: Let plain text message is.**

$$M = \begin{pmatrix} m_1 & m_2 \\ m_3 & m_4 \end{pmatrix}$$

Where $m_i > 0$; i = 1, 2, 3, 4,...

**Step 2: Choose n = 6 and p = 1 such that.**

$$Q^6 = \begin{pmatrix} 13 & 8 \\ 8 & 5 \end{pmatrix} = \begin{pmatrix} 1101 & 1000 \\ 1000 & 0101 \end{pmatrix}$$

**Step 3:**

$$M * K = \begin{pmatrix} m_1 & m_2 \\ m_3 & m_4 \end{pmatrix} * \begin{pmatrix} 1101 & 1000 \\ 1000 & 0101 \end{pmatrix} =$$

$$\begin{pmatrix} 1101*m_1 + 1000*m_2 & 1000*m_1 + 0101*m_2 \\ 1101*m_3 + 1000*m_4 & 1000*m_3 + 0101*m_4 \end{pmatrix}$$

$$
\begin{aligned}
e_1 &= 1101 * m_1 + 1000 * m_2 \\
e_2 &= 1000 * m_1 + 0101 * m_2 \\
e_3 &= 1101 * m_3 + 1000 * m_4 \\
e_4 &= 1000 * m_3 + 0101 * m_4
\end{aligned}
\tag{8}
$$

## 3.2   Decryption Process

Stakhov [10] explained the decryption process as follows:

**Step 1: Received encoded message is represented in the matrix form.**

$$
E = \begin{pmatrix} e_1 & e_2 \\ e_3 & e_4 \end{pmatrix}
$$

**Step 2: Compute the reversible decryption function.**

$$
Q^{-6} = \begin{pmatrix} 5 & -8 \\ -8 & 13 \end{pmatrix} = \begin{pmatrix} 0101 & -1000 \\ -1000 & 1101 \end{pmatrix}
$$

**Step 3: Recover plain text.**

$$
M = \begin{pmatrix} m_1 & m_2 \\ m_3 & m_4 \end{pmatrix} = \begin{pmatrix} e_1 & e_2 \\ e_3 & e_4 \end{pmatrix} * \begin{pmatrix} 0101 & -1000 \\ -1000 & 1101 \end{pmatrix}
$$

Above process has been analyzed, implemented and tested in [15] and they concluded that the algorithm works faster than symmetric algorithms (including DES, 3DES, AES and Blowfish). Current work is extension of this proposed work.

## 4     Proposed Algorithm

The concept of Fibonacci Q-matrices [2], allows us to develop a symmetric key cryptographic algorithm. This algorithm assumes an initial message in the form of square matrix M of size (p+1) x (p+1) where p = 0, 1, 2, 3,.... Now choose the Fibonacci Q-matrix, Q p n, of size (p+1) * (p+1) as a encryption (key) matrix and it's inverse matrix, Q p–n , of the same size as decryption (key) matrix. Therefore, the encryption and decryption are defined by parameters n and p.

## 4.1    Encryption Algorithm: Algorithm Encrypt(M)

//This algorithm accepts plain text and produced cipher text.
// n and p are used as parameter to derive Encryption Key

a.  Choose n
b.  Choose p
c.  Compute $Q_p^n$
d.  $E \leftarrow M * Q_p^n$
    // Computation of Cipher text over's here, now transmit it.
e.  End of algorithm

The working of above symmetric key encryption algorithm based on classical Q-matrix is beautifully illustrated in [10] and [11].

**Step 1: Let plain text message is:**

$$M = \begin{pmatrix} m_1 & m_2 \\ m_3 & m_4 \end{pmatrix}$$

Where $m_i > 0$; $i = 1, 2, 3, 4$.

**Step 2: Choose n = 6 and p = 1 such that:**

$$Q^6 = \begin{pmatrix} 13 & 8 \\ 8 & 5 \end{pmatrix}$$

**Step 3:**

$$M * Q^6 = \begin{pmatrix} m_1 & m_2 \\ m_3 & m_4 \end{pmatrix} * \begin{pmatrix} 13 & 8 \\ 8 & 5 \end{pmatrix} =$$

$$\begin{pmatrix} 1101*m_1 + 1000*m_2 & 1000*m_1 + 01001*m_2 \\ 1101*m_3 + 1000*m_4 & 1000*m_3 + 0101*m_4 \end{pmatrix}$$

The generated cipher text will be:

$$m_1 = 0101*e_1 - 1000*e_2$$
$$m_2 = -1000*e_1 + 1101*e_2 \tag{9}$$
$$m_3 = 0101*e_3 - 1000*e_4$$
$$m_4 = -1000*e_3 + 1101*e_4$$

## 4.2    Decryption Algorithm: Algorithm Fibodecrypt (n, p, E)

This 3-step algorithm takes plain encrypted text, parameters n and p and produces original plain text.

//This algorithm accepts Cipher text E and produced cipher text.
// n and p are used as parameter to derive Decryption Key

    a.   Compute $Q_p^{-n}$
    b.   $M \leftarrow E \times Q_p^{-n}$
       // Regeneration of plain text over here, now it is ready for use by recipient
    c.   End of algorithm

The decryption process can be explained with the help of an example as follows:

**Step 1: Received encoded message is represented in the matrix form as:**

$$E = \begin{pmatrix} e_1 & e_2 \\ e_3 & e_4 \end{pmatrix}$$

**Step 2: Compute the reversible decryption function:**

$$Q^{-6} = \begin{pmatrix} 5 & -8 \\ -8 & 13 \end{pmatrix} = \begin{pmatrix} 0101 & -1000 \\ -1000 & 1101 \end{pmatrix}$$

**Step 3: Recover plain text:**

$$M = \begin{pmatrix} m_1 & m_2 \\ m_3 & m_4 \end{pmatrix} = \begin{pmatrix} e_1 & e_2 \\ e_3 & e_4 \end{pmatrix} * \begin{pmatrix} 0101 & -1000 \\ -1000 & 1101 \end{pmatrix}$$

This reveals the original plain text message.

## 5    Experimental Setup

For implementation the scheme of section 4 using Python is illustrated in [15], this code-snippet of encryption and decryption was tested for large values of n. For clarity of code, a necessary comment has been introduced in proper place.

## 6    Analysis

Fibonacci sequence $\{f_n\}$ defined by

$$f_n = f_{n-1} + f_{n-2} \tag{10}$$

For $n \geq 2$ with initial conditions $f_0 = 1$ and $f_1 = 1$. Using Shift operator above equation can be re-written as:

$$E^2 f_n = E f_n + f_n \tag{11}$$

$$(E^2 - E - 1) f_n = 0 \tag{12}$$

The characteristic equation corresponding to this recurrence equation would be

$$r^2 - r - 1 = 0 \tag{13}$$

This equation can be solved by two approaches:

## 6.1    Approach-1

Using Iterative methods like Bisection, Method of false position, Secant and Newton Raphson approach.

Table 1. Iterative methods for solution of auxiliary equation

| Iteration No. | Bisection | False Position | Secant | Newton Raphson |
|---|---|---|---|---|
| 1 | 1.500000 | 1.500000 | 1.500000 | 1.666667 |
| 2 | 1.750000 | 1.600000 | 1.600000 | 1.619048 |
| 3 | 1.625000 | 1.615385 | 1.619048 | **1.618034** |
| 4 | 1.562500 | 1.617647 | **1.618026** | |
| 5 | 1.593750 | 1.617978 | | |
| 6 | 1.609375 | **1.618026** | | |
| 7 | 1.617188 | | | |
| 8 | 1.621094 | | | |
| 9 | 1.619141 | | | |
| 10 | 1.618164 | | | |
| 11 | 1.617676 | | | |
| 12 | 1.617920 | | | |
| 13 | **1.618042** | | | |

The root of this equation is shown in following plot , methods starts converging with different rates and approaches towards final solution within finite number of iteration, Bisection, False position both converges to the 1.6180 with full guarantee, But among all the method first one is slow and has gain of one bit per iteration, False position shows linear. Secant and raphson approach converges fast with the rate of 1.62 and 2 respectively. But depends on the choice of initial value and may not reach to the root

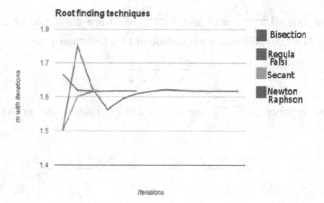

**Fig. 1.** Solving Auxiliary equation of Fibonacci sequence using various iterative methods

in case of wrong choice. This behavior can be observed very well from Fig. 1. Finally r=1.6180 with the initial guess [1. 2] similarly for the guess [0,1] it concludes to 0.618.

## 6.2    Approach-2

Using Analytical method of solving auxiliary equation,

Solving it, we get $c_1 = \frac{1+\sqrt{5}}{2}$ and $c_2 = \frac{1-\sqrt{5}}{2}$ as the characteristic roots. These will make the general solution as

$$f_n = A_1 C_1^m + A_2 C_2^m \qquad (14)$$

Where $A_1$ and $A_2$ are constants. Since $f_0 = 1$ and $f_1 = 1$ we get

$$A_1 + A_2 = 0 \qquad (15)$$

$$A_1 C_1 + A_2 C_2 = 1 \qquad (16)$$

$$A_1 \left(\frac{1+\sqrt{5}}{2}\right) + A_2 \left(\frac{1-\sqrt{5}}{2}\right) = 1 \qquad (17)$$

After simplification of 15 and 17, we get

$$A_1 = \frac{1}{\sqrt{5}} \text{ and } A_2 = \frac{-1}{\sqrt{5}} \qquad (18)$$

Substituting these values in equation 14 Hence, the $n^{th}$ term of Fibonacci sequence would be

$$f_n = \frac{1}{\sqrt{5}}\left[\left(\frac{1+\sqrt{5}}{2}\right)^n - \left(\frac{1-\sqrt{5}}{2}\right)^n\right] = \frac{\emptyset^n - \emptyset'^n}{\sqrt{5}} \qquad (19)$$

Here, the ratio $\left(\frac{1+\sqrt{5}}{2}\right)^{n} = \emptyset$ and $\left(\frac{1-\sqrt{5}}{2}\right)^{n} = \acute{\emptyset}$ are conjugate of each other. $\emptyset =$ 1.61803 and $\acute{\emptyset} = 0.61803$. Since magnitude of $\acute{\emptyset}$ is less than 1, so

$$\frac{\acute{\emptyset}^{n}}{\sqrt{5}} < \frac{1}{\sqrt{5}} < \frac{1}{2} \tag{20}$$

$n^{th}$ Fibonacci number $f_n = \frac{\emptyset^n}{\sqrt{5}}$ rounded to nearest integer and grows exponentially, see fig.2.

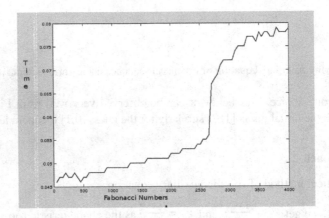

**Fig. 2.** Computation of Fibonacci terms v/s time

From fig.2[5], it is obvious that, for n>2500, computation time for guessing 3 alterative terms would be quite high, some time may be much higher then the length of communication over channel, so higher guessing time would add another security level for the proposed system, specially for computing probable keys.

In encryption approach based on AVK approach at any instance Alice or Bob uses key for secure transmission using following matrix, this has three possibilities $A = F_{n-1}$, $B = F_n$ and $C = F_{n+1}$.

Let $X_n$ represents the nth step of transmission. The state space is [A, B,C]. This is a Markov chain since Alice or Bob using the key for a particular session is not influenced by those sessions who previously had process key. Let the t. p. m. the

$$\text{Markov Chain is: } P = \begin{array}{c} A \\ B \\ C \end{array} \begin{pmatrix} 0.75 & 0.25 & 0 \\ 0.25 & 0.5 & 0.25 \\ 0 & 0.75 & 0.25 \end{pmatrix}$$

and the initial state distribution of the chain is P[ $X_0=i$]=1/3, i=0,1,2. We analyze the Fibonacci(Q) matrix approach under following situations:

**Situation-1: Hacker is interested to future key say predict P[X2=2].**
In this case, he/she can use Chapman-Kolmogorov theorem , according to which if P
is the t. p.m. of a homogeneous Markov chain, then n-step t. p. m. P $^{(n)}$ is equal to
P $^{(n)}$. i.e. $[P_{ij}{}^{(n)}]= [ [P_{ij}]$. So,

$$P^{(2)} = P^2 = \begin{pmatrix} 0.625 & 0.3125 & 0.0625 \\ 0.3125 & 0.5 & 0.1875 \\ 0.1875 & 0.5625 & 0.25 \end{pmatrix}$$

Now using the definition of conditional probability:

$$P[X_2=2] = \sum P[X_2 = 2 \mid X_0 = i] P[X_0 = i]$$

$$P[X_2 = 2 \mid X_0 = 0] * P[X_0 = 0]$$
$$+ P[X_2 = 2 \mid X_0 = 1] * P[X_0 = 1]$$
$$+ P[X_2 = 2 \mid X_0 = 2] *$$
$$= P[X_0 = 2]$$

(21)

Since,$(X_n$

$$\begin{matrix} & & 0 & 1 & 2 \\ p = (X_{n-1}) & \begin{matrix}0\\1\\2\end{matrix} & \begin{pmatrix} 0.625 & 0.3125 & 0.0625 \\ 0.3125 & 0.5 & 0.1875 \\ 0.1875 & 0.5625 & 0.25 \end{pmatrix} \end{matrix}$$

So, $P[X_2 = 2 ] = P_{02}{}^{(2)} * P[X_0 = 0] + P_{12}{}^{(2)} * P[X_0 = 1] + P_{22}{}^{(2)}$
$P[X_0 = 2 ]= (0.0625 + 0.1875 + 0.25) * (1/3) = 0.16667$

**Situation-2:** Hacker is interested in prediction of sequence P[X 3 = 1, X2 = 2, X1 = 1,
X0 = 2] Since

$$P = \begin{matrix}0\\1\\2\end{matrix} \begin{pmatrix} 0.75 & 0.25 & 0 \\ 0.25 & 0.5 & 0.25 \\ 0 & 0.75 & 0.25 \end{pmatrix}$$

So, $P[X_3 = 1, X_2 = 2, X_1 = 1, X_0 = 2 ] = P [X_3 = 1 \mid X_2 = 2 ] * P[ X_2 = 2 \mid X_1 = 1] *$
$P[ X_1=1 \mid X_0 = 2] * P[ X_0 = 2 ]$
$= P_{21}{}^{(1)} * P_{12}{}^{(1)} * P_{21}{}^{(1)} * P[X_0 = 2 ] = 0.75 * 0.25 * 0.75 * (1/3) = 0.046875,$

(22)

this situation is less likely to occur.

**Situation-3:** Hacker assumes that Alice or Bob will never use present key. In this situation, Alice or Bob have only two choices either to use $F_{n-1}$ or $F_{n+1}$ as key, Let prior choice is represented by state $q_0$ and later one be $q_1$, these two states is depicted in following diagram

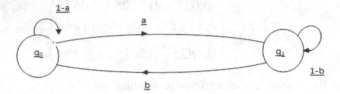

**Fig. 3.** Transition Graph for next choosing key

In this transition diagram, Alice and Bob are working properly in state $q_0$ and system will undergo in state $q_1$, to vary the key. The transition probability matrix of this Markov Chain is. $P = \begin{pmatrix} 1-a & a \\ b & 1-b \end{pmatrix}$ The steady state probability vector for this Markov Chain is

$$V = \left( \frac{b}{a+b} \quad \frac{a}{a+b} \right). \tag{23}$$

Now We consider following situations from hacker's perspective

**Situation-4:** Alice either uses $F_{n-1}$ or $F_{n+1}$ in such a manner that Alice never uses $F_{n+1}$ in row. But if he uses Fn-1 in one session then the next session he is just likely to use Fn-1 as he is to use $F_{n+1}$.

In this situation we suppose that at the first session of communication, alice tossed a die and uses Fn-1 for encryption if a 6 appeared then hacker may be interested in computing the chances that he uses Fn+1 in the third session or the chances where he uses Fn-1 to encrypt in long run .In order to achieve this the transmission pattern of Markov Chain with state space=$(F_{n+1},F_{n-1})$ .The transition probability matrix (t.p.m) would be

$$F_{(n+1)}F_{(n-1)} = \begin{matrix} F(n+1) \\ F(n-1) \end{matrix} \begin{pmatrix} 0 & 1 \\ 0.5 & 0.5 \end{pmatrix}$$

Probability of using Fn-1 = P[getting 6 in the toss of die]=(1/6).
Probability of using $F_{n+1}$ = (5/6).
As the initial state probability distribution is $P^{(1)}$=(5/6  1/6).

$$P^{(2)} = P^{(1)} * P \left( 5/6 \quad 1/6 \right) * \begin{pmatrix} 0 & 1 \\ 0.5 & 0.5 \end{pmatrix} = (1/12 \quad 11/12)$$

$$P^{(3)} = P^{(2)} * P(1/12 \quad 11/12) * \begin{pmatrix} 0 & 1 \\ 0.5 & 0.5 \end{pmatrix} = (11/24 \quad 13/24)$$

So the probability P[The Alice/Bob uses $F_{n+1}$ in the third session] = 11/24= 0.458333.

The steady state probability distribution of this Markov Chain would be:

Let $S = (S_1, S_2)$ be the steady state vector, Then $S*P = S$ So,

$$(S_1, S_2) * \begin{pmatrix} 0 & 1 \\ 0.5 & 0.5 \end{pmatrix}$$

will lead to

$$S_1 = S_2 / 2 \text{ and } S_2 = S_1 + S_2 / 2 \tag{24}$$

Further $S_1 + S_2 = 1$, and $2*S_1 = S_2$ gives $S_1 = 1/3$ and $S_2 = 2/3$. Hence the Steady state probability vector $S = (1/3 \quad 2/3)$. This concludes that hacker can predict that the Alice will use $F_{n-1}$ in long run with probability of 67%.

# 7    Conclusion

This work presents implementation of novel approach for the investigation of alternative binary operator like XOR. The paper presents recurrence relation solution in terms of golden ratio and it's conjugate together with Markov approach. The Analysis and deductions presented in this paper shows the vulnerability effects, under different conditioned as described above .The Situation 4 points out that extra care is to taken care off. With fixing up Fibonacci terms of Six character length, the optimum security with time variant scheme can be extended for further improvement for performance gain for low power and hand held devices.

**Acknowledgement.** This work is supported by research project under Fast Track Scheme for Young Scientist from DST, New Delhi, India. Scheme 2011-12, No. SR/FTP/ETA-121/ 2011 (SERB), dated 18/12/2012.

# References

1. http://technet.microsoft.com/en-s/library/cc961628.aspx
2. Prajapat, S., Jain, A., Thakur, R.S.: A Novel Approach For Information Security With Automatic Variable Key Using Fibonacci Q-Matrix. IJCCT **3**(3), 54–57 (2012)
3. Nalli, A.: On the Hadamard Product of Fibonacci Q n matrix and Fibonacci Q–n matrix. International Journal of Contemporary Mathematical Sciences **1**(16), 753–761 (2006)
4. Chakrabarti, P., Bhuyan, B., Chowdhuri, A., Bhunia, C.: A novel approach towards realizing optimum data transfer and Automatic Variable Key (AVK) in cryptography. IJCSNS **8**(5), 241 (2008)

5. Prajapat, S., Thakur, R.S.: Time variant approach towards symmetric key, SAI 2013, pp. 398–405. IEEE
6. Diffe, W., Hellman, M.: Exhaustive Cryptanalysis of the NBS Data encryption standard. IEEE Computer, 74–84, June 1977
7. Chakrabarti, P.: Application of Automatic Variable Key (AVK) in RSA. International Journal HIT Transactions on ECCN 2(5)
8. Bhunia, C.T.: Application of AVK and selective encryption in improving performance of quantum cryptography and network (2006)
9. Blaze, M., Diffie, W., Rivest, R.L., Schneier, B., Shimomura, T., Thompson, E., Wiener, M.: Minimal Key Lengths for Symmetric Ciphers to Provide Adequate Commercial Security. A Report by an Ad Hoc Group of Cryptographers and Computer Scientists (1996)
10. Stakhov, A.P.: Fibonacci matrices, a generalization of the 'Cassini formula', and a new coding theory. Chaos, Solutions & Fractals 30(1), 56–66 (2006)
11. Fernández1, M., Diaz1, G., Cosme1, A., Negrón, I., Negrón, P., Alfredo: Cryptography: algorithms and security applications. The IEEE Computer Society's Student 8(2), Fall 2008
12. Nadeem, A., Javed, M.Y.: A performance comparison of data encryption algorithms. Information and Communication Technologies, ICICT 2005, pp. 84–89 (2005)
13. Elminaam, D.S.A., Hadhoud, M.M., Abdul Kader, H.M.: Performance Evaluation of Symmetric Encryption Algorithm. IJCSMS 8(12), 58–64 (2008)
14. Elminaam, D., Kader, H., Hadhoud, M.: Performance Evaluation of Symmetric Encryption Algorithms on Power Consumption for Wireless Devices. International Journal of Computer Theory and Engineering 1(4), 1793–8201 (2009)
15. Prajapat, S., Saxena, S., Jain, A., Sharma, P.: Implementation of information security with Fibonacci-Q matrix. In: Proceeding of International Conference ICICIS-2012 and International Journal of Electronics Communication and Computer Engineering, ISSN(Online): 2249-071 62575/BPL/CE/2012
16. Prajapat, S., Thakur, R.S.: Sparse approach for realizing AVK for Symmetric Key Encryption. IJRDET & proceeding of International Research Conference on Engineering, Science and Management (IRCESM 2014), vol. 2, pp. 15–18
17. Ross, S.M.: Introduction to Probability models, 10th edn., pp. 191–268

# Comparison of Static Analysis Tools for Quality Measurement of RPG Programs

Zoltán Tóth[1], László Vidács[2]($\boxtimes$), and Rudolf Ferenc[1]

[1] Department of Software Engineering, University of Szeged, Szeged, Hungary
{zizo,ferenc}@inf.u-szeged.hu
[2] MTA-SZTE Research Group on Artificial Intelligence, Szeged, Hungary
lac@inf.u-szeged.hu

**Abstract.** The RPG programming language is a popular language employed widely in IBM i mainframes nowadays. Legacy mainframe systems that evolved and survived the past decades usually data intensive and even business critical applications. Recent, state of the art quality assurance tools are mostly focused on popular languages like Java, C++ or Python. In this work we compare two source code based quality management tools for the RPG language. The study is focused on the data obtained using static analysis, which is then aggregated to higher level quality attributes. SourceMeter is a command line tool-chain capable to measure various source attributes like metrics and coding rule violations. SonarQube is a quality management platform with RPG language support. To facilitate the objective comparison, we used the SourceMeter for RPG plugin for SonarQube, which seamlessly integrates into the framework extending its capabilities. The evaluation is built on analysis success and depth, source code metrics, coding rules and code duplications. We found that SourceMeter is more advanced in analysis depth, product metrics and finding duplications, while their performance of coding rules and analysis success is rather balanced. Since both tools were presented recently on the market of quality assurance tools, we expect additional releases in the future with more mature analyzers.

**Keywords:** Static analysis · Software quality · SonarQube · SourceMeter · IBM RPG · Metrics · Coding rules

## 1 Introduction

Rapid development life cycles provided by 4GL languages resulted in a number of large software systems decades ago, that are mostly considered legacy systems nowadays. On the other hand, the role of quality assurance of these data intensive and often business critical systems is increasingly important. The IBM i platform – initially called AS/400 platform – became very popular to the end of the last century. Business applications developed for the IBM i platform usually use the RPG high-level programming language (Reporting Program Generator), which is still widely employed, supported and evolving. In the early days of the appearance of

© Springer International Publishing Switzerland 2015
O. Gervasi et al. (Eds.): ICCSA 2015, Part V, LNCS 9159, pp. 177–192, 2015.
DOI: 10.1007/978-3-319-21413-9_13

4GL (like RPG), several studies were published in favour of their use. The topics of these studies are mostly focused on predicting the size of a 4GL project and its development effort, for instance by calculating function points [22] or by combining 4GL metrics with metrics for database systems [17]. In the literature only few papers are available considering the software quality of these languages [12], [8], [16], while the main focus of current QA tools and techniques is on the more popular object-oriented languages.

In this paper we compare two state of the art tools for RPG quality measurements by analyzing the capabilities of static analyzers. Several measurable aspects of the source code may affect higher level quality attributes, however this comparison is based on five important aspects: analysis success, analysis depth, source code metrics, coding rule violations and code duplications.

This paper is organized as follows. Related research is outlined in Section 2. Section 3 briefly introduces the RPG 4GL language, our subject analyzer tools capable of RPG quality measurements. In depth comparison of the tools in terms of source code metrics, coding rule violations and clones is presented in Section 4, while our findings are discussed in Section 5. Finally, we conclude our work and present ideas for further research in the last section.

## 2    Related Work

Numerous studies have been published in the last decades focusing on different software metrics. Chidamber and Kemerer introduced a number of object oriented metric definitions [4]. Basili et al. validated these metrics by applying them on early software defect prediction [1]. A revalidation were done by Gyimothy et al. [7] to present fault prediction technique results of the open source Web and e-mail suite called Mozzila. Despite RPG is not located in OO domain, these studies are cornerstones for further investigations on software metrics.

At present, RPG and other early programming languages like COBOL are used by a narrowed set of developers since RPG programs cannot be run on personal computers (only via remote connection) and mainly newly constructed languages are tutored. Thus, many effort was put into researches dealing with effective migration mechanisms to transform RPG legacy programs into an object oriented environment [3]. The migration process presented stands from six sequential phases, however it transforms RPG II and RPG III into RPG IV. A migration technology was also proposed to handle COBOL legacy systems as Web-based applications[5] by applying wrapping techniques on them.

Research papers dealing with software metrics are commonly applied on widely used programming languages like C, C++ [23],[6], Java[2], C#[13],[9]. The Magic 4GL language has similar attributes to RPG, with similar need for quality assurance solutions ([15], [14]). Only a few study focuses on software metrics specialized for RPG. Hartman focused on McCabe and Halstead metrics [8] because of their usefulness in identifying modules containing a high number of errors. Another early research paper also focuses on the characteristic of programs written in RPG [16]. Naib conducted an experiment using environmental (varying with time) and internal (McCabe, Halstead, LOC that do not vary with time) factors and constructed a regression model to predict errors in the given systems. Bakker and Hirdes analyzed mainly legacy systems with more than 10 million lines of code written in COBOL, PL/I, C, C++, and RPG (1.8 million lines of code). They found that maintenance problems are highly correlates with design issues. They recommended to re-design and re-structure rather than re-build applications since it better worth it. Further maintenance difficulties including improvement and extension can occur, so a flowchart extraction method was made by Suntiparakoo and Limpiyakorn [21] that can serve as a quality assurance item supporting the understanding of RPG legacy code in maintenance process. One can see that many approach use software metrics as a low level component to produce or model a higher level characteristic (e.g fault-prone modules) describing a given system. Low level metrics can be applied for a wide variety of software quality purposes such as using quality models to characterize whole systems (often includes benchmarking). Different models have been proposed based on ISO/IEC 25010 [11], and its ancestor called ISO/IEC 9126 [10] to serve these purposes. Ladanyi et al. built a quality model[12] for especially RPG programs. They used software metrics, rule violations, and code duplications to estimate the maintainability of RPG software systems. A case study was also introduced on how such a quality assurance tool can be integrated into a given development cycle with the less interference.

Due to focusing on high level characteristics the above mentioned studies pay little attention on different low level software metrics and rules. In the following sections we will propose two state of the art RPG static source code analyzers and compare their functionalities from the view of the users.

## 3  Static Analysis of RPG Programs

### 3.1  The RPG Language from the Analysis Perspective

RPG has a long history in view of the fact that IBM has been developing the language since 1959. Originally it was designed as the Report Program Generator (RPG) with a purpose to replicate punched card processing on an IBM 1401. However, RPG quickly evolved into a high-level programming language (HLL) equivalent to COBOL and PL/I. At present, the language itself participates in IBM's Integrated Language Environment (ILE), represents a dramatic step forward in RPG's evolution. Despite the fact that in our static analysis method we support programs written in RPG III, predecessor of RPG IV, we only will focus solely on the latter since a well-defined conversion process is provided by IBM to transform the programs of old, furthermore, SonarQube is also not able to handle them. A great enhancement was presented by announcing RPG IV, since it has contained a new function type beside subroutines called procedures. They can have parameters and can be called from other programs, contrary to subroutines that do not support flexibilities like this. A new era has opened with the appearance of free-form blocks providing modern programming approach by omitting column-sensitive layout requirements.

```
....1....+....2....+....3....+....4....+....5....+....6....+....7....+..
*=====================================================================*
* Convert String to UpperCase
*=====================================================================*
p UCase           b
d UCase           pi          256
d  inString                   256
d  outString      s           256
d Up              c                       const('ABCDEFGHIJKLMNOPQRSTUVWXYZ')
d Lo              c                       const('abcdefghijklmnopqrstuvwxyz')
 /free
   outString = %trim(%xlate(lo:up:inString));
   return outString;
 /end-free
p UCase           e
```

**Fig. 1.** Sample RPG IV code

The purpose of this paper is not to introduce the language possibilities but to compare the available static analysis tools. Only a short sample RPG IV program is shown in Figure 1 converting a given string to uppercase form.

### 3.2  RPG Program Analyzer Tools

In this paper we provide in depth comparison of two tool-chains for quality centric static analysis of RPG programs. Source code based quality measurements usually consider several aspects of the code, form which the most popular ones are architecture & design, comments, coding rules, potential bugs, complexity, duplications, and unit tests. The RPG language is not provided with so extensive free tool support as in the case of object-oriented languages. In our comparison we selected two recently announced and partially free / low cost software quality tools: SourceMeter for RPG version 7.0 and SonarQube RPG (version 4.5.4). Although the categorization of quality attributes are different in these tools, we found them comparable as the results of the SourceMeter toolchain are integrated into the SonarQube framework.

**SourceMeter for RPG.** SourceMeter [20] is an innovative tool built for the precise static source code analysis of projects implemented in languages like Java, C/C++, Python or RPG [12]. This tool makes it possible to find the weak spots of a system under development from the source code itself without the need of simulating live conditions.

SourceMeter can analyze source code conforming to RPG III and RPG IV versions, including free-form as well. The input of the analysis can be specified as a raw source code file or a compiler listing. In case of using raw source code as an input, the analyzer could not calculate some code metrics, and detect various rule violations because the raw source contains less information than the compiler listing (for instance, cross references are detected using compiler listing entries). As it is recommended, we used compiler listing inputs in our work. For constructing RPG compiler listing files, we use RPG compiler with version V6R1M0.

SourceMeter is essentially a command line tool-chain to flexibly produce raw results of static analysis. Visualization and further processing of these results are done in other tools like the QualityGate [18] software quality management platform and the SourceMeter plugin to integrate data into the SonarQube framework.

**SonarQube RPG.** SonarQube [19] is an open source quality management platform with several extensibility possibilities. In this platform the concrete static analyzers of various programming languages are implemented as plugins as well. As it supports several languages, the depth and type of analysis results depend on the actual tool-chain. The main starting point of the user interface is the so called Dashboard, however the interface can also be highly extended and customized. Figure 2 shows the SonarQube RPG dashboard, where all aspects of quality are represented. The SonarQube RPG analyzer is a commercial plugin, however trial licence is available. The plugin supports the RPG IV language. However, no possibility is present to perform an analysis on RPG III programs or to handle free-form code blocks in RPG IV.

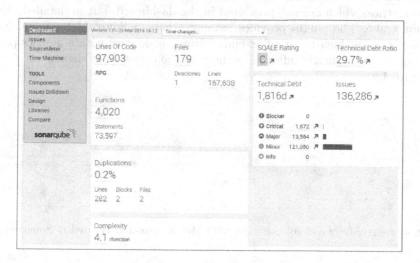

**Fig. 2.** SonarQube dashboard

**SourceMeter for SonarQube Plugin.** SourceMeter is bundled with a free SonarQube RPG analyzer plugin. The plugin conforms to the analysis process of the SonarQube and provides necessary data for the integration. This way, analysis results (metrics, code clones, rule violations) of SourceMeter can be re-used and the SonarQube framework provides the user interface and the management layer of quality data.

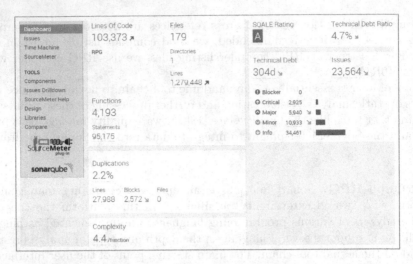

**Fig. 3.** SourceMeter for RPG SonarQube plugin dashboard

In Figure 3, the dashboard can be seen with SourceMeter data. Results are different from the ones shown in Figure 2, for example much more coding rules and clones are found by the SourceMeter. In addition, there are several additional metrics, which are not presented in the dashboard, but in detailed views of SonarQube. The plugin provides, among others, a SourceMeter menu item with custom dashboard and an extended source code view with a metrics panel showing hands on metric information next to the actual code element as shown in Figure 4.

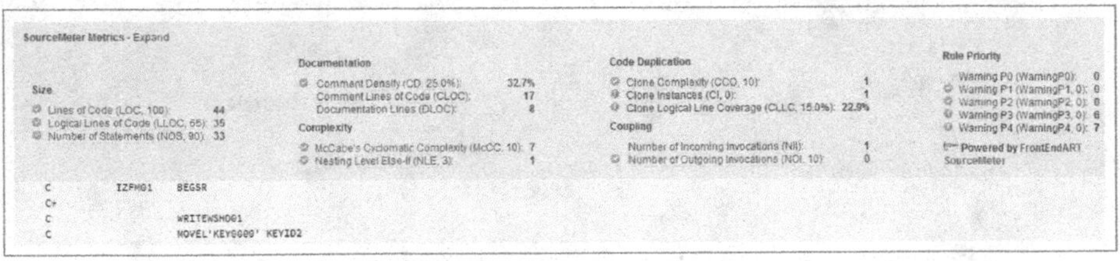

**Fig. 4.** SourceMeter source code view with Metrics panel integrated in SonarQube

## 4    Comparative Evaluation

We conducted experiments using 179 RPG programs containing around 100k lines of code. These programs belong to a software development company specialized for IBM i mainframe applications. While these programs are considered typical in their purpose, they are influenced by the coding style and tradition of the company.

## 4.1    Comparison of Source Code Metrics

The SourceMeter tool provides a large variety of metrics at four levels of program elements: system, program, procedure, and subroutine levels. The SonarQube model is restricted to file level metrics, which we treat as program level metrics. In addition, system level summary is also available. On the other hand, the extensibility mechanism of SonarQube makes it possible to incorporate additional metrics into the user interface (as shown in Figure 4).

**Table 1.** System level metric values

|  | Files | LOC | Functions | Statements | Duplications | Complexity |
|---|---|---|---|---|---|---|
| SonarQube | 179 | 97,903 | 4,020 | 73,597 | 0.2% | 16,667 |
| SourceMeter | 179 | 103,373 | 4,193 | 95,175 | 2.2% | 18,296 |
| Difference | 0 | 5,470 | 173 | 21,578 | - | 1,629 |
| 3 files | 0 | 5,289 | 173 | 5054 | - | 1,113 |
| Abs. Diff. | 0 | 2 (181-179) | 0 | 16524 | - | 516 |

Table 1 shows high level metric values of the analyzed system, which metrics are available in both tools. Each tool is able to calculate the number of files, lines of code, number of functions, number of statements, the percentage of duplicated code portion, and the global complexity. Considering the indicated values, one can see that many of them are not the same. Further investigations showed that SonarQube could not analyze three files, thus the metric values are also not calculated and aggregated. Metric values that are calculated for these three files by SourceMeter is also showed in the table. SourceMeter counts the last empty line into the LOC metric, so absolute difference can be caused by the distinct calculating methods used by each tool, moreover it is based on that no previous baselines were unified when dealing with software metrics related to RPG programming language. Different operations can be taken into consideration when calculating complexity or number of statements.

We summarized the available metrics of both tools in Table 2. SourceMeter definitely operates with a more comprehensive set of metrics. SourceMeter handles subroutines as basic code elements and propagates the calculated metric values to higher levels (procedures and programs can contain subroutines). SonarQube focuses only on file (program) and system levels and also works with a narrowed set of metrics. For detailed descriptions of the computed metrics we refer to the websites and users guides of the tools.

## 4.2    Comparison of Coding Rules

The lists of coding rules of the two analysis tools have a significant common part. Figure 5 shows the distribution of coding rules between each of the following categories: common rules checked by both tools, SourceMeter-only rules, SonarQube-only rules. SourceMeter also provides a set of rules for validating metric values by specifying an upper or lower bound for each metric shown in Table 2.

**Table 2.** Defined Metrics

| Level | Category | SourceMeter for RPG | SonarQube RPG |
|---|---|---|---|
| System | Coupling | TNF | TNF |
| | Documentation | TCD, TCLOC, TDLOC | |
| | Complexity | - | McCC |
| | Size | TLLOC, TLOC, TNOS, TNPC, TNPG, TNSR, TNNC, TNDS | TNOS, TNSR, TLOC, TLLOC |
| Program/File | Coupling | TNOI, NF, TNF, NIR, NOR | |
| | Documentation | CD, CLOC, DLOC, TCD, TCLOC, TDLOC | CLOC, CD |
| | Complexity | NL, NLE | McCC |
| | Size | LLOC, LOC, NOS, NUMPAR, TLLOC, TLOC, TNOS, TNPC, TNSR, NNC, TNNC, NDS, TNDS | TNSR, TNOS, LOC, LLOC |
| Procedure | Coupling | NOI, TNOI, NF | |
| | Documentation | CD, CLOC, DLOC, TCD, TCLOC, TDLOC | |
| | Complexity | McCC, NL, NLE | |
| | Size | LLOC, LOC, NOS, NUMPAR, TLLOC, TLOC, TNOS, TNSR, NNC, NDS | |
| Subroutine | Coupling | NII, NOI | |
| | Documentation | CD, CLOC, DLOC | |
| | Complexity | McCC, NL, NLE | |
| | Size | LLOC, LOC, NOS | |

Precisely set values can help developers to focus on code segments that are possible weak spots. SonarQube does not support rules like this. Many rules are implemented in both tools ($31\% \approx 30$ rules), that confirms that a similar set of rules are considered important by the developers of each tools.

Table 3 shows a comparison of the implemented rules in both tools. Based on the different implementation, many rule violation trigger numbers are not equal. In the following, we mainly wanted to focus on the rule violation occurrence values that differs. The rule dealing with comment density is not the same in these tools since SourceMeter desires comment lines after x lines, where x is an upper threshold, contrary SonarQube only examines the Comment density metric (CD) for a program. None of the tools found any subroutine without documentation. The reason for this is that the RPG sources are generated from compiler listing files that contain comments since the compiler automatically places comments before subroutines. In the given source files no naming convention was applied on subroutine names, so SourceMeter detects all the 4193 subroutines (found) except one which name starts with SR. However, the list contains the *INZSR subroutines (172) which is not correct, since the initialization subroutine must be named exactly like that. The remaining 173 rule violation comes from the three unanalyzed files. Copyright checks are not sufficient by SonarQube (found no violation), contrary to SourceMeter that found 28 case when copyright is not located in the source code. Some random case were validated manually and

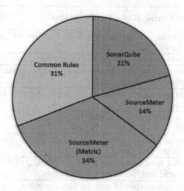

**Fig. 5.** Distribution of common and unique coding rules

SourceMeter triggers correctly. Nesting different control flow statements like do, do-while, if, select too deeply may result in complexity problems. SonarQube located 264 deep nesting case, while SourceMeter detected 352. A possible reason for this can be the different parameter setting for the maximum nesting level. SourceMeter should use a better default parameter for subroutine complexity since it detects numerous subroutines with high complexity. "/EJECT" compiler directive should be used after F, D, and C specification sections. Empirically validated the fact that SonarQube do not detects all the possibilities (after C specifications it does not require an /EJECT directive). When dealing with unused subroutines SonarQube counts the initialization subroutine as one of the never called ones although it is called automatically (there are cases when explicit call is used). SonarQube detects commented out code sections, but SourceMeter locates commented out statements. SourceMeter explores avoid ("testn" – occurred 4 times) and forbidden ("leave" operation – occurred once) operations (only the priority differs) and does not use a particular rule only for GOTO operation. SonarQube handles all of the occurrence of '0' or '1' as a possible rule violation and asks the developer to change it for *ON or *OFF (not only for indicators, that causes many false positive violations). SourceMeter desires the presence of *NODEBUGIO option too in the header not only the *SRCSTMT keyword. Missing error handling rule violations differs only because of the three unanalyzed files. "Code blocks (IF, DO, Files, WHEN clauses) containing too many lines" rules possibly have different occurrence values since the default parameter differs. SourceMeter has a similar rule for limiting the usage of the /COPY compiler directive but it operates with a nesting level limit (currently one level of copy is allowed). However, SonarQube does not detect the forbidden copy operations.

Table 4 presents the SourceMeter-only rules. There can be found rules like subroutine circular call detection, different naming conventions, constantly false

**Table 3.** Rules implemented in both tools

| Group by SonarQube | SC Occ. | SonarQube Rule Description | SM Occ. | Group by SourceMeter |
|---|---|---|---|---|
| | 0 | Source files should have a sufficient density of comment lines | 185 | Documentation |
| | 0 | Subroutines should be documented | 0 | Documentation |
| convention | 3847 | Subroutine names should comply with a naming convention | 4192 | Naming |
| | 0 | Copyright and license headers should be defined | 28 | Security |
| | 0 | "E" (externally described) indicator should be found in F spec lines | 0 | Design |
| | 7 | Numeric fields should be defined as odd length packed fields | 7 | Design |
| brain-overload | 264 | Control flow statements "IF", "FOR", "DO", ... should not be nested too deeply | 352 | Design |
| brain-overload | 31 | Subroutines should not be too complex | 1,454 | Design |
| | 1 | Line count data should be retrieved from the file information data structure | 1 | Design |
| convention | 334 | "/EJECT" should be used after "F", "D" and "C" specification sections | 520 | Basic |
| | 305 | The first parameter of a "CHAIN/READx" statement should be a "KLIST" | 315 | Design |
| unused | 227 | Unused subroutines should be removed | 80 | Unused Code |
| unused | 130 | Sections of code should not be "commented out" | 185 | Unused Code |
| | 0 | Certain operation codes should not be used | 4 + 1 | Basic |
| brain-overload | 0 | "GOTO" statement should not be used | 0 | Basic |
| cwe, security | 0 | Debugging statements "DEBUG(*YES)" and "DUMP" should not be used | 0 | Basic |
| | 0 | The correct "ENDxx" statement should always be used | 0 | Basic |
| | 0 | "IF" statements should not be conditioned on Indicators | 0 | Basic |
| cwe | 0 | All opened "USROPN" files should be explicitly closed | 0 | Basic |
| | 3 | An indicator should be used on a "CHAIN" statement | 1 | Basic |
| | 1111 | Standard figurative constants *ON, *OFF and *BLANK should be used in place of '1', '0' and ' ' | 17 | Basic |
| | 0 | The "*SRCSTMT" header option should be used | 4 | Basic |
| error-handling | 699 | Error handling should be defined in F spec | 749 | Basic |
| cert | 0 | "IF ELSEIF" constructs shall be terminated with an "ELSE" clause | 0 | Basic |
| brain-overload | 643 | "WHEN" clauses should not have too many lines | 308 | Size |
| brain-overload | 58 | Files should not have too many lines | 120 | Size |
| brain-overload | 55 | "DO" blocks should not have too many lines | 17 | Size |
| brain-overload | 145 | "IF" blocks should not have too many lines | 151 | Size |
| | 0 | "/COPY" should be avoided | 1 | Design |
| brain-overload | 0 | Subroutines should not have too many lines | 0 | Size |

conditional statements (like 1 equals to 2). A bad programming practice when a variable is given as an operand of call operation since it hardens the debugging process.

**Table 4.** Rules implemented only in SourceMeter (without metrics-based rules)

| SourceMeter Rule Description | Group by SourceMeter | Occ. |
|---|---|---|
| Uncommented conditional operation | Documentation | 4,629 |
| File uses prefixed name | Naming | 0 |
| Too short name | Naming | 22 |
| Too long name | Naming | 271 |
| Character variable names should begin with '$'. | Naming | 0 |
| Numeric variable names should begin with '#'. | Naming | 0 |
| Lower case letter in the name of called program or procedure | Naming | 0 |
| Large static array | Design | 33 |
| Circular reference between subroutines | Design | 1 |
| Variable only referenced from an unused subroutine | Unused Code | 18 |
| Conditional expression is always false | Unused Code | 1 |
| Numeric operands of MOVE(L) are not compatible | Type | 2 |
| Call operand is a variable | Basic | 3 |
| Complete Conditional Operation Needed | Basic | 179 |

Table 5 shows the list of SonarQube-only rules and the number of triggers. Some rules have a very high trigger value. Uppercase form was not used in 107,768 cases that can seriously distort the technical dept information.

## 4.3 Comparison of Duplicated Code Sections

SonarQube contains a rule for noting suspicious duplicated code sections. Sonar can show duplicated lines in the source files, however no grouping can be obtained that makes it hard to understand code clones. Sonar only deals with Type-1 clones that means the traditional copy-paste programming habit, so, every character must be the same in the clone instances. A clone class encapsulates the same code portions (clone instances) from different source locations into a group. SonarCube considered 0.2% of the whole RPG code as code duplication (2 duplicated section with 141 lines). SourceMeter has a poor display technique in sonar environment, namely no highlighting on affected lines are done. In a different context, SourceMeter supports a kind of well-defined format for marking various clone classes and the relevant code instances. The tool is also capable to find Type-2 clones (e.g variable names may differ) that is confirmed by the found 2.2% of code that play a role in code duplications. Its clone detection algorithm tries to match similar code sections (syntax-based) based on source code elements (subroutine, procedure). Contrary, SonarQube only uses textual similarities to detect clones, but no structural information is used in clone detection. For example, clone instances containing one and a half subroutines may be produced, however they should be splitted into two clone instances (holds more information when considering refactoring). Another advantage of SourceMeter is that it accepts parameters such as the minimum lines of code contained by a clone instance.

While SonarQube shows duplicated code locally in the inspected program, SourceMeter extends its capabilities with a separate code duplication view, where clone instances belonging to the same clone class can be investigated easily.

**Table 5.** Rules implemented only in SonarQube

| SonarQube Rule Description | Group by SonarQube | Occ. |
|---|---|---|
| Variables used in only one subprocedure should not be global | pitfall | 0 |
| "/COPY" statements should include specification letters | convention | 185 |
| "CONST" should be used for parameters that are not modified | | 2 |
| Columns to be read with a SELECT statement should be clearly defined | sql | 0 |
| Comment lines should not be too long | convention | 9,891 |
| Expressions should not be too complex | brain-overload | 86 |
| LIKE keyword should be used to define work fields | | 703 |
| Nested blocks of code should not be left empty | bug | 32 |
| Operation codes and reserved words should be in upper case | convention | 107,768 |
| Prototypes should be used | convention, obsolete | 1,423 |
| Record formats should be cleared before each use | bug | 973 |
| Source files should not have any duplicated blocks | | 2 |
| SQL statements should not join too many tables | performance, sql | 0 |
| Subprocedures should be used instead of subroutines | obsolete | 4,019 |
| Subprocedures should not reference global variables | brain-overload | 0 |
| The data area structure for "IN" should be defined in D spec lines. | | 148 |
| The parameters of a "CALL" or "CALLB" statement should be defined as a "PLIST" | | 68 |
| Non-input files should be accessed with the no lock option | | 0 |
| Unused variables should be removed | unused | 14 |
| String literals should not be duplicated | | 2872 |

# 5    Discussion

## 5.1    Summary of Results

We summarize our findings along five viewpoints as follows.

*Analysis success and depth* The program analysis went almost without problems with both tools. While SourceMeter successfully analyzed all source files, SonarQube RPG failed to analyze three of them. Although this is not considered as a blocker problem in its use. On the other hand, SonarQube works at file level, while SourceMeter analyzer works at finer levels of details (like procedure, subroutine level), which provides a more detailed view of the analyzed system.

*Source code metrics* SourceMeter provides wider range of metrics and works even at procedure and subroutine levels. SonarQube provides a limited set of metrics, which restricts the quality model that can be built upon it.

*Coding rules* A large portion of analyzed coding rules are common or very similar in both tools. SonarQube has slightly more unique rules implemented, but SourceMeter provides a wide set for validating metric rules. Generally, the two tools provide balanced functionality.

*Code duplications* SourceMeter found significantly more duplicated code fragments with better granularity. SourceMeter detects Type-2 clones (syntax-based), SonarQube only deals with copy-paste clones. SourceMeter extends SonarQube with improved display of code clones.

**Table 6.** Overall comparison results

| Aspect | Result | Note |
|---|---|---|
| Analysis success | Balanced | SonarQube failed to analyze some input files |
| Analysis depth | SourceMeter | SourceMeter provides statistics in lower levels |
| Code metrics | SourceMeter | SourceMeter provides much more metrics |
| Coding rules | Balanced | Large common set, balanced rule-sets |
| Code duplications | SourceMeter | SourceMeter found more duplicated code blocks |

Table 6 summarizes our findings with a short explanation of the result of our experiments. During the comparison of our subject tools, we experienced that coding rules for the RPG language in general need to be evolved, compared to similar solutions of other popular languages. Given that both tools appeared recently on the market, we foresee extended versions in the coming years.

## 5.2 Effect on Quality Indexes

Low level, measurable attributes such as code metrics, rule violations and code duplications contribute to higher level code quality indexes. Such quality indexes give an overall picture of the analyzed project, helping stakeholders to take actions in case of low or decreasing quality. SonarQube operates with two concepts to assess higher level quality: technical debt and SQALE rating.

Technical debt is a metaphor of doing things in a quick but dirty way, which makes future maintenance harder. If the debt is not paid back (e.g. software quality is not considered as an important aim in the development), it will keep accumulating interest – similarly to a financial debt. In case of SonarQube, the technical debt is measured purely based on coding rule violations. Each coding rule has an estimated time to correct it. The overall technical debt is the sum of the estimated correction time of all rule violation instances. The SQALE rating is based on the technical debt, as such, it is based on coding rules as well. Hence, other quality attributes, like various metrics (e.g. complexity, coupling) and code duplications do not affect these quality indexes. We provide dashboard data of quality indexes computed in case of all rules checked (Figure 6) and the dashboard for an analysis when only the common rules were active (Figure 7). On the other hand, we recommend quality models that relies on more quality attributes, like the QualityGate [18] models.

## 5.3 Threats to Validity

We identified several threats to validity of our study. The validation of the results was done manually on selected metrics/rules. The initial plan was to export the whole list of rule violations and filter automatically at least the common results. While SourceMeter is a command line tool-chain that can produce csv outputs, we did not manage to obtain the full list from SonarQube. It is possible to

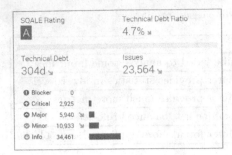

**Fig. 6.** Quality indexes based on SourceMeter for RPG analyzer (left) and SonarQube RPG analyzer (right) – computed using all coding rules

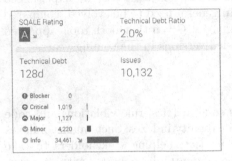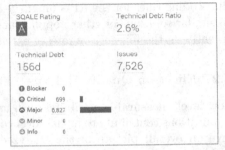

**Fig. 7.** Quality indexes based on SourceMeter for RPG analyzer (left) and SonarQube RPG analyzer (right) – computed using common coding rules only

obtain a report from SonarQube, but that is not a complete list of rule violations. Although exhautive manual validation is not feasible, the current study involves three aspects of quality measurements. We believe these three aspects are of high importance (technical debt is computed based on only one aspect), however adding other viewpoints or even dynamic analysis results would increase the validity of the results. The measured RPG programs belongs to the same domain and implemented by developers of the same software house who followed coding policies of the company. Further experiments are needed with larger RPG codebase from various domains and developers. Although we identified this threat, we note that the measured RPG programs are part of legacy, data intensive applications typical in IBM i mainframes.

## 6    Conclusion and Future Work

In this study we experimented with the static analyzers of quality management tools for the RPG programming language employed on the IBM i mainframe. We compared the SourceMeter for RPG command line tool-chain together with its SonarQube plugin to the RPG analyzer of the SonarQube framework. Five important aspects of quality measurements were examined: analysis success, analysis depth, source code metrics, coding rules, and code duplications. Sonar-Qube can not handle some source files, moreover the depth of analysis is limited to system and file level. SourceMeter can perform analysis in finer granularity (procedures, subroutines). We found that from the metrics point of view the SourceMeter tool provides much wider range of possibilities, while handling of coding rules is balanced since the common set of coding rules is relatively large. SonarQube detects clones using copy-paste (Type-1) clones, SourceMeter can detect Type-2 clones since it uses a syntax-based mechanism and also take into consideration the bounds of source code elements (subroutines, procedures).

In the future we need to conduct further experiments on larger and more diverse set of programs. We plan to extend the investigated aspects to other lower level quality attributes, and to automate the validation process to increase the confidence and generalizability of the comparison.

**Acknowledgments.** The publication is partially supported by the European Union FP7 project "REPARA – Reengineering and Enabling Performance And poweR of Applications" (project number: 609666) and by the Hungarian national grant GOP-1.1.1-11-2012-0323.

# References

1. Basili, V.R., Briand, L.C., Melo, W.L.: A validation of object-oriented design metrics as quality indicators. IEEE Transactions on Software Engineering **22**(10), 751–761 (1996)
2. Bruntink, M., Van Deursen, A.: Predicting class testability using object-oriented metrics. In: Fourth IEEE International Workshop on Source Code Analysis and Manipulation, 2004, pp. 136–145. IEEE (2004)
3. Canfora, G., De Lucia, A., Di Lucca, G.A.: An incremental object-oriented migration strategy for rpg legacy systems. International Journal of Software Engineering and Knowledge Engineering **9**(01), 5–25 (1999)
4. Chidamber, S.R., Kemerer, C.F.: A metrics suite for object oriented design. IEEE Transactions on Software Engineering **20**(6), 476–493 (1994)
5. De Lucia, A., Francese, R., Scanniello, G., Tortora, G.: Developing legacy system migration methods and tools for technology transfer. Softw. Pract. Exper **38**, 1333–1364 (2008)
6. Ferenc, R., Siket, I., Gyimothy, T.: Extracting facts from open source software. In : Proceedings 20th IEEE International Conference on Software Maintenance, 2004, pp. 60–69, september 2004
7. Gyimothy, T., Ferenc, R., Siket, I.: Empirical validation of object-oriented metrics on open source software for fault prediction. IEEE Transactions on Software Engineering **31**(10), 897–910 (2005)
8. Hartman, S.D.: A counting tool for rpg. In: ACM SIGMETRICS Performance Evaluation Review, vol. 11, pp. 86–100. ACM (1982)
9. Hegedűs, P.: A probabilistic quality model for c#-an industrial case study. Acta Cybern. **21**(1), 135–147 (2013)
10. ISO/IEC. ISO/IEC 9126. Software Engineering - Product quality. ISO/IEC (2001)
11. ISO/IEC. ISO/IEC 25000:2005. Software Engineering - Software product Quality Requirements and Evaluation (SQuaRE) - Guide to SQuaRE. ISO/IEC (2005)

12. Ladányi, G., Tóth, Z., Ferenc, R., Keresztesi, T.: A software quality model for rpg. In: Proceedings of the 22nd IEEE International Conference on Software Analysis, Evolution, and Reengineering (SANER 2015), pp. 91–100. IEEE, March 2015
13. Lanza, M., Marinescu, R.: Object-oriented metrics in practice: using software metrics to characterize, evaluate, and improve the design of object-oriented systems. Springer Science & Business Media (2007)
14. Nagy, C., Vidács, L., Ferenc, R., Gyimóthy, T., Kocsis, F., Kovács, I.: Magister: Quality assurance of magic applications for software developers and end users. In: IEEE International Conference on Software Maintenance (ICSM 2010), pp. 1–6. IEEE, September 2010
15. Nagy, C., Vidács, L., Ferenc, R., Gyimóthy, T., Kocsis, F., Kovács, I.: Complexity measures in 4GL environment. In: Murgante, B., Gervasi, O., Iglesias, A., Taniar, D., Apduhan, B.O. (eds.) ICCSA 2011, Part V. LNCS, vol. 6786, pp. 293–309. Springer, Heidelberg (2011)
16. Naib, F.A.: An application of software science to the quantitative measurement of code quality. In: ACM SIGMETRICS Performance Evaluation Review, vol. 11, pp. 101–128. ACM (1982)
17. Piattini, M., Calero, C., Genero, M.: Table oriented metrics for relational databases. Software Quality Control 9(2), 79–97 (2001)
18. QualityGate quality management platform http://www.quality-gate.com (2015)
19. SonarQube quality management platform (2015). http://www.sonarqube.org
20. SourceMeter (2015). http://www.sourcemeter.com
21. Suntiparakoo, K., Limpiyakorn, Y.: Flowchart knowledge extraction on rpg legacy code. Advanced Science and Technology Letters 29, 258–263 (2013)
22. Witting, G.E., Finnie, G.R.: Using artificial neural networks and function points to estimate 4GL software development effort. Australasian Journal of Information Systems 1(2) (1994)
23. Xenos, M., Stavrinoudis, D., Zikouli, K., Christodoulakis, D.: Object-oriented metrics-a survey. In: Proceedings of the FESMA, pp. 1–10 (2000)

# Novel Software Engineering Attitudes for Bussiness-Oriented Information Systems

Jaroslav Král[1] and Michal Žemlička[2]([⊠])

[1] Faculty of Informatics, Masaryk University,
Botanická 68a, 60200 Brno, Czech Republic
kral@fi.muni.cz
[2] The University of Finance and Administration,
Estonská 500, 10100 Praha 10, Czech Republic
michal.zemlicka@post.cz

**Abstract.** Modern ICT technologies should and can be more business-
and businessmen-friendly than they are now. We discuss a variant of
service-oriented architecture solving this challenge. The architecture is a
network of the analogues of real-life services communicating by exchange of
business documents via a network of organizational (architecture) services.
The architecture is especially useful in small-to-medium enterprises – generally in the cases when sourcing and dynamic business processes are used.
It is also useful for e-government systems. It has many business as well as
technical/engineering advantages. The architecture enables a smooth collection of transparent data for management and research. It further enables
business document oriented communication of business services and simplifies the collaboration of users with developers.

**Keywords:** Business-oriented information systems · Document-oriented
communication · Software confederations

## 1 Introduction

Contemporary business is ICT driven. The contribution of ICT to the quality
of business and economical processes is enormous. Substantial issues and challenges are actual for decades. Contemporary business is simultaneously more
open, globalized, and more dynamic. Its dynamics must be supported online by
different professionals (businessmen, managers, or bookkeepers). We will call the
professionals agents.

Some agents must define business processes being networks of economic (business) actions. The processes must be agile, i.e., they must be online changed by
the interventions of the agents. The agents are often no IT professionals. The
agents must be responsible for their actions. Some agents fix the business rules;
others supervise business cases, operations, or whole business processes.

The dynamics of the business processes is a consequence of the dynamics of
(global) market, failures of business relations, changing production conditions,
and changing customer needs, etc. The most frequent reasons of the changes are:

© Springer International Publishing Switzerland 2015
O. Gervasi et al. (Eds.): ICCSA 2015, Part V, LNCS 9159, pp. 193–205, 2015.
DOI: 10.1007/978-3-319-21413-9_14

- Inability of business partners to meet crucial points of business agreements: prices, terms, quality;
- inevitable accidents (black outs, floods; transport problems);
- changes of business partners;
- business failures.

From the point of view of the agents the collaboration of the business partners is given by business documents. IT helps the agents to do their actions correctly and easier; it also simplifies the reconstruction of the previous activities and the detection of reasons why something had gone wrong.

The business partners can be themselves networks of business (sub)units. The supporting ICT system (usually an information system) should reflect the properties of contemporary business. A straightforward solution is to design the system as a business document driven service-oriented architecture (SOA). The SOA integrates coarse-grained applications (services) exchanging coarse-grained messages (documents) [1]. A specific implementation of such SOA is called *software confederation* [2] or *confederation* for short. We will discuss here the advantages of confederations driven by digitalized real-life documents or their collections.

Such confederations offer many business as well as technological (software engineering) opportunities. They enable business transparency and business agility in the large – e.g. in global business processes, in simplifying the solution of business issues (court trials, changes of business partners). Technical opportunities are of big importance – new methods of development like prototyping or incremental development in the large.

The runtime advantages include easy maintenance [3], smooth sourcing, optimization, easy integration of commonly used tools like spreadsheets. Our philosophy is well applicable in e-government and open new possibilities for small-to-medium vendors. It holds especially for small-to-medium enterprises (SME) and surprisingly for state administration.

The paper is structured as follows: Section 2 describes desirable properties of business processes. Sections 3 and 4 discuss the cons and pros the top-down (cathedral) and bottom-up (bazaar) construction of business-oriented software systems. The business requirements are analyzed in Section 5. Sections 6–8 are dedicated to document-oriented communication. Section 9 focuses on architectural services, especially gateways. Section 10 mentions further research. Finally, Section 11 presents the conclusions.

## 2    The Desirable Properties of Business Processes

The desirable properties of modern (global) business processes are:

1. Business processes often control the collaboration of rather autonomous business subjects. The collaboration is as a rule based on business agreements and on exchange of business documents (like orders and invoices).

2. The business processes are supervised and possibly on-line modified by authorized people, businessmen inclusive.
3. The processes and their supervisors must be able to respond to business as well as technical issues, threats, and opportunities.
4. It is desirable, sometimes necessary, to generate business data to simplify the solution of business controversies that could have the form of court trials.
5. The processes should provide transparent data enabling optimizations of business processes and hopefully business or economic research.

The requirements can be met if the business processes are supported by information systems having a specific service-oriented architecture (SOA) called confederation. The components of confederations communicate by the exchange of (electronic) business documents in appropriate formats.

Such business documents could be more like plain text (many business documents have relatively fixed structure of the text, they can be therefore relatively simply parsed), like forms that are filled during individual steps of the business process, or in a complex formats based on XML as described e.g. in [4].

The application components can be analogs of real-life business applications or services (bookkeeping) or complete information systems of independent business bodies (enterprises, offices). It is advantageous to use similar philosophy and tools for the whole system, its autonomous parts and information systems of business partners.

The requirements 4 and 5 can be met if the transported documents are logged for future use.

## 3   Top-Down or Bottom-Up?

Large business information systems usually contain autonomous components (subsystems) and are developed and operated by autonomous, often distributed, teams. They very often support not only business processes in the global economy. The components can work in different local ecosystems communicating with changing business partners and their systems and must cope with local restrictions and legislative peculiarities. Their development, operation, management, and maintenance require specific attitudes. It implies the use of specific software architectures.

The development of large information systems is usually based on one of the following two strategies:

– The top-down development working as one large entity starting from overall detailed definition (vision). It is designing the system as one monolithic whole. It is somewhat like building a cathedral, see [5]. It is often combined with holistic approach (understanding the system as a whole). It is especially popular in the case of Enterprise information systems – compare [6]. This approach is advantageous for building systems from scratch – compare e.g. object-oriented methodologies [7] or some service-oriented methodologies like the recommendations of OASIS [8] or OpenGroup [9]. It has the drawbacks discussed below.

– The bottom-up development based on an integration of existing legacy or third-party products or newly developed parts into a new system. The parts are in fact integrated as black boxes. It reminds to some degree the bazaar attitude from [5]. In the case of the methodology discussed below the top-down process reminds the development of a city built according an urban plan. This approach is typical for Software Confederations [10], useful hints can be found in OpenGroup Reference Architecture [9].

A very important advantage of the bottom-up attitude is that it usually enables preservation of the existing structures of enterprises and business processes. It prevents undesirable business process restructuring. It moreover substantially simplifies the user involvement into software development, operation, and maintenance. There are further advantages discussed below. The bottom-up (integration oriented) approach is well suited for the integration of existing software entities and can preserve their autonomy – compare the integration of local and central offices that are governed by different authorities or integration of companies into temporal cooperating business alliances.

The points are especially important in the case when a system has (ought to have) a peer to peer architecture.

## 4    Pros, Cons, and Traps of Top-Down Approach

It is known from software engineering practice and also from the studies of software metrics (see e.g. the notion "inaccessible area" and "effort consumption dynamics" in [11,12]) that complex systems cannot be neither developed nor substantially changed quickly enough. The changed or redeveloped systems are either obsolete or erroneous or both unless the systems are small or they have a proper architecture – are properly decomposed – [12] and the terms are not too tight.

If a top-down cathedral attitude is used, the development time often cannot be shortened enough. During system maintenance any system becomes after some time obsolete and error prone and must be completely redeveloped. It is a challenge as it implies that the system cannot be redeveloped in time and that usually two systems (old and new ones) must be operated in parallel for a very long time (almost permanently). Many contemporary tools and standards are better applicable in the framework of top-down approach. It has snags. The models used during systems development and system operation are not user-friendly. Even worse. The operation is then not transparent for users. It implies substantial limitations for the agile business. It is often accompanied with the pressure of some software vendors to induce users to modify their business processes more than necessary, see the Business Process Restructuring Antipattern.

The cathedral philosophy is not good for incremental development and, what is more important, for incremental maintenance. It is especially unacceptable for small-to-medium user organizations as big systems are developed by big vendors for big users. Such systems are therefore good for the organizations having the management culture of big enterprises. SME cannot develop such a system or to integrate their products with them. The use of bottom-up philosophy is now technically easy as there is enough infrastructure services like web or cloud simplifying the implementation of a document-oriented communication.

Information systems must work in dynamically changing environments. Users (businessmen) must be able to understand and control the communication between subsystems. It is often required to add new abilities to a system being used as a black box. All the requirements are difficult to meet in the top-down framework. As software systems grow they gradually become unmaintainable even for the large software vendors.

## 5    Business Requirements Summary

The above challenges and issues can be avoided if the following measures are applied:

- The system should be divided into autonomous parts supporting activities processed by autonomous or even independent individual people or organizational units (agents).
- The communication of parts (respective the software entities supporting them) should be declarative and problem oriented – preferably very similar to the ones used by people. It can be implemented if the communication is based on an exchange of digitalized business documents.
- Well-working subsystems should be reused. If there is a satisfactory IT system supporting an agenda, it should be possible to use (insource) it.
- The system should be easily and quickly reconfigurable if necessary – e.g., in the case of emergency or unexpected situation(s). The (sub)systems should be easily replaceable. It is crucial for easy sourcing.
- The system should enable or even support an easy user involvement during development, operation, and maintenance.

We will show that these requirements can be met in systems having a specific service-oriented architecture. These systems use the exchange of business documents between their constituent parts. The communication between the parts is powered by specific fine-grained services called architectural services. The concept of architectural services is very flexible. Services having very different capabilities can have very similar technical structure and the structure can be changed very easily.

## 6    Towards Document-Oriented Interfaces

The individual functionality providers (people or organizational units) should be supported by proper IT entities having interfaces covering needs of the communication partners. The software supporting a business subject should communicate with software systems of its business partners. The communication should be similar to the one already in use by the real-world "processors" (individual people or institutional units). There are multiple reasons for it:

- Many interfaces of human-provided services are in use without substantial changes for a very long time. The interfaces are therefore in some sense optimal and well tested. Existing know-how, skills, and tools can be reused.
- If the communication is business oriented, it is transparent for businessmen and other economic agents. It is a crucial advantage. The communicated documents can be logged together with time and routing data. The logged data are business transparent and can be easily used in management, research, or in business trials. The business document oriented systems can use up the capabilities of document management systems. It is very good for business problem identification and testing.
- It is possible to communicate with human partners instead of the software ones, if necessary. Such situations can appear during development (some parts are already developed but their partner parts are not) or in emergency situations.
- Business document oriented communication is usually based on business standards and practices. It is preferable especially for global business processes and e-government.

## 7    Advantages of Document-Oriented Interfaces

The solution proposed by us below solves the problem of the document-oriented communication between autonomous bodies in a real world as well as in business information systems. It enables flexible adaptation of message formats designed according agreements between business people. It enables not to follow into the trap of too complex standards (see the experience with EDIFACT and with some applications of Web Services). Our solution enables agile (online) involvement of various business professionals (managers inclusive) into activities during entire systems' lifetime include their operation and continuous enhancements. It simplifies the response to technical as well as business failures of partners as well as the responses to their tricky cooperation. Architectural services moreover enable the use of gateways in the sense of BPMN and collect data necessary for the process optimization.

It is very important that the policy enabled by our solution practically orthogonal to coding. Small changes of gateway code substantially change the properties of the entire system to react properly on external stimuli. Examples are the choice of business partners as well as of various components of the system to perform an elementary business action.

Surprisingly enough a very important property is a smooth and changeable combination of manual or computerized activities. It enables to avoid often neglected antipattern Overautomatization being dangerous especially in small-to-medium business.

Technological advantages are very important like the enhancement of the principle (paradigm) of information hiding recommended by Parnas [13]. Multi-level connectors enable on the level of gateways to achieve the situation that the application components are used absolutely black boxes. We have tools enabling to handle the black boxes as grey ones. It is, we can see and change some implementation details if necessary. It is an example of technological aspects of our solution.

There are many technological advantages having substantial influence on the capabilities of the system and its parts. The most important advantages are:

- There is almost no difference between development and maintenance. It enables to do incremental enhancements and changes and to avoid the antipattern permanent obsolescence described in [14].
- The significant autonomy or even practical independence of the components of a confederation enables in- and out-sourcing of the system parts. It can also result into an avoidance of Vendor Lock-In Antipattern [15].

The permanent obsolescence is especially important for very large and complex systems like e-government. Accidentally, it is often neglected that something like that could happen.

According [11] and others, optimal system development duration grows with system size. The duration cannot be substantially reduced (e.g. halved).

Yet another important effect is that it enables small-to-medium firms to add, replace, or change the components of large systems developed by large vendors. Small-to-medium enterprises can then enter the global business – independently or in cooperating groups (consortia).

The above mentioned principles allow broader application of the rule that quite every newly developed service can be (sooner or later) part of another created service. It is possible to create practically unbounded hierarchies of services. The services can, if necessary, define and control their local business processes.

Last but not least, the fact that the communication is business document based opens the way to implement the network of architectural services (gateways) using the capabilities of document management systems.

It together enables extremely flexible and powerful prototypes. The prototypes could be used during system operation as emergency solutions.

# 8   Implementation of Business Document Oriented Interfaces

Let us discuss some technical issues. A crucial condition, especially in small-to-medium enterprises, is the ability to reuse/integrate existing business software – e.g. the ones being third-party products. Such artifacts must be used as black boxes as their code is as a rule not open.

The software has often no appropriate business-document-oriented interface. It can be e.g. too fine-grained or it can have an improper format.

We propose to implement a tool allowing the use of communication based on exchange of business documents (like orders or invoices) in an appropriate format modifiable according the needs of cooperating partners.

The partners are business applications A. They can be developed using various paradigms. We can expect that any A is encapsulated using a component G so that the tuple (G, A) can be used as a service in a peer-to-peer network (e.g. as a service in a service-oriented system).

It is possible and desirable to encapsulate portals in the same way as the applications to gain the same interface as the other services have. Such service is than a tuple (G,U) where G is a wrapper and U a user interface (portal).

There are various reasons why G usually does not support any business-document-oriented coarse-grained interface. The interface usually is, according object-oriented paradigm, fine-grained using remote procedure call or remote method invocation. The relation between the fine-grained interface provided by G and the business-document interface required by the service partners reminds the relation between an assembler and a high-level programming language.

We propose a solution being applicable and advantageous generally. We propose to wrap the application by a service (i.e. to use wrapper as a service). We call it front-end gate (FEG). A FEG compiles business documents from sequences of fine-grained messages produced by G and transforms the incoming business documents into sequences (tuples) of fine-grained messages for G. A FEG has the following structure (compare Fig. 1):

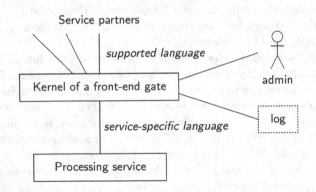

- *The processing service can be either an application service or a composite service connected through its head.*
- *The service specific language is given by the available interface of the service; it can be implementation-based (e.g. RPC/RMI), declarative, or even document-based.*
- *The supported language is usually declarative, problem oriented. It usually supports problem-oriented messages or business documents.*

**Fig. 1.** Front-end gate

1. FEG logic performing necessary transformations of messages. Its actions can be changed or supervised by an administrator.
2. Distinguished channel for the communication with wrapped application (service). It need not provide any transfer of business documents.
3. Channels transporting business documents to (from) partner services (i.e. their FEGs)

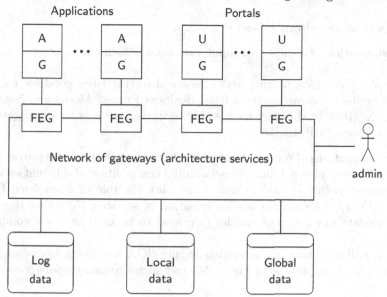

**Fig. 2.** Structure of a confederation

4. Memory for logged messages. There is a channel enabling (on-line) involvement/administration of a supervisor or administrator.

Some countries, e.g. Czech Republic, have digitalized filling rooms of offices as a part of the development of their e-government systems. The digitalized filling rooms can be quite easily modified to be used as front-end gates in our sense. It is, however, necessary to gain political support for it.

# 9   Architectural Services (Gateway as a Service)

Front-end gates can be easily modified to be architectural services. Architectural services communicate with other architectural services and front-end gates. Architectural services can be used as intelligent nodes of a virtual communication network between front-end gates. There are several types of architectural services.

The simplest variant of architectural service is a router. It enables dynamic routing of messages (documents). A more advanced variant of this function is provided by general architecture services (gateways).

The capabilities of a router can be on-line modified by its administrator. Examples of low-level capabilities of routers:

- explicit destination change,
- destination change by a remote procedure,
- advanced logging.

Examples of advanced capabilities of gateways:

- transformation of n-tuples of input messages into m-tuples of output messages,
- support of complex waiting and message directing rules good for e.g. the implementation some concepts from Business Process Model and Notation (BPMN, [16]). In this way we are able to implement a generalized gateway in the sense of BPMN [16].

Various combinations of routers, front-end gates and related policies provide very flexible and powerful capabilities like unlimited composition of different services.

If a proper policy is used, a router can play the role of generalized front-end gate (FEG). An example is a composition of services. Note that this turn can encapsulate any group of entities (services) to be used as a new composed service.

The overall structure of the system having SOA and using front-end gates according to our proposal is in Fig. 2. We call such systems *confederations*.

## 10    Other Applications

Complex systems contain as a rule parts supporting different problem domains. If these parts can use the logic, habits and turns typical for their domains, the cooperation of developers with users can be tighter. Application of front-end gates and gateways can integrate such parts without breaking the above mentioned advantage.

Current software systems are significantly influenced by various standards and recommendations. Complex systems must cope with standards and recommendations for various domains. It can therefore happen that these standards are based on different, sometimes even contradictory, assumptions.

Using all such standards at the same time can significantly complicate the work. If individual subsystems supporting different problem domains use only corresponding subsets of the standards, it can reduce the work (even reading and understanding of tens of standards having hundreds of pages each is a hard task) and eliminate many requirements clashes. It is, applying the information hiding (and communication conversion) we can reduce the system complexity or can even enable solution of such complex projects.

We have successfully applied this approach in our industrial projects from manufacturing, administration, and business.

## 11    Further Research

Although there is a quite long practical experience with confederations there are open questions:

- For which kind of enterprises is the bottom-up construction more advantageous than the top-down one.

- Conditions for the applicability of document-driven confederations in e-government.
- Usability and usefulness of document management systems (DMS) in our solutions. There are promising examples of the applications of DMS. There are cases when modern document management systems were successfully used in small firms to enhance the power of their information systems.
- Confederations use document-oriented communication. The use business documents need not be preferable under certain circumstances.
  The documents can have various formats. Very popular are formats of sheets (cleversheets) and various applications of spreadsheets.

A proper use of our proposal simplifies the collaboration between enterprises and authorities. It offers a method for integration of products of small and large software vendors. It seems that technically our proposals are mature enough. The issue is the lack of support by large vendors. The vendors could be afraid that they will not ensure their future business based on top-down attitude requiring too many changes of business processes of system users.

## 12    Conclusions

We have proposed a service-oriented architecture (confederation) for distributed and complex information systems. It narrows the logical gap between the structure of an organization and the structure of the supporting information system. The structure if the information system can mirror the organizational structure of the supported institution. It simplifies emergency use and prototyping of the system, its development and operation. The system can be very flexible.

Confederations simplify incremental development and maintenance. It enables solution of the issue of permanent obsolescence. It enhances flexibility and reduces expenses of development, operation, and maintenance. It is especially important for small and medium-sized companies.

A main long-term contribution can be the fact that powerful and dynamic logging can provide detailed snapshots collected and used for optimization and for formulation of economic hypotheses and for their testing.

In software development and programming there is a new paradigm every about 10 years. We believe that properly applied business document driven architecture is an extremely powerful tool and a kernel of a new paradigm well applicable also in cloud systems. It is indirectly indicated by, besides cloud tools themselves, the explosive growth of the number of document management systems (like SharePoint, alfresco, RunMyProcess) and their applications, compare also the steeply growing number of pages mentioning document management systems.

Document oriented communication promises many variants of further application. A very perspective application of it is the use of complicated formats like data sheets or e-governmental structured forms.

There are, however, indications that some people again tend to apply the "What's New Antipattern" from [17].

**Acknowledgments.** The paper was supported by The institutional funding of long-term conceptual development of research organization, the University of Finance and Administration in the year 2015.

# References

1. Žemlička, M., Král, J.: Flexible business-oriented service interfaces in information systems. In: Filipe, J., Maciaszek, L. (eds.) Proceedings of Enase 2014–9th International Conference on Evaluation of Novel Approaches to Software Engineering, SciTePress, pp. 164–171 (2014)
2. Král, J., Žemlička, M.: Electronic government and software confederations. In: Tjoa, A.M., Wagner, R.R. (eds.) Twelfth International Workshop on Database and Experts System Application, pp. 383–387. IEEE Computer Society, Los Alamitos, CA (2001)
3. Král, J., Žemlička, M.: Simplifying maintenance by application of architectural services. In: Murgante, B., Misra, S., Rocha, A.M.A.C., Torre, C., Rocha, J.G., Falcão, M.I., Taniar, D., Apduhan, B.O., Gervasi, O. (eds.) ICCSA 2014, Part V. LNCS, vol. 8583, pp. 476–491. Springer, Heidelberg (2014)
4. Molnár, B., Benczúr, A.: Issues of modeling web information systems proposal for a document-centric approach. Procedia Technology **9**(0), 340–350 (2013). CENTERIS 2013 - Conference on ENTERprise Information Systems / ProjMAN 2013 - International Conference on Project MANagement / HCIST 2013 - International Conference on Health and Social Care Information Systems and Technologies
5. Raymond, E.S.: The cathedral and the bazaar. First Monday 3(3) (1998)
6. Op't Land, M., Proper, E., Waage, M., Cloo, J., Steghuis, C.: Enterprise Architecture. The Enterprise Engineering Series. Springer (2009)
7. Rumbaugh, J., Blaha, M., Premerlani, W., Eddy, F., Lorensen, W.: Object-Oriented Modeling and Design. Prentice-Hall, Englewood Cliffs, New Jersey, 07632 (1991)
8. MacKenzie, C.M., Laskey, K., McCabe, F., Brown, P.F., Metz, R.: Reference model for service-oriented architecture 1.0, OASIS standard, 12 October 2006 (2006)
9. Open Group: Open Group standard SOA reference architecture (2011)
10. Král, J., Žemlička, M.: Towards design rationales of software confederations. In: Seruca, I., Filipe, J., Hammoudi, S., Cordeiro, J., eds.: ICEIS 2004: Proceedings of the Sixth International Conference on Enterprise Information Systems, vol. 1., Setúbal, Portugal, EST Setúbal pp. 105–112 (2004). Also available in Seruca, I., Cordeiro, J., Hammoudi, S., Filipe, J. (eds.) Enterprise Information Systems VI, pp. 89–96, Springer, Dordrecht, The Netherlands (2006)
11. Boehm, B.W.: Software Engineering Economics. Prentice-Hall, Upper Saddle River (1981)
12. Král, J., Žemlička, M.: Inaccessible area and effort consumption dynamics. In: Dosch, W., Lee, R., Tuma, P., Coupaye, T. (eds.) Proceedings of 6th International Conference on Software Engineering Research, Management and Applications (SERA 2008), pp. 229–234. IEEE CS Press, Los Alamitos (2008)
13. Parnas, D.L.: Designing software for ease of extension and contraction. IEEE Transactions on Software Engineering **5**(2), 128–138 (1979)
14. Armour, P.: The reorg cycle. Communications of the ACM **46**, 19–22 (2003)

15. Brown, W.J., Malveau, R.C., McCormick, H.W., Mowbray, T.J.: AntiPatterns: Refactoring Software, Architectures, and Projects in Crisis. John Wiley & Sons, New York (1998)
16. Object Management Group: Business process model and notation (BPMN) (2011) [Online:] http://www.omg.org/spec/BPMN/2.0/
17. Ang, J., Cherbakov, L., Ibrahim, M.: SOA antipatterns (2005) http://www-128. ibm.com/developerworks/webservices/library/ws-antipatterns/

# Data Consistency:
# Toward a Terminological Clarification

Hendrik Decker[1], Francesc D. Muñoz-Escoí[1], and Sanjay Misra[2(✉)]

[1] Instituto Tecnológico de Informática, Universidad Politécnica de Valencia,
Valencia, Spain
[2] Covenant University, Ota, Nigeria
sanjay.misra@covenantuniversity.edu.ng

**Abstract.** 'Consistency' is an 'inconsistency' are ubiquitous term in data engineering. Its relevance to quality is obvious, since 'consistency' is a commonplace dimension of data quality. However, connotations are vague or ambiguous. In this paper, we address semantic consistency, transaction consistency, replication consistency, eventual consistency and the new notion of partial consistency in databases. We characterize their distinguishing properties, and also address their differences, interactions and interdependencies. Partial consistency is an entry door to living with inconsistency, which is an ineludible necessity in the age of big data.

## 1 Introduction

In the field of databases, the meaning of the word 'consistency' is overloaded with multiple, often unclear meanings, as revealed by googling or looking it up in Wikipedia [59]. The terminological disarray becomes even more discomforting when trying to clarify the meaning of 'consistency' in terms of 'data quality'. Consistency is unanimously considered one of the most important 'aspects' or 'dimensions' of data quality [61] [21]. Occasionally, both terms even are identified [15]. In general, however, different people (authors as well as readers) may mean different things with 'consistency', without being explicit or clear about the differences.

Another frequently used synonym for 'consistency' is 'correctness'. But, without further explanation, that just replaces one unclear term by another. So, two questions arise that beg for an answer: What precisely is meant by 'consistency', and how to avoid, eliminate or at least contain inconsistency. In this paper, we propose some prolegomena for answering the first one. The second is for future work, but first steps for answering it are taken already in Sections 6 and 7.

Necessary conditions for data consistency can be expressed by an *integrity theory*, i.e., a finite set of *integrity constraints* (sometimes also called *assertions*, or, in database normalization theory, *dependencies*). These are sentences in the database language that are supposed to evaluate to *true* (i.e., to be *satisfied*) in each committed state of the database. We are going to clarify the meaning

H. Decker and F.D. Muñoz—supported by the Spanish MINECO grant TIN 2012-37719-C03-01.

O. Gervasi et al. (Eds.): ICCSA 2015, Part V, LNCS 9159, pp. 206–220, 2015.
DOI: 10.1007/978-3-319-21413-9_15

of 'consistency' in connection with updates that may affect the truth value of integrity constraints associated to the given database.

Four situations or reasons can be distinguished, why or when updates may lead to consistency violation. One is simply that a bad update directly contradicts what is required by some constraint, e.g., the insertion of *married(fred, fred)* which goes against the constraint that nobody can marry herself or himself. The second situation which may lead to constraint violations are the well-known 'update anomalies' that are due to an insufficient observance of data redundancies or dependencies that have not been eliminated by schema normalization [5]. The third kind of occurrence of integrity violation is due to equally well-known 'update phenomena' caused by deficient concurrency control [10]. The fourth is due to bad management of data distribution or replication [55].

The mentioned reasons of inconsistency are conductive to the different kinds of consistency studied in the remainder. In particular, we are going to address *semantic consistency, transaction consistency, distribution consistency, replication consistency* and *eventual consistency*. Also, we propose the new notion of *partial consistency*, which alleviates the abrasiveness of inconsistency. We conclude with an outlook on coping with big data inconsistency and more future work.

We are not going to deal with normalization except to emphasize that there is no normal form that could guarantee any of the consistency properties we are going to deal with. In fact, normalization can be very helpful, but is neither sufficient nor necessary for consistency. Moreover, we are not going to deal with physical consistency (a.k.a. disk consistency, or file integrity i.e., the inviolacy of binaries) [52], nor with network consistency [43], nor with the consistency or integrity of data transmission or communication (a topic of coding theory and cryptography) [51], nor with any other integrity issue of concern to database security, such as fraudulent tampering, authenticity or trustworthiness of data [50]. Nor is 'relational consistency' (an issue in constraint programming [16]) of interest in this paper. Also, we do not discuss issues related to the provenance, accuracy or truthfulness of data, concerning the consistency between the (real or imaginary) world that the database is supposed to model and the database content itself. For instance, if the *born* attribute $b$ in a database entry about a person $p$ is not $p$'s actual birth date, then we do not count that mistake as an inconsistency, as long as there is no other stored information which would contradict $b$. Such issues are studied in fields such as data lineage [38], truth finding [40], truth discovery [62] and data fusion [26].

## 2 Semantic Consistency

Semantic consistency is a property of database states. Thus, for a given database schema, semantic consistency can be identified with a subset of all possible states. Ideally, the predicate that corresponds to the characteristic function of that subset is described by an integrity theory.

In general, however, semantic consistency is not expressed completely by explicit declarative integrity constraints in the database schema. In fact, the integrity conditions of an application even may not all be cleanly documented. Worse, not all developers of applications may be aware of every semantic constraint. Hence, semantic consistency tends to be hard to guarantee, in general.

Instead of an automated enforcement of declarative constraints, semantic consistency is often realized by nested subtransactions, or firing triggers, or running stored procedures, or executing inline code of application programs, or compensating transactions, the latter in case an update is detected post-factum to have violated integrity. Such procedural ways of integrity enforcement are known to be error-prone and hard to maintain.

In this paper, we identify, for simplicity, semantic consistency with what is expressed (or is expressible) in the integrity theory that is (implicitly or explicitly) associated to the database. Thus, semantic consistency means that all integrity constraints in the given integrity theory are satisfied.

Semantic consistency is also known under the name of *data integrity* (which, however, also suffers from a fuzzy overload of different meanings, in- and outside of the literature on data quality [53] [60]; occasionally, semantic integrity is even identified with data quality [54]).

Ideally, semantic consistency is enforced automatically, by the DBMS or some module on top of its core, so that transaction designers, application programmers and users need not be asked to pay attention to the preservation of integrity. Automatic integrity enforcement usually sanctions an update only if it does not produce any constraint violation. The common built-in enforcement of some specific forms of declarative consistency conditions such as primary and foreign key constraints is an imperfect realization of that ideal. Imperfect because, among others, more general forms of constraints usually are not supported by the systems on the market. For query optimization or mere documentation, some do support a declarative assertion of more complex constraints (e.g., Oracle's RELY construct [46]), but not their enforcement. More comprehensive methods to support an automated checking of declarative integrity constraints, such as described, e.g., in [44] [17], have not yet found their way into marketed products.

## 3   Transaction Consistency

Simply put, transaction consistency means concurrency-transparent semantic consistency. That is, the database system ensures that each complete history $H$ of concurrent transactions preserves semantic consistency if each transaction $T$ in $H$ preserves semantic consistency whenever $T$ is executed alone [10].

In Section 2, semantic consistency has been characterized as a property of database states. Transaction consistency is a property of state transitions. More precisely, each successful execution of a transaction (which may consist of a single update or comprise a partially ordered set of updates) effects a state change that is supposed to preserve semantic consistency. Thus, for non-concurrent transactions, all of what has been said in Section 2 applies, without further worries.

However, concurrent transactions may violate the semantic consistency in a way that is not to blame on possible integrity violations by individual transactions, but on some harmful interleaving of actions of different transactions. Such interleavings may lead to well-known problems of concurrency such as dirty or non-repeatable reads, lost updates and other anomalies. These are not necessarily problematic by themselves, but only if they lead to the violation of semantic consistency. To avoid such anomalies, *serializability* was invented [34].

Transaction consistency sometimes is wrongly identified with serializability, i.e., the equivalence of the execution of a history $H$ of transactions with a one-by-one execution of all transactions in $H$. However, neither semantic consistency nor transaction consistency are guaranteed by serializability alone, nor does a history $H$ the transactions of which preserve semantic consistency necessarily entail its serializability [39]. Yet, given a consistent input state, it is plausible that $H$ and in particular each transaction $T$ in $H$ teminate consistently if $H$ is serializable and each *solitary* execution of $T$ (i.e., running $T$ alone) preserves semantic consistency [10]. That can be expressed schematically as follows.

(*)   serializability + solitary consistency $\Rightarrow$ transaction consistency

In Section 2, we have pointed out practical problems of warranting semantic consistency. Hence, the solitary consistency preservation of each transaction in a history is an Achilles heel of (*).

In 3.1 – 3.4, we relate transaction consistency to the popular *ACID* properties [35], by discussing them with regard to serializability and semantic consistency.

## 3.1   ACID

Brewer is quoted saying that the acronym *ACID* is "contrived [...] much more than people realize", that Jim Gray "admitted that *ACID* was a stretch", and "*A* and *D* have high overlap and the *C* is ill-defined" [14]. Nevertheless, many authors identify *ACID* with either serializability or (*) or transaction consistency.

The following brief explanation of *ACID* leans on [10]: *A* stands for *atomicity*, *C* for *consistency*, *I* for *isolation*, and *D* for *durability*. Atomicity means that each transaction in $H$ terminates either by committing or aborting. Consistency means semantic consistency as characterized in Section 2. Isolation means that the interleaved execution of transactions in $H$ does not harm semantic consistency. Durability means that, once committed, the effects of a transaction will remain persistent, or will be recovered after temporary unavailability, until they are modified, undone or overridden by any subsequent transaction.

Without concurrency, $I$ is trivially satisfied, both $A$ and $D$ can be ensured fairly straightforwardly, and $C$ can be handled as indicated in Section 2. For concurrent transactions, however, $A, C, I$ and $D$ are interrelated much more than without concurrency. For instance, imagine a transaction that infringes atomicity by terminating with only half of its writes done (e.g., a subtraction from credit but no corresponding addition to debit), or a DBMS that violates durability by an incomplete recovery from a crashed history that would replay not all of the writes of a committed transaction. In both cases, semantic inconsistencies (e.g., a faulty balance of accounts, or an incomplete and thus potentially inconsistent state) are likely to happen, too. In general, each of $A, C, D$ is complicated by concurrency, i.e., by the need to cater for $I$. Conversely, $I$ is complicated whenever $A$, $C$ or $D$ has to be catered for by the DBMS as well.

## 3.2  Isolation or the Transparency of Concurrency

The $I$ of $ACID$ is also called "concurrency transparency" [56], i.e., the designer or programmer or user of a transaction $T$ does not have to worry about possibly concurrent executions of $T$, and the output of $T$ in such executions is the same as if $T$ had been executed alone, so that users and applications do not take notice of concurrency.

As already indicated in 3.1, a history $H$ of transactions is *concurrency-transparent*, i.e., satisfies $I$, if it has the same effect as (or, is equivalent to) a solitary one-by-one execution of the transactions in $H$. To formalize this notion, it is necessary to precisely define what is "the effect of", i.e., the equivalence relation between, concurrent and sequential histoires.

Usually, each of the two most well-known variants of serializability (called conflict-serializability and, resp., view serializability [10]) is taken to ensure the sameness of effects or execution equivalence, thus guaranteeing the transparency of concurrency. Yet, serializability is only a sufficient, though not a necessary condition for isolation. And, similar to serializability, also isolation is neither sufficient nor necessary for transaction consistency. Since serializability is not always necessary for isolation, there are several similar, more or less exigent definitions of serializability [57], as well as a large amount of various weakenings, e.g., [30] [47], each of which determines its own degree or modality of isolation [6] [39] [8]. In fact, each weakening of serializability may debilitate isolation and hence transaction consistency, i.e., (*) above may no longer hold.

Instead of discussing specific definitions of serializability, we leave it here with the intuition that, for each of them, transaction designers, application programmers and users are supposed to be not bothered by concurrency, whenever transactions are guaranteed to always be executed serializably by the DBMS.

## 3.3  Consistency of Final and Commited States

The $C$ of $ACID$ means semantic consistency in the context of concurrency. Independent of the given variant of serializability, there are two significantly different notions of semantic consistency of concurrent transactions in a history $H$: either each transaction's commit in $H$ contributes to a state that satisfies all constraints – we call that *committed state consistency* –, or only the final state at the end of $H$ is required to satisfy integrity – we call that *final-state consistency*. This difference had been made in [48], and was further discussed in [25].

To deal with committed-state consistency is more difficult than with final-state consistency, since the committed state at the end of any transaction $T$, except the one that commits last, in any history $H$, is not necessarily *quiescent*, i.e., is not a committed state at any time at which no transaction is in course.

In [10], a state of a database is given by "the values of the data items at any one time", while a committed state is defined with respect to some execution, "to be the state in which each data item contains its last committed value". Since non-committed concurrent transactions may be in course at any time, the committed state of a transaction $T$ may never materialize physically at any one time. However, if all non-committed values of a transaction are protected from being accessed by other transactions, then each transaction only "sees" (parts of) committed input states. In general, the difficulties of pinpointing non-quiescent states that are committed or "seen" by users or applications tend to increase with the degree of relaxing serializability, and so do the intricacies of characterizing the consistency of such states.

Many papers about concurrent transaction processing only deal with final state consistency, and ignore what may happen with the states reached at commit time of transactions that terminate before a history comes to its end. However, transactions usually are issued without concern for the potential concurrency of their execution, and tend to have a vital interest in the consistency of their own individual outcome, rather than in the consistency of the state reached after the execution of all transactions that accidentally have been running concurrently. Hence, committed-state consistency is at least as relevant as final-state consistency, if not more.

### 3.4 Atomicity and Durability

The $A$, i.e., atomicity (also often spelled out adroitly as 'all or nothing') means that each transaction terminates either by comitting all of its updates, or by aborting without leaving behind any changes. Atomicity is a fairly straightforward standard in centralized database systems [27] [42], even if there are multiple, possibly remote users. We come back on atomicity in Sections 4 and 5.

The $D$, i.e., durability, has to do with the persistence of stored data beyond terminated transactions, user sessions and application program runs, and with the recovery from failures. While durability is essential for the reliability of database systems, we deliberately exclude it from further discussion in this paper, except to mention that properties of histories that define or ensure recoverability overlap but do not coincide with serializability, as pictured nicely in [10], pages 36 and 46; in particular, technologies that cater for recovery also may be beneficial for atomicity and isolation, and vice-versa.

## 4  Distribution Consistency and Replication Consistency

Distributed consistency means transaction consistency in distributed databases, where distribution is transparent to the designers, programmers, users and applications of transactions. (A more refined characterization of transparency is given in [56].) Similarly, replication consistency means distribution consistency in replicated databases, i.e., replication does not harm the transparency of distribution and concurrency.

For obtaining such transparency, some amount of system-level communication between the server nodes of a distributed database network is due, since data items must be accessible from each node but are either not stored at each one, or are replicated at several nodes, in which case changes of their values must be synchronized or at least coordinated to some extent.

The communication needed for data access and node coordination may suffer network latency, transmission delay and failures of nodes or network links. To cope with that, special attention has to be paid to the coordiation of concurrent accesses to replicated data items and of the atomic commit or abort of transactions. The ensurance that all commit and abort actions are carried out "consistently" is characterized in [10] as "the only non-trivial problem" caused by possible failures in distributed database systems without replication. Here, atomic commitment is a property for achieving distribution consistency. Together with transaction consistency, it is sufficient for distribution consistency. Thus, the latter can be schematically described as follows:

(**)   transaction consistency + atomic commit ⇒ distribution consistency

Depending on how it is realized, atomic commitment is not categorically orthogonal to transaction consistency. For instance, if it is realized as a two- or three-phase commit [10], then the respective protocol already covers part of the work necessary to ensure isolation.

In replicated databases, the situation is further complicated by the need to make replication transparent to designers and programmers of transactions, as well as to human and programmed agents that use transactions. Thus, not only the actions of commit and abort, but also some or all actions that read or write data need to be coordinated transparently. That is usually achieved by some system protocol [28,45]. In analogy to serializability properties as mentioned in Section 3, the one-copy serializability property (1SR) [10] is a sufficient, though not a necessary property for obtaining replication consistency. That is described schematically as follows.

(***)   distribution consistency + 1SR protocol ⇒ replication consistency

Usually, 1SR protocols are not orthogonal to distribution consistency. Typically, such protocols are meant to cater for part or all of distribution consistency, and also part or all of recoverability [1] [29] [2].

## 5   Eventual Consistency

A striking example of the babel (and sometimes babble) around 'consistency' is the ongoing discussion about 'eventual consistency' in distributed systems. In databases, it is a form of 'lazy' ('optimistic') replication consistency [49], which weakens the guarantees made by serializability, born out of the urge to scale. Eventual consistency is often mixed up with related but different issues such as availability, semantic consistency, transaction consistency, atomicity, recoverability or other consistency aspects associated to concurrency, shared memory coherence, distribution and cloud computing; see, e.g., [13] [58] [37] [63] [7] [41] [9] [3] [32].

According to [47], eventual consistency requires that replicated copies are consistent at certain times – usually in states that are quiescent for a sufficiently long period of time (some milliseconds may suffice, but maybe more) – but may be inconsistent in the interim intervals. In other words, eventual consistency means that violations of replication consistency will disappear after some indefinite delay, usually when states that are quiescent for a sufficiently long period of time are reached.

Stronger variants of eventual consistency require each violation of replication consistency to be repaired after some indefinite time, but before any write access to violated data items. Some even require that inconsistent values should be repaired by the time they are accessed, no matter if read or written, and if that is not feasible, replication inconsistency must be repaired asynchronously (typically by some compensating transactions), i.e., any consistency guarantee may be suspended indefinitely.

In summary, we can say that, in databases, eventual consistency is a compromised form of replication consistency, at the cost of strict consistency requirements. So, the question is, which kind of consistency is compromised. The answer is that each of the four kinds of consistency as addressed in Sections 2–4 can be violated in eventually consistent databases. However, what is violated in the first place by eventual consistency is the "consistency" of atomic commitment, as broached in Section 4.

Semantic consistency, transaction consistency, distribution consistency and replication consistency as characterized in Sections 2–4 are defined for whole database states, in terms of the satisfaction of integrity constraints in those states. As opposed to that, atomic commit consistency is defined for individual data items. Since eventual consistency does not comply with the strict atomic commit requirements of the $1SR$ property of replication consistency, also distribution consistency, transaction consistency and semantic consistency guarantees are at risk.

Instead of requiring a coordinated order of committed updates, eventual consistency tolerates a local and immediate commit at the node where a transaction is executed. Later, the locally committed updates are propagated to the remaining database replicas, in a lazy FIFO way. Lazy propagation may be the source of consistency conflicts among concurrent transactions. Such conflicts can be dealt with by one of the two following general approaches [49].

- Each data item has a manager node. That manager uses a deterministic criterion for deciding which will be the surviving update value in case of conflicts, using compensating transactions in nodes where those values were not the latest ones being applied. That usually entails the *lost update* phenomenon, but ensures eventual value convergence in all copies.

  This solution may demand that all data items accessed by each of the concurrent transactions share the same manager node. That demand may be difficult to maintain in general.

– *Semantic scheduling.* Suppose that the operations in each transaction of a given history is commutative, and that all transactions are applied in each replica node. Then, the value of the copies of each data item will converge.

In this case, the database system should ensure that all updates are eventually propagated to every node of the distributed network.

The data-item-based concept of atomic commit consistency is the $C$ in the widely discussed $CAP$ theorem [31] [14] [1]. According to [36], the $C$ of $CAP$ is a special case of strict serializability, where transactions are restricted to consist of a single operation applied to a single object.

Originally, $CAP$ had not been formulated particularly for databases, but for networked shared data systems in general. According to [14], $CAP$ says that a distributed system cannot have at a time the three properties of "consistency" (all nodes see the same data at the same time), availability (each request receives a response of success or failure) and partition tolerance (system continues operation despite loss of network connectivity).

In large-scale distributed systems (e.g., data-related cloud services deployed in multiple datacenters), network partitions may appear. In order to maintain system responsiveness to users, such systems prioritize availability and partition tolerance (i.e., $A$ and $P$) over the $C$ of $CAP$. In other words, the $C$ property is being sacrificed, and that is one of the reasons for using eventual consistency instead of "strong" replication consistency in such kind of systems.

## 6   Partial Consistency

Partial consistency means that a given committed state may violate some constraint. Traditionally, a state that violates integrity is called inconsistent, but, arguably, calling it partially consistent is more adequate, being more suggestive of the positive potential of such states inspite of their compromised integrity, than the negative connotations associated to inconsistency.

By the usual understanding of semantic and transactional consistency, inconsistency should definitely be avoided, because inconsistencies may be severely harmful. In distributed and replicated databases, transitory inconsistency is accepted as inevitable, but even the eventual consistency paradigm insists on a convergence of the states seen by distributed users and applications toward consistency. In general, inconsistency is taken to be uncontrollable by means of classical logic, due to its *ex falso quodlibet* rule which invalidates each and every answer given to any query by an inconsistent database.

Classical logic is the acknowledged fundament of database theory and practice. Hence, inconsistency is ill-reputed, if not considered monstrous. On the other hand, inconsistency is ubiquitous in practice, to the extent that an insistence on total consistency is illusory. Moreover, practical experience shows that the majority of answers given by database systems that contain some inconsistencies are not nonsensical but valuable.

In fact, all kinds of consistency violations may easily manifest themselves in databases. For instance, legacy data that are not checked for compliance with newly introduced constraints may violate them. Or, updates that directly violate what is required by some integrity constraint will not be rejected whenever integrity control is switched off for boosting performance. Or, transaction consistency is violated by running some transaction that is not programmed according to the rules in concurrence with other transactions. Or, distributed consistency is violated by a prolongued laziness of protocols that lead to the violation of some constraint. Or, replication consistency is violated by giving priority to availability requirements and thereby weakening semantic consistency guarantees.

The predictions made by conventional transaction processing theory for transaction consistency, and hence also for distribution and replication consistency of committed database states, all depend on the fulfillment of the promise of solitary consistency preservation of each transaction. In other words, the control of transaction consistency is not in the hands of any single transaction programmer or user, but is a collective achievement of all authors of the transactions that accidentally run concurrently in the same history. However, it is naïve to trust that each transaction is written such that each of its solitary executions will preserve consistency. After all, by (*) as described in Section 3, none of the transactions in any history $H$ is guaranteed to produce consistent outcome if there is just a single transaction in $H$ such that a solitary execution of it fails to preserve integrity. Then, (*) does not make any consistency guarantee at all, not even for those transactions in $H$ the solitary executions of which are perfectly consistency-preserving. Or, to put it differently: Would you bet that each transaction, programmed by a possibly unkown colleague, maybe before the latest changes in the application, that happens to run concurrently with your own (no matter how carefully written) transaction, is correctly taking into account all integrity constraints (including those that you possibly might not even know of)? If not, then the usual transaction consistency guarantees are not for you.

However, integrity violations are losing large portions of their theoretical sting, by recent advances in database research. Already since early last century, the absolute intolerance of inconsistency in logic had been questioned [12], and the ramifications of that movement have arrived in database research [18]. Although inconsistency thereby has not lost all of its potential calamities, logically sound ways to work consistently in inconsistent databases have been devised [11] [25] [24] [19] [23], so that database inconsistency has become tolerable, not only pragmatically speaking, but also from an austere theoretical point of view.

The theories presented in the preceding references provide a logical justification of database reasoning in the presence of semantic inconsistency, no matter if inconsistency has resulted from direct violations of constraints by updates, or because of any failures of controlling concurrency, distribution or replication. Concretely, [11] contains articles that describe how consistent answers to queries in inconsistent databases can be computed. In [24], it is shown that conventional integrity checking methods can be soundly applied also in inconsistent databases, even if their integrity theories are not satisfiable by any state whatsover. As shown in [23] [20], integrity violation becomes controllable and repairable by quantifying it with inconsistency measures [33]. In [19] [22], it is shown that query answers the causes of which are independent of any causes of integrity violation have the same integrity as answers in perfectly consistent databases. The authors of [25] outline how concurrent transaction consistency guarantees can be extended to inconsistent database states and histories containing transactions that do not comply with the solitary integrity preservation requirement of (*). We expect that such theories will be further developed toward a logically sound processing of transactions in situations where consistency is harmed by abandoned precautions in terms of eventual consistency, including applications of big data processing.

# 7    Conclusion

For clarifying the meaning of the frequently used technical term 'consistency' in the sense of stored data quality, we have characterized semantic consistency, transactional consistency, distribution consistency, replication consistency, eventual consistency and partial consistency. Semantic consistency means the satisfaction of all integrity constraints associated to a given database. Transaction consistency means concurrency-transparent semantic consistency. Distribution consistency means distribution-transparent transaction consistency. Replication consistency means replication-transparent distribution concurrency. Eventual consistency means lazy replication consistency. Partial consistency means that not necessarily all integrity constraints are satisfied.

Partial consistency is more general than all forms of consistency addressed in Sections 2–5: the latter all are oriented toward the ideal of total consistency (even though eventual consistency only promises to reach total consistency after an indefinite time), while partial consistency is prepared to always work consistenly in the presence of inconsistency, no matter if integrity violations will ever disappear or not.

The preceding lineage of database consistency that ascends from the total satisfaction of constraints for semantic consistency to a relaxation of integrity in the sense of partial consistency can be reversed to a description that descends from user-friedly partial consistency to the satisfaction of schema-level constraints, as follows. Integrity violations that are possible with partial consistency and eventual consistency are tolerable as long as they do not interfere with computing answers to queries. From a human or programmed agent's point of view, replication consistency is the same as distributed consistency, which, from the same point of view, is the same as transaction consistency, i.e., concurrency-transparent semantic consistency preservation.

We have recalled that serializability entails isolation, and also that neither serializability nor isolation alone is neither necessary nor sufficient for semantic consistency, nor for transaction consistency. Moreover, we have scrutinized the theorem that serializability in conjunction with solitary integrity preservation by all transactions in a history is sufficient (though not necessary) for transacion consistency. We have argued that the applicability of this theorem is severely limited in two ways. Firstly, total semantic consistency, and hence totally consistent states before or after the execution of transactions, are rarely given in practice, at least not in large (let alone in big) databases. Secondly, to trust in that theorem is very risky, since it actually makes no consistency guarantees unless each contingently concurrent transaction is bug-free in the sense that it will never violate integrity when run solitarily.

We have introduced the notion of partial consistency, which serves to alleviate the deficiencies of the theorem that serializability plus solitary consistency preservation entail transaction consistency. More generally, partial consistency is meant to replace the notion of inconsisteny as the natural complement of consistency. The main theoretical advantage of partial consistency over inconsistency is that classical logic throws up whenever inconsistency is encountered, while partial consistency provides a *modus vivendi* with integrity violations. The main practical advantage of partial consistency is that the consistency guarantees described schematically by (*), (**), (***) in Sections 3 and 4 are useless in databases that are not totally consistent, but become useful when relaxing the requirement of total consistency by admitting partial consistency. The main challenge of partial consistency is to apply it systematically for transaction processing in modern database systems such as column store, main-memory, NoSQL ('not only SQL') and NewSQL architectures, and to use it for reasoning with big data.

In this context, it is useful to recall that partial consistency generalizes eventual consistency (by not insisting on a convergence to total consistency), while its applications as mentioned in Section 6 (consistent query answering, inconsistency-tolerant integrity checking and repairing, inconsistency-tolerant concurrency control) do not at all forfeit strong consistency guarantees. In a similar spirit, eventual consistency is fortified in [4] by explicitly stating integrity constraints ("invariants" that define consistency conditions for a given application) as logical properties that have to hold in each state of a history ("a given set of transactions"). Thus, eventual consistency is directly linked back to semantic consistency, so that a reliable form of eventual consistency can be enforced by using inconsistency-tolerant integrity checking [24], which is an application of partial consistency. Future work may further explore this idea.

More work lies ahead also for a clarification of the meaning of 'consistency' in the fields of data science and software engineering, where 'consistency' is subsumed under 'quality', which in fact is a wider and even more fuzzy term than 'consistency'. Thus, it is a challenge to embark on a terminological study of 'data quality', similar to what we have done in this paper for 'consistency'.

# References

1. Abadi, D.: Consistency tradeoffs in modern distributed database system design: Cap is only part of the story. Computer **45**(2), 37–42 (2012)
2. Bailis, P. (2015). http://www.bailis.org/blog/
3. Bailis, P., Ghodsi, A.: Eventual consistency today: limitations, extensions, and beyond. ACM Queue, **11**(3) (2013)
4. Balegas, V., Duarte, S., Ferreira, C., Rodrigues, R., Preguica, N., Najafzadeh, M., Shapiro, M.: Putting consistency back into eventual consistency. In: 10th EuroSys. ACM (2015). http://dl.acm.org/citation.cfm?doid=2741948.2741972

5. Beeri, C., Bernstein, P., Goodman, N.: A sophisticate's introduction to database normalization theory. In: VLDB, pp. 113–124 (1978)
6. Berenson, H., Bernstein, P., Gray, J., Melton, J., O'Neil, E., O'Neil, P.: A critique of ansi sql isolation levels. SIGMoD Record **24**(2), 1–10 (1995)
7. Bermbach, D., Tai, S.: Eventual consistency: how soon is eventual? In: 6th MW4SOC. ACM (2011)
8. Bernabé-Gisbert, J., Muñoz-Escoí, F.: Supporting multiple isolation levels in replicated environments. Data & Knowledge Engineering **7980**, 1–16 (2012)
9. Bernstein, P., Das, S.. Rethinking eventual consistency. In: SIGMOD 2013, pp. 923–928. ACM (2013)
10. Bernstein, P., Hadzilacos, V., Goodman, N.: Concurrency Control and Recovery in Database Systems. Addison-Wesley (1987)
11. Bertossi, L., Hunter, A., Schaub, T.: Inconsistency Tolerance. In: Bertossi, L., Hunter, A., Schaub, T. (eds.) Inconsistency Tolerance. LNCS, vol. 3300, pp. 1–14. Springer, Heidelberg (2005)
12. Bobenrieth, A.: Inconsistencias por qué no? Un estudio filosófico sobre la lógica paraconsistente. Premios Nacionales Colcultura. Tercer Mundo Editores. Magister Thesis, Universidad de los Andes, Santafé de Bogotá, Columbia (1995)
13. Bosneag, A.-M., Brockmeyer, M.: A formal model for eventual consistency semantics. In: PDCS 2002, pp. 204–209. IASTED (2001)
14. Browne, J.: Brewer's cap theorem (2009). http://www.julianbrowne.com/article/viewer/brewers-cap-theorem
15. Cong, G., Fan, W., Geerts, F., Jia, X., Ma, S.: Improving data quality: consistency and accuracy. In: Proc. 33rd VLDB, pp. 315–326. ACM (2007)
16. Dechter, R., van Beek, P.: Local and global relational consistency. Theor. Comput. Sci. **173**(1), 283–308 (1997)
17. Decker, H.: Translating advanced integrity checking technology to SQL. In: Doorn, J., Rivero, L. (eds.) Database integrity: challenges and solutions, pp. 203–249. Idea Group (2002)
18. Decker, H.: Historical and computational aspects of paraconsistency in view of the logic foundation of databases. In: Bertossi, L., Katona, G.O.H., Schewe, K.-D., Thalheim, B. (eds.) Semantics in Databases 2001. LNCS, vol. 2582, pp. 63–81. Springer, Heidelberg (2003)
19. Decker, H.: Answers that have integrity. In: Schewe, K.-D., Thalheim, B. (eds.) SDKB 2010. LNCS, vol. 6834, pp. 54–72. Springer, Heidelberg (2011)
20. Decker, H.: New measures for maintaining the quality of databases. In: Murgante, B., Gervasi, O., Misra, S., Nedjah, N., Rocha, A.M.A.C., Taniar, D., Apduhan, B.O. (eds.) ICCSA 2012, Part IV. LNCS, vol. 7336, pp. 170–185. Springer, Heidelberg (2012)
21. Decker, H.: A pragmatic approach to model, measure and maintain the quality of information in databases (2012).
www.iti.upv.es/~hendrik/papers/ahrc-workshop_quality-of-data.pdf,
www.iti.upv.es/~hendrik/papers/ahrc-workshop_quality-of-data_comments.pdf.
Slides and comments presented at the Workshop on Information Quality. Univ, Hertfordshire, UK
22. Decker, H.: Answers that have quality. In: Murgante, B., Misra, S., Carlini, M., Torre, C.M., Nguyen, H.-Q., Taniar, D., Apduhan, B.O., Gervasi, O. (eds.) ICCSA 2013, Part II. LNCS, vol. 7972, pp. 543–558. Springer, Heidelberg (2013)
23. Decker, H.: Measure-based inconsistency-tolerant maintenance of database integrity. In: Schewe, K.-D., Thalheim, B. (eds.) SDKB 2013. LNCS, vol. 7693, pp. 149–173. Springer, Heidelberg (2013)

24. Decker, H., Martinenghi, D.: Inconsistency-tolerant integrity checking. IEEE Transactions of Knowledge and Data Engineering **23**(2), 218–234 (2011)
25. Decker, H., Muñoz-Escoí, F.D.: Revisiting and improving a result on integrity preservation by concurrent transactions. In: Meersman, R., Dillon, T., Herrero, P. (eds.) OTM 2010. LNCS, vol. 6428, pp. 297–306. Springer, Heidelberg (2010)
26. Dong, X.L., Berti-Equille, L., Srivastava, D.: Data fusion: resolving conflicts from multiple sources (2015). http://arxiv.org/abs/1503.00310
27. Eswaran, K., Gray, J., Lorie, R., Traiger, I.: The notions of consistency and predicate locks in a database system. CACM **19**(11), 624–633 (1976)
28. Muñoz-Escoí, F.D., Ruiz-Fuertes, M.I., Decker, H., Armendáriz-Íñigo, J.E., de Mendívil, J.R.G.: Extending middleware protocols for database replication with integrity support. In: Meersman, R., Tari, Z. (eds.) OTM 2008, Part I. LNCS, vol. 5331, pp. 607–624. Springer, Heidelberg (2008)
29. Fekete, A.: Consistency models for replicated data. In: Encyclopedia of Database Systems, pp. 450–451. Springer (2009)
30. Fekete, A., Gupta, D., Lynch, V., Luchangco, N., Shvartsman, A.: Eventually-serializable data services. In: 15th PoDC, pp. 300–309. ACM (1996)
31. Gilbert, S., Lynch, N.: Brewer's conjecture and the feasibility of consistent, available, partition-tolerant web services. SIGACT News **33**(2), 51–59 (2002)
32. Golab, W., Rahman, M., Auyoung, A., Keeton, K., Li, X.: Eventually consistent: Not what you were expecting? ACM Queue, **12**(1) (2014)
33. Grant, J., Hunter, A.: Measuring inconsistency in knowledgebases. Journal of Intelligent Information Systems **27**(2), 159–184 (2006)
34. Gray, J., Lorie, R., Putzolu, G., Traiger, I.: Granularity of locks and degrees of consistency in a shared data base. In: Nijssen, G. (ed.) Modelling in Data Base Management Systems. North Holland (1976)
35. Haerder, T., Reuter, A.: Principles of transaction-oriented database recovery. Computing Surveys **15**(4), 287–317 (1983)
36. Herlihy, M., Wing, J.: Linearizability: a correctness condition for concurrent objects. TOPLAS **12**(3), 463–492 (1990)
37. R. Ho. Design pattern for eventual consistency (2009). http://horicky.blogspot.com.es/2009/01/design-pattern-for-eventual-consistency.html
38. Ikeda, R., Park, H., Widom, J.: Provenance for generalized map and reduce workflows. In: CIDR (2011)
39. Kempster, T., Stirling, C., Thanisch, P.: Diluting acid. SIGMoD Record **28**(4), 17–23 (1999)
40. Li, X., Dong, X.L., Meng, W., Srivastava, D.: Truth finding on the deep web: Is the problem solved? VLDB Endowment **6**(2), 97–108 (2012)
41. Lloyd, W., Freedman, M., Kaminsky, M., Andersen, D.: Don't settle for eventual: scalable causal consistency for wide-area storage with cops. In: 23rd SOPS, pp. 401–416 (2011)
42. Lomet, D.: Transactions: from local atomicity to atomicity in the cloud. In: Jones, C.B., Lloyd, J.L. (eds.) Dependable and Historic Computing. LNCS, vol. 6875, pp. 38–52. Springer, Heidelberg (2011)
43. Monge, P., Contractor, N.: Theory of Communication Networks. Oxford University Press (2003)
44. Nicolas, J.-M.: Logic for improving integrity checking in relational data bases. Acta Informatica **18**, 227–253 (1982)
45. Muñoz-Escoí, F.D., Irún, L., H. Decker: Database replication protocols. In: Encyclopedia of Database Technologies and Applications, pp. 153–157. IGI Global (2005)

46. Oracle: Constraints. http://docs.oracle.com/cd/B19306_01/server.102/b14223/constra.htm (May 1, 2015)
47. Ouzzani, M., Medjahed, B., Elmagarmid, A.: Correctness criteria beyond serializability. In: Encyclopedia of Database Systems, pp. 501–506. Springer (2009)
48. Rosenkrantz, D., Stearns, R., Lewis, P.: Consistency and serializability in concurrent datanbase systems. SIAM J. Comput. **13**(3), 508–530 (1984)
49. Saito, Y., Shapiro, M.: Optimistic replication. JACM **37**(1), 42–81 (2005)
50. Sandhu, R.: On five definitions of data integrity. In: Proc. IFIP WG11.3 Workshop on Database Security, pp. 257–267. North-Holland (1994)
51. Simmons, G.: Contemporary Cryptology: The Science of Information Integrity. IEEE Press (1992)
52. Sivathanu, G., Wright, C., Zadok, E.: Ensuring data integrity in storage: techniques and applications. In: Proc. 12th Conf. on Computer and Communications Security, p. 26. ACM (2005)
53. Svanks, M.: Integrity analysis: Methods for automating data quality assurance. Information and Software Technology **30**(10), 595–605 (1988)
54. Technet, M.: Data integrity. https://technet.microsoft.com/en-us/library/aa933058 (May 1, 2015)
55. Terry, D.: Replicated data consistency explained through baseball. Technical report, Microsoft. MSR Technical Report (2011)
56. Traiger, I., Gray, J., Galtieri, C., Lindsay, B.: Transactions and consistency in distributed database systems. ACM Trans. Database Syst. **7**(3), 323–342 (1982)
57. Vidyasankar, K.: Serializability. In: Encyclopedia of Database Systems, pp. 2626–2632. Springer (2009)
58. Vogels, W.: Eventually consistent (2007). http://www.allthingsdistributed.com/2007/12/eventually_consistent.html. Other versions in ACM Queue **6**(6), 14–19. http://queue.acm.org/detail.cfm?id=1466448 (2008) and CACM **52**(1), 40–44 (2009)
59. Wikipedia: Consistency model. http://en.wikipedia.org/wiki/Consistency_model (May 1, 2015)
60. Wikipedia: Data integrity. http://en.wikipedia.org/wiki/Data_integrity (May 1, 2015)
61. Wikipedia: Data quality. http://en.wikipedia.org/wiki/Data_quality (May 1, 2015)
62. Yin, X., Han, J., Yu, P.: Truth discovery with multiple conflicting information providers on the web. IEEE Transactions of Knowledge and Data Engineering **20**(6), 796–808 (2008)
63. Young, G.: Quick thoughts on eventual consistency (2010). http://codebetter.com/gregyoung/2010/04/14/quick-thoughts-on-eventual-consistency/ (May 1, 2015)

# Workshop on Virtual Reality and its Applications (VRA 2015)

# Challenges and Possibilities
# of Use of Augmented Reality in Education
## Case Study in Music Education

Valéria Farinazzo Martins[1], Letícia Gomes[1], and Marcelo de Paiva Guimarães[2,3(✉)]

[1] Faculdade de Computação e Informática,
Universidade Presbiteriana Mackenzie, São Paulo, SP, Brazil
valfarinazzo@gmail.com, leticiagomez01@hotmail.com
[2] Universidade Aberta do Brasil, UNIFESP, São Paulo, SP, Brazil
marcelodepaiva@gmail.com
[3] Programa de Mestrado em Ciência da Computação,
FACCAMP, Campo Limpo Paulista, SP, Brazil

**Abstract.** This paper aims to discuss the difficulties and possibilities of using augmented reality in education, especially for musical education. Among the difficulties addressed are the following types of issues: physical, technological, sociocultural, pedagogical and managerial. The possible solutions presented involve the use of authoring tools that are easily usable by teachers. An augmented reality application to teach musical perception was developed using an authoring tool, and tests with children are presented and discussed.

## 1 Introduction

In recent years, computational resources have been increasingly present in the teaching–learning process. New technologies have provided advances in traditional teaching methods, which may make it easier for students to learn and also change the way that teachers share knowledge. In this new context, where educational issues are compelling and supported by technology in a basic and almost invisible way, educational applications should be simple to set up or adapt. They must also provide flexibility in their configuration, depending on the content generated or manipulated by the user. They must function with the minimal training that the potential user already has or that is in the domain of her/his social group, for instance, creating videos with mobile devices, manipulating videos in repositories such as YouTube, manipulating images and videos on social networks, using text editors, etc. These technological tools are dominated by users and can be mastered with ease, when those users have been exposed to environments where these technologies are naturally relevant [1].

The increasing use of these technologies in education has occurred primarily due to the lower costs of computers and the emergence of new software tools. However, some computer technologies, due to their peculiarities such as those involving Augmented Reality (AR), are not yet widely used. In a general analysis, we note that this is due to the disparity between the state of the art of these technologies and the time of

O. Gervasi et al. (Eds.): ICCSA 2015, Part V, LNCS 9159, pp. 223–233, 2015.
DOI: 10.1007/978-3-319-21413-9_16

maturity at which they can be implemented effectively, that is, made available easily and affordably [2].

According to Azuma et al. [3] and Billinghurst [4], AR is a technology that enables the user to see the real world, with virtual objects superimposed upon or composited with the real world. An AR system has the following three characteristics: it combines the real and the virtual, it is interactive in real time and it is registered in 3D.

AR originated from another technology called virtual reality (VR). While VR can be defined as an advanced interface with computer applications, where the user can navigate and interact in real time in a 3D synthesized environment, using multisensory devices, AR simplifies its use. A nonconventional device is not required; just a webcam and markers are needed [5].

AR was indicated by the Horizon Report [6] as one of the technologies that will revolutionize education in the coming years. This is due to its fun and interactive features, among other reasons. The use of AR education applications is made possible by the cheapening and improvement of the hardware and the need for more user-friendly interfaces for interaction with nonexpert users along with the indispensability of working with other ways of teaching, using active practices. In this sense, using computers in the classroom may allow the simulation of situations not previously possible or imagined.

This article aims to discuss the difficulties and possibilities of using AR in education. A case study was developed in the musical education area, using an authoring tool. This choice was made because since 2008, music education has again become mandatory content in Brazilian schools (Law No. 11,769 of August 18, 2008); therefore, greater attention must now be paid to the use of technology in this area of knowledge, motivating students and allowing learning beyond the classroom. Another relevant factor is the small number of studies using AR and musical education [7] [8].

This paper is organized as follows. In sections 2 and 3 are discussed some of the challenges and possibilities of the effective use of AR in education. Section 4 presents the Music-AR application for music education using AR. Section 5 shows tests performed with the application, as well as the results obtained. Finally, the implications of this work are discussed in section 6.

## 2     Challenges for the Effective Use of Augmented Reality

The effective deployment of emerging technologies such as AR is still a challenge, because it requires the overcoming of various barriers. The first relates to physical and technological issues, dealing with the gap between the process of development proposed in software engineering for interactive applications [9, 10] and how these projects are being developed. The second relates to sociocultural issues. The last barrier is related to pedagogical and management issues. Although one of the areas most cited for the potential use of emerging technologies is education, very few projects are in fact implemented in schools to support learning in an effective way.

## 2.1  Physical and Technological Issues

Educational applications supported by computers are usually developed by computer experts consulting with education professionals. In many cases, the applications are much more influenced by the computer specialists than by the education professionals; this is the traditional use of technology in education.

The evolution of AR environment development tools in the last decade has been considerable; today there is a range of solutions available. Nevertheless, these solutions still require a high technical knowledge and/or considerable time to generate content, which makes it a challenge to create AR educational environment and to generate content in an easy and effective way.

## 2.2  Sociocultural Issues

Sociocultural issues have also been an obstacle to the use of new technologies such as AR. The first obstacle is training teachers in the use of this technology. In this context, the teacher must be a student using AR. Maintaining the awareness of the use of these tools becomes essential in order to stay focused on one's purpose; otherwise there is a movement of some teachers to the use of technologies, as stated by [11].

It is also necessary to promote technology usage that is safe, healthy and responsible, that is, it is necessary to pay attention so that students do not become dependent on technology in the classroom, hampering their learning.

In the Brazilian socioeconomic environment, equal access to technology should be considered. It is therefore necessary to ensure that everyone in the classroom has the same rights, and if there are extra activities, it must be ensured that everyone has equal access to educational tools.

## 2.3  Pedagogical and Management Issues

Content creation for AR applications demands considerable time and effort. To develop such content, not only computational technical knowledge is needed, but also a knowledge of the subject as well as teaching skills. Because most development tools are not high-level or easy to use, teachers feel unable to generate applications. The solution found by some teachers is to search for support staff, but this restricts the technology to only a few institutions and people.

For educational content to be broadly developed, the tools have to be easily usable even for complex tasks, allowing teachers to focus on the content of the lessons. That way, teachers do not need to have a thorough knowledge of the underlying system (e.g., knowledge of computer graphics, programming languages, interface with the operating system, etc.). It is also necessary to create educational methodologies for the use of such technologies. In addition, all created material should be conceived to function within a dynamic new type of classroom.

On the other hand, teachers also need to be trained so that they feel safe and capable in using the technology in the classroom. Open basic courses and training in the creation and manipulation of images, videos, stories, etc. are needed in order to resolve or alleviate this problem [1].

Technology should improve communication and education management. The technology in education should not be a barrier but should be a means of stimulating communication between parents, teachers and students.

## 3   Possibilities

There are many people without programming skills, such as teachers, who are interested in developing AR applications. To work around this problem, there are some authoring tools to provide resources for nonprogrammers, so that they can create their own applications [12]. It is important to mention that these applications must be independent of compilers, operational systems and programing languages, but should, instead, use text editing, configuration procedures, visual interface and tangible actions.

An example of a high-level tool is Flaras [13]. This tool is free open-source software available on the Internet at http://ckirner.com/flaras2. It includes support material such as downloadable versions of the tool; tutorials based on texts and videos; FAQs; an e-book; a repository with various online applications, accompanied by their respective projects, open and licensed for the adaptation and creation of derivative works; and information about the software developers. Flaras can be used with a single marker (a drawn card with a frame and a symbol therein) and is based on the concept of spatial points with stacked virtual scenes containing images, sounds, videos and 3D objects.

However, it is important to mention other tools that can be differentiated by their degree of ease in creating material for nonspecialist users and also by their limitations. Among these, the following can be highlighted.

- ARToolKit [14] is one of the first AR tools that used markers and computer vision. To use it to create AR applications, developers need to have skills in programming in C/C++. Its content design tool eliminates the dependence on the programming language, replacing it with the description of the virtual objects and their relationships with the real environment.
- DART [15] is a tool implemented on Macromedia Director, using a model of a visual type of authorship "drag-and-drop" and an interpreted scripting language (textual descriptions).
- AMIRE [16] is based on a structure that uses oriented components technology, providing a minimum set of components for demonstration, a collection of reusable resources and a visual authoring tool.
- ComposAR [12] is an extensible tool for authoring AR applications for nonprogramming users. It supports a scripting and interface type of "drag-and-drop" and interpreted input in real time as well as features added by users.
- ARSFG [17] is a software tool for the rapid prototyping of AR applications, based on scene graphs, which enables remote collaboration through an XML-based protocol.
- BasAR [18] is an authoring tool where the authoring layers are separated into infrastructure, structure, content, performance and behavior.

- FlarToolKit [19] is a library developed in the ActionScript language. It runs on a majority of web browsers that support the technology of the Adobe Flash Player. Therefore, the development of FlarToolKit applications enables applications to run AR on the Web, providing greater flexibility on the issue of access to applications.

Another important issue for the development of these tools involves the fact that an educational application, involving various media, can be designed or planned using the approach of the problem of division, also known as a "top-down" approach [9]. In this approach, a complex problem is divided into smaller and smaller parts until all of them can be resolved.

On the other hand, most applications developed by professionals in the technological area are presented to users as a whole, integrating the structure and content. This makes it difficult to make any adjustments, due to system complexity, since programming changes are required in most cases. The application of a division in structure and content enables making changes of substance in a more simple way, allowing the participation of teachers and students in changing the applications.

This approach to problem-solving of dividing into smaller parts and the implementation of a division in structure and content, along with a simple tool, creates favorable conditions for the development of educational applications for teachers and students. These simple tools should be based on visual actions, should be easier to use and should allow adaptations to the project, enabling the creation of applications from which others can be derived through changing the content and possibly the structure [1].

# 4    Music-AR

According to Schafer [20], music is an organization of sounds (rhythm, intensity, melody and timbre), with the intention of being heard. In this regard, the vision of musical art transcends the orthodox definitions of classical, popular and folk music and personal preference. Music can also be described as an imitation of nature or of daily sounds. The awareness of environmental sounds occurs through much training, and musical education provides training so that people become able to preserve familiar sounds and create new ones. Thus, the leaners should listen to, analyze and make sounds. For the authors of this paper, musical education should be constantly linked to experiences and discovery, just so efficient results can be achieved. These are the experiences and discoveries that happen very early in childhood.

Music should elicit different emotional responses in listeners. It is assumed that these responses are generated because different extremes related to sound characteristics (e.g., high and low, strong and smooth, short and long, fast and slow) must have some power over emotions. This fact can be used to create a composition with a specific character that may affect the listeners' emotional states [21, 22].

Music-AR is a set of short games that was designed based on interviews with a music teaching expert, the Arts Coordinator of a private teaching school located in the city of São Paulo, Brazil. Information gathered through this interview revealed shortages of software for teaching musical perception to children. According to the person interviewed, few software aim to teach trivial details such as musical notes. However, it is necessary to teach sound properties that precede musical teaching, such as pitch, rhythm and timbre. Thus, the goal of the researchers was not to teach music or a musical instrument; rather, the core focus was to develop an application that could teach musical perception in children.

The Flaras tool was used to develop the application. Although it is possible to import ready-made 3D objects, for reasons of ease and the adequacy of the objects, they were created in this case. As the target audience was made up of children, Music-AR needed to meet the requirements of being attractive, playful, easy to use, easy to learn, fun, intuitive and visually pleasing. The development took about two weeks and involved a first-semester student of computer science without a deep knowledge of application development.

To achieve these goals, short and intuitive games (exercises) were generated. All games allowed the child to hear/see explanations about them and then use the concepts learned to solve the exercises. These exercises encompassed the following sound features:

- Pitch: This refers to the bass, midrange and treble, that is, how high or low a sound is. Figure 1(a-e) shows a sequence in which the user performs one of the games related to the topic, "Pitch." To explain the concept of pitch, the application uses the idea of the stretching and loosening of a rope to represent a high or a low sound, respectively. This is what really happens with stringed musical instruments.

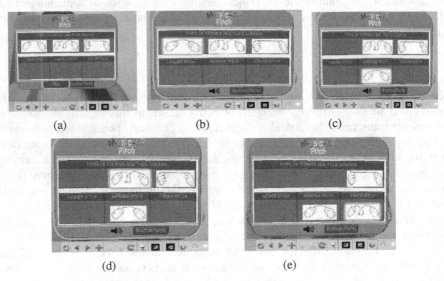

Fig. 1. Main screens running the game on the Pitch

- Intensity: This is the strength at which the sound occurs. Figure 2(a-d) shows the main screen for the application relating to sound intensity, the strength at which the sound occurs. In this exercise, after hearing an explanation about "Sound Intensity," the sounds of three bees buzzing are presented to the child, giving the feeling of distance (the sound of the closest or the farthest bee) and the sound with greater or lesser intensity.

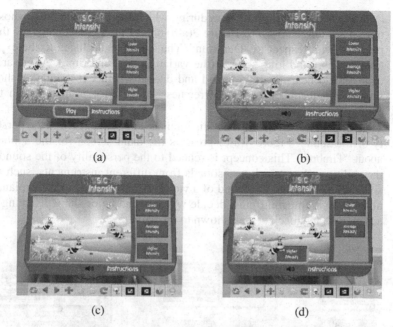

**Fig. 2.** Main screens running the game on the Intensity

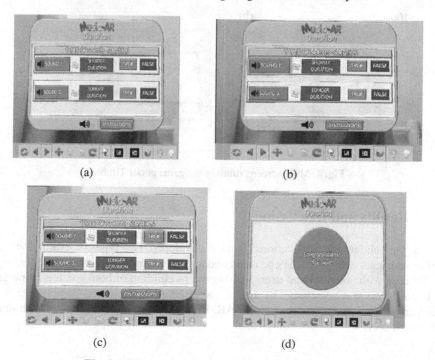

**Fig. 3.** Main screens running the game about Duration

- Duration is the variation of time during which the sound endures, whose extremes are long and short. Figure 3(a-d) shows the main screens for the application with respect to "Duration." The child should listen to the explanations of short and long sounds (the variation of sound time). Afterward, the child should hear a certain sound and decide whether it is long or short. A screen with feedback on the correctness or the error is then shown to the child.

- Timbre is the sound quality that distinguishes instruments. It is the personality of the sound. Figure 4(a-e) presents the main screen for the application about "Timbre." This concept is related to the personality of the sound; it is the ability to distinguish two sounds from different instruments, such as the sound of a piano and the sound of a violin. After listening to the explanations about timbre, the child must decide which sound he or she is listening to. A screen with feedback is then shown to the child.

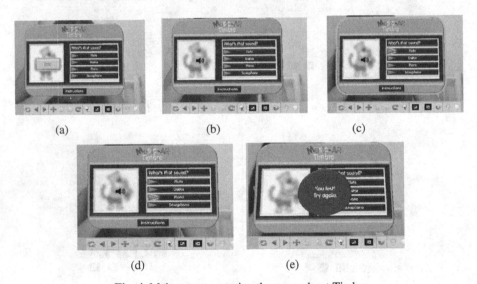

(a)                    (b)                    (c)

(d)                    (e)

**Fig. 4.** Main screens running the game about Timbre

## 5     Results and Discussion

In order to validate the applications (games) developed, two types of evaluation were performed: tests with potential end users and an evaluation with a music teacher.

For tests with potential end users, a group of 14 children of both genders (nine girls and five boys) was selected. The age of the children is presented in Figure 5. Five children had had previous contact with AR. Eight children had had contact with musical education.

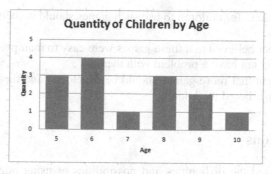

**Fig. 5.** Quantity of Children by Age

The following methodology was applied for the tests with potential end users. First, we collected parental permission for the children to participate in testing. Some questions about their profile were asked. For each child who did not know AR, its principles were explained, and a small demonstration was performed. Subsequently, each child was asked to carry out the execution of the applications about sound properties. Finally, some questions were asked about usability.

The physical environment used for the tests was not a controlled environment. The tests took place in the users' homes in order not to cause a nuisance to the children's usual environment. The users were then asked to run the applications.

During the test sessions, the tester, by observation techniques, noted some information about the use of the applications:

- Five children had some difficulty performing the activities;
- This difficulty was not related to the lack of musical education nor to a lack of interest;
- This difficulty was related to problems of usability of the technology itself, such as those caused by inadequate light or a lack of detection of the marker within a few moments;
- 100% of the children could understand the musical perception using the applications.

These were the children's responses to the usability questions:

- 100% of the children believed that the layout of the four applications was appropriate and attractive;
- 11 children considered the mouse interaction to be good, and 13 considered the marker interaction to be good;
- 100% of the children liked the sounds of the applications;
- 100% of the children liked using the applications and would like to use this type of technology frequently.

In interviewing the music teacher, the following results were obtained:

- The teacher did not k0now the technology of AR but considered it important to use for music education;
- The teacher considered the applications very interesting and fun and felt that they could really help in teaching music;

- The age range for children to use such games should be between four and six years;
- The teacher believed that these games were easy to manipulate and that many kids would not have a problem with them;
- She thought that more games should be developed and then organized into a kit and distributed to children.

## 6    Conclusions

This paper discussed the difficulties and possibilities of using augmented reality in education, especially in musical education. Physical, technological, sociocultural, pedagogical and management issues are among the difficulties inherent in the use of this technology, and the possible solutions presented involve the use of authoring tools that are easy for teachers to use.

We also presented a case study that uses AR for teaching musical perception to Brazilian children, especially the topics of pitch, sound intensity, duration and timbre. From the observation results, we can conclude that the children could understand, through using AR, the concepts of pitch, sound intensity, duration and timbre, sound properties that must be learned before teaching music. All children could solve the musical perception exercises.

According to the questionnaire results, the children felt motivated to use the technology. Another important observation was that children seemed to understand how to use the AR as soon as they began to interact with the technology.

Through an interview with a music teacher, it was concluded that AR could be used for music education; applications developed can motivate children to learn music.

AR seems to be adequate for use with children, due to the playful nature of its technology. AR can be used as a tool in the learning–teaching process with respect to musical perception.

Flaras is a very easy tool to use for nonspecialist people. On the other hand, it is a little limited, for exactly the same reason. Some types of exercises on musical perception could not be performed due to a lack of resources in the tool. For example, it is impossible to include a selection command (if-then and if-then-else). This command is used by programmers to determine what will happen if a certain event occurs. In the case of the "if" statement, if a certain event is true, then the command will be executed.

However, the results show that we can use AR as an educational tool. This technology has a considerable potential impact on the teaching–learning process. In future work, we aim to investigate tools and techniques for creating educational AR learning objects that can be easily proposed to suit different learning contexts.

## References

1. Kirner, C.: Educação permeando a Tecnologia em Aplicações Educacionais Abertas baseadas em Hipermídia e Realidade Aumentada. Revista Contemporaneidade, Educação e Tecnologia 1, 75–87 (2013)

2. Martins, V.F., Guimarães, M.P.: Desafios para o uso de realidade virtual e aumentada de maneira efetiva no ensino. In: Anais do Workshop de Desafios da Computação Aplicada à Educação, pp. 100–109 (2012)
3. Azuma, R., Baillot, Y., Behringer, R., Feiner, S., Julier, S., MacIntyre, B.: Recent advances in augmented reality. Computer Graphics and Applications, IEEE 21(6), 34–47 (2001)
4. Billinghurst, M., Cheok, A., Prince, S., Kato, H.: Real world teleconferencing. Computer Graphics and Applications, IEEE 22(6), 11–13 (2002)
5. Kirner, C., Kirner, T.G.: Virtual reality and augmented reality applied to simulation visualization. Simulation and Modeling: Current Technologies and Applications 1, 391–419 (2007)
6. Johnson, L., Adams, S., Cummins, M., Estrada, V., Freeman, A., Ludgate, H.: The NMC horizon report: 2013 higher education edition (2013)
7. Corrêa, A.G.D., Assis, G.A., Nascimento, M., Lopes, R.D.: Genvirtual: Um Jogo Musical Para Reabilitação De Indivíduos Com Necessidades Especiais. Revista Brasileira de Informática na Educação 16(01) (2008)
8. Zorzal, E.R., Buccioli, A.A.B., Kirner, C.: O uso da realidade aumentada no aprendizado musical. In: Workshop de Aplicações de Realidade Virtual, Minas Gerais (2005)
9. Sommerville, I.: Software Engineering, 6th edn. Addison Wesley (2001). ISBN 0-39815-X
10. Pressman, R.S.: Software engineering: a practitioner's approach. McGraw-hill, pp. 466–472 (2001)
11. Martins, V.F., de Oliveira, A.J.G., Guimarães, M.P.: Implementação de um laboratório de realidade virtual de baixo custo: estudo de caso de montagem de um laboratório para o ensino de Matemática. Revista Brasileira de Computação Aplicada 5(1), 98–112 (2013)
12. Seichter, H., Looser, J., Billinghurst, M.: ComposAR: an intuitive tool for authoring AR applications. In: Proceedings of the 7th IEEE/ACM International Symposium on Mixed and Augmented Reality, pp. 177–178. IEEE Computer Society (2008)
13. Souza, R.C., Moreira, H.D.F., Kirner, C.: FLARAS – Flash Augmented Reality Authoring System (2013). http://ckirner.com/flaras2 (accessed on January 28, 2015)
14. Kato, H.: ARToolKit (1999). http://www.hitl.washington.edu/artoolkit/ (accessed on February 31, 2015)
15. MacIntyre, B., Gandy, M., Dow, S., Bolter, J.D.: DART: a toolkit for rapid design exploration of augmented reality experiences. In: Proceedings of the 17th Annual ACM Symposium on User Interface Software and Technology, pp. 197–206. ACM (2004)
16. Grimm, P., Haller, M., Paelke, V., Reinhold, S., Reimann, C., Zauner, R.: AMIRE-authoring mixed reality. In: Augmented Reality Toolkit, The First IEEE International Workshop, p. 2 (2002)
17. Di Wu, Y. Y., Liu, Y.: Collaborative Education UI in Augmented Reality. ETCS 9, 670–673 (2009)
18. Cerqueira, C.S., Kirner, C.: basAR: ferramenta de autoria de realidade aumentada com comportamento. In: VIII Workshop de Realidade Virtual e Aumentada. Uberaba–MG (2011)
19. Saquoosha, T.K.A.: FlarToolKit (2011). http://www.libspark.org/wiki/saqoosha/FLARToolKit /en (accessed on February 31, 2015)
20. Schafer, R.M.: The soundscape: Our sonic environment and the tuning of the world. Inner Traditions/Bear & Company, p. 320 (1993). ISBN: 9780892814558
21. Zatorre, R.J.: Musical perception and cerebral function: a critical review. Music Perception: An Interdisciplinary Journal 2, 196–221 (1984)
22. Wiggins, J.: Teaching for musical understanding, 1st edn., p. 319. McGraw-Hill Humanities/Social Sciences/Languages (2001)

# A Relational Database for Human Motion Data

Qaiser Riaz[✉], Björn Krüger, and Andreas Weber

Insitute for Computer Science II, University of Bonn, Bonn, Germany
qaiser.riaz@uni-bonn.de, {kruegerb,weber}@cs.uni-bonn.de

**Abstract.** Motion capture data have been widely used in applications ranging from video games and animations to simulations and virtual environments. Moreover, all data-driven approaches for analysis and synthesis of motions are depending on motion capture data. Although multiple large motion capture data sets are freely available for research, there is no system which can provide a centralized access to all of them in an organized manner. In this paper we show that using a relational database management system (RDBMS) to store data does not only provide such a centralized access to the data, but also allows to include other sensor modalities (e.g. accelerometer data) and various semantic annotations. We present two applications for our system: A motion capture player where motions sequences can be retrieved from large datasets using SQL queries and the automatic construction of statistical models which can further be used for complex motion analysis and motions synthesis tasks.

## 1  Introduction

Motion Capturing (mocap) has become a standard technique in the last decades. Various data-driven applications like capturing with low cost sensors (e.g. Kinect), action recognition and motion synthesis have been developed. All these data-driven methods require high quality motion capture data.

For research purposes multiple datasets [1–4] containing tens of gigabytes of mocap data are published and available for free download. These so called *databases* are usually available as collections of files in different formats, like C3D, BVH and ASF/AMC. Currently, the two largest, well established, and freely available collections of mocap data are the CMU [1] and HDM05 [2] databases where the data is organized in flat files. The file names do not include any information about the nature of the event stored in the file. An indication of the content of each file is given as rough textual description only.

For the CMU data collection, textual descriptions are given for each motion file on the web page. In addition, links to motion files are sorted by subjects and categories, such as *Human Interaction, Locomotion, Physical Activities & Sports*. Each file name contains the subject number and an event number only. For the HDM05 data collection, a technical report is available, where a script with the description of the tasks the subjects had performed is published. Here, the file-name refers to the subject, a chapter in this script and thus, gives an indirect indication on the actual content. Additionally a frame rate is stored in the HDM05 file names.

© Springer International Publishing Switzerland 2015
O. Gervasi et al. (Eds.): ICCSA 2015, Part V, LNCS 9159, pp. 234–249, 2015.
DOI: 10.1007/978-3-319-21413-9_17

On the one hand there exists no centralized policy for frame accurate annotations of motion capture data. Annotations on the frame level are only available for few example motion sequences, as shown for comparison in the context of motion segmentation [5,6]. This makes it difficult to search the available data for contents of interest. It is not directly possible to extract a sub part of a relevant information e.g. extracting those frames where subject is in T-pose. To this end, whole files have to be parsed and relevant contents have to be extracted by hand. Another disadvantage of flat files is an inability to query, search, and retrieve information in a structured manner. On the other hand a huge amount of data-driven techniques were developed (see [7–10] for some examples) that make use of carefully selected motion data. Our goal is to provide tools that simplify and speed-up the work flow of data selection and model building for such data-driven approaches.

The main contributions of this paper are:

- We present a flexible Entity-Relationship (ER) model that is capable to handle motion capture data from various public available datasets.
- We come up with solutions to incorporate additional sensor data and data that can be derived from the original measurements.
- We show that hierarchical annotations can be handled to describe the content of the motion data.
- We show that default applications can be supported by relational databases.

The remainder of this work is organized as follows: We give an overview on previous and related work in Section 2. The ER model of our database system is explained in detail in Section 3, while we show some basic operations that are available in Section 4. In Section 5 we consider details of the performance optimization of the database scheme. Exemplary applications are shown in Section 6 and the work is concluded in Section 7.

## 2   Related Work

To handle the growing amount of mocap data, many alternative methods for fast search and retrieval have been proposed by the research community. Based on frame by frame comparison, *Match Webs* [11] were introduced to come up with a efficient index structure. Due to the quadratic complexity it is not possible to build index structures on all currently available mocap data with this method. To avoid quadratic complexity of frame by frame comparisons, segmentation based methods were developed [12]. Cluster based methods classify motion database into small clusters or groups. The classification is either based on similar frames in different motions [13] or on clustering similar poses [14]. Binary geometric features [15] can be used to tackle this kind of segmentation problems. While methods based on boolean features can not describe close numerical similarity, this is the case when low dimensional representations of the motion data are used for comparison. To compute low-dimensional representations of mocap data principal component analysis (PCA) is a well established method [16,17].

Another way to come up with low dimensional representations is to compute feature sets in a normalized pose space [18]. Dimensionality reduction is achieved by using a subset of joint positions only. For the authors, a feature set consisting of hand, feet and the head positions is the one of choice. They also show, that it is possible to search these feature sets efficiently by employing a kd-tree as index structure.

This retrieval technique is adopted for the database architecture proposed by [19]. In this work the authors focus on both, data compression and retrieval of motion capture data. A compression ratio of 25% of the original database size is achieved with their method. To this end a curve simplification algorithm is applied to reduce the number of frames to 20% of the original ones. An adaptive k-means clustering algorithm is used to group similar motion clips together. In addition, a three-step motion retrieval method is used which accelerates the retrieval process by filtering irrelevant motions. A database architecture consisting of data representation, data retrieval and selection modules has been proposed [20]. An in-memory internal representation of the motion data consisting of a collection of poses is created via a data representation module. The data retrieval and selection module queries this in-memory representation using PhaseQuery and results are expressed as sequences of segments. The search and retrieval time is almost real-time for small data sets but it increases dramatically for larger data sets. The in-memory condition requires enough physical memory to load large data sets.

For annotation purposes, semi-supervised techniques were employed. Based on a support vector machine (SVM), a database of football motions was annotated [17]. Such kind of annotations are transferred from motion capture data to video sequences in [21]. To visualize and explore huge data sets hierarchical clustering on normalized pose features [22] was used to obtain an overview on huge data collections. In opposite, Lin [23] presents a system where Kinect queries are used to search and explore mocap datasets that are modeled as sets of 3-attribute strings.

## 3   Database Architecture

The Entity-Relationship (ER) Model of the proposed database architecture is shown in Figure 1. The core schema is divided into four different categories:

1. **Controller Entity:** The heart of the proposed schema, which controls the flow of information.
2. **Sensor-specific Entities:** To handle sensor-specific data for each sensor.
3. **Annotations Control Entities:** Control annotation policy mechanism.
4. **Derived Entities:** To entertain non-mocap data which is computed from the mocap data.

We will briefly explain each of these categories in the following subsections.

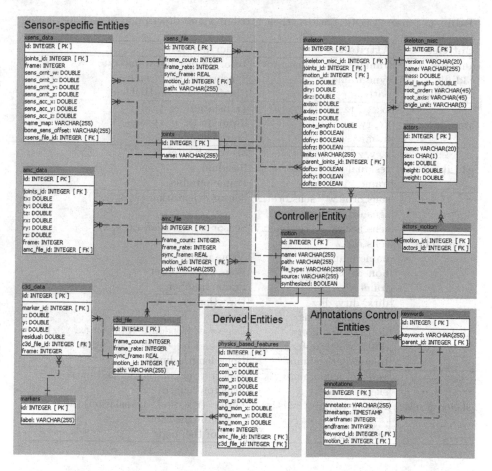

**Fig. 1.** Entity Relationship (ER) model of the proposed database architecture. The core schema is divided into four different categories, each of which handles an aspect of the proposed schema.

## 3.1   Controller Entity

The purpose of the controller entity is to control the flow of information in a logical manner. It functions as the heart of the schema providing a logical bounding among different entities. All vital information passes through the controller entity. In our proposed database architecture, the *motion* entity acts as the controller entity (Figure 1). The controller entity has a simple schema with certain general attributes of a file such as 1) *name:* stores actual name of the file, 2) *file_type:* stores type of the file e.g. amc, c3d, 3) *source:* stores the source of the file e.g. HDM05, CMU, and 4) *synthesized:* a boolean flag to indicate if the motion data is an actual recording or an artificial motion obtained by some motion synthesis procedure.

**Table 1.** Sensor-specific Entities, their attributes and description of each attribute

| Entities | Attributes | Description |
|---|---|---|
| amc_file, | frame_count, frame_rate | Total frames and frame rate |
| c3d_file, | sync_frame | Synchronization frame |
| xsens_file | path | Physical path on HDD |
| amc_data | tx, ty, tz | Translation (x, y, z) |
| | rx, ry, rz | Rotation (x, y, z) |
| | frame | Frame number |
| c3d_data | x, y, z | 3D coordinates |
| | residual | Residual |
| | frame | Frame number |
| xsens_data | sens_ornt_w, sens_ornt_x, sens_ornt_y, sens_ornt_z | Sensor orientation (w, x, y, z) |
| | sens_acc_x, sens_acc_y, sens_acc_z | Sensor acceleration (x, y, z) |
| | name_map | Name map |
| | bone_sens_offset | Bone sensor offset |
| | frame | Frame number |
| skeleton | dirx, diry, dirz | Direction (x, y, z) |
| | axisx, axisy, axisz | Axis (x, y, z) |
| | dofrx, dofry, dofrz | Degree of freedom - rotation (x, y, z) |
| | doftx, dofty, doftz | Degree of freedom - translation (x, y, z) |
| | bone_length, limits | Bone length, limits |
| skeleton_misc | version, name, mass | Sensor's version, name, and mass |
| | skel_length | Skeleton length |
| | root_order, root_axis | Order (rotation, translation) and Axis (x,y,z) of the root |
| | angle_unit | Unit of the angle |

## 3.2   Sensor-Specific Entities

Most of the entities in our proposed database schema are sensor-specific. Sensor-specific entities, as the name indicates, are used to store sensor specific information in the database. In order to achieve a flexible design of the database schema, general properties of each recording are stored in separate entities (name of the entity in a format: *sensor name + an underscore + 'file'* e.g. *c3d_file*) and actual data is stored in separate entities (name of the entity in a format: *sensor name + an underscore + 'data'* e.g. *c3d_data*). Each sensor can have any additional supporting entities. For example to store AMC data, the general properties are stored into *amc_file* table and the actual data is stored into *amc_data* table. The supporting entity in this case is *joints* table. We have processed and stored data from different sensors in our proposed database, which include:

**ASF Data:** The ASF skeleton data is distributed into two entities namely *skeleton_misc* and *skeleton*. The *skeleton_misc* entity stores the general attributes of skeleton while the *skeleton* entity stores specific skeleton data of each joint in each frame. The attributes of both entities are described in Table 1.

**Table 2.** Annotations Control Entities, their attributes and description of each attribute

| Entities | Attributes | Description |
|---|---|---|
| annotations | annotator | Name of the annotator |
| | timestamp | Record creation timestamp |
| | startframe | Starting frame number of the motion |
| | endframe | Ending frame number of the motion |
| keywords | keyword | Keyword |
| | parent_id | A self relation, *null* if no parents |

**AMC Data:** The AMC motion data is stored into two mandatory entities *amc_file* and *amc_data* and a supporting entity *joints*. The *amc_file* entity stores general information about the data such as *frame count, frame rate, synchronization frame* etc. A synchronization frame is used to overcome synchronization problem amongst different sensor systems, which occurs when a single motion is simultaneously recorded by multiple motion capture devices. The *amc_data* stores rotation and translation data for each joint in each frame. The *joints* entity has a *one-to-many* relationship with the *amc_data* and provides an easy mechanism of joint-based data search using standard SQL statements. The attributes of AMC data entities are described in Table 1.

**C3D Data:** The C3D data is stored into two mandatory entities *c3d_file* and *c3d_data* and a supporting entity *markers*. Like the *amc_file*, the *c3d_file* entity also stores general information about the data such as *frame count, frame rate, synchronization frame* etc. The *c3d_data* entity stores 3D information of each marker in each frame. The *markers* entity has a *one-to-many* relationship with the *c3d_data*. The database can be queried based on markers to fetch data of a particular marker using standard SQL statements. The attributes of C3D data entities are explained in Table 1.

**Accelerometer Data (Xsens):** The accelerometer data is not available in CMU or HDM05 mocap data sets. However, some recordings of accelerometer data were captured later on and we have included these data sets in our database schema. This shows the flexibility and extensibility of our database architecture that any new sensor can be easily integrated within the existing schema. The data has been recorded using Xsens's MTi accelerometer [24]. In order to store data; two mandatory entities *xsens_file* and *xsens_data* and a supporting entity *joints* are used. The *xsens_file* has same attributes as *amc_file* and *c3d_file*. The *xsens_data* entity stores orientation and acceleration data of each joint in each frame. The *joints* entity has a *one-to-many* relationship with the *xsens_data*. The attributes of accelerometer data entities are explained in Table 1.

**Table 3.** Derived entities - physics based features. Physics based features are derived from mocap data sets and do not exit on their own.

| Entity | Attributes | Description |
|---|---|---|
| physics based features | com_x, com_y, com_z | Center of Mass (x, y, z) |
| | zmp_x, zmp_y, zmp_z | Zero Moment Point (x, y, z) |
| | ang_mom_x, ang_mom_y, ang_mom_z | Angular Momentum (x, y, z) |
| | frame | Frame number |

### 3.3 Annotations Control Entities

Annotations control entities are one of the important entities in the proposed database architecture. These entities define an annotation policy mechanism and provide an easy way to query the database based on an event keyword. In the proposed database architecture, two entities have been introduced to handle annotations. The *annotations* entity stores the general attributes of an annotation such as *start frame, end frame, timestamp* etc. the *keywords* entity serves as a dictionary of annotations and has a *one-to-many* relationship with the *annotations* entity. It also has a *self-relation* to maintain a *hierarchical relationship* between different keywords. The *hierarchical relationships* are parent- child relationships and define annotations from high-level general terms to low-level more specific terms. For example, in HDM05, a *'jumping jack'* motion is a child of the *'workout'* event and a grand child of the *'sports'* event. So the *parent_id* of the *'jumping jack'* will be the *id* of the *'workout'* and the *parent_id* of the *'workout'* will be the *id* of the *'sports'*. The attributes of annotations control entities are expressed in Table 2.

### 3.4 Derived Entities

In the proposed database architecture, there are certain entities which are derived from the existing mocap data sets. These entities do not exist in any freely available mocap data set. However, they are required in many research activities and researchers have to manually compute them whenever required. A good example of derived entities is physics based features such as center of mass, zero moment point, and angular momentum, which can be computed through kinematics [25]. To entertain these features, a separate entity namely *physics_based_features* has been created. The attributes of the *physics_based_features* are explained in Table 3. This table has no real data at the moment and computing and dumping physics based features into the database will be carried out in near future.

## 4   Basic Database Operations

### 4.1 Processing and Storing Mocap Data into Database

We have used PostgreSQL, version 9.0 - 64 bit, to store extracted data from mocap data sets. PostgreSQL is a widely used object-relational database management system which is freely available under PostgreSQL license. A database has been created using our proposed database schema. In order to extract data from different formats and store into database, special scripts are written in Visual C++ and Matlab. These scripts read each file of a particular format, extract data of interest, and generate text files with structured queries. These queries are then executed under PostgreSQL environment to store data into different tables. In order to optimize insertion process by minimizing data insertion time, concepts of bulk data insertion are used.

## 4.2   Retrieving Collections

Collections can be retrieved using standard SQL statements. In the upcoming subsections, we will give some examples of retrieving collections using standard SQL queries.

**Retrieving All Actors:** This is a simple example of retrieving data of all *actors*. Each event in mocap data is performed by one or more actors and motions can be retrieved based on actor information.

```
select * from actors;
```

**Retrieving All Annotation Keywords:** This is another simple example of retrieving all *keywords*. Each event in mocap data is annotated through a keyword, which explains the nature of the event.

```
select * from keywords;
```

**Retrieving Annotations of a Specific Event Group:** Sometimes one is interested to find all annotations of a specific group of events e.g. finding all '*sports*' annotations. In this example we show how one can retrieve annotations of a specific event group. The event group of interest, in this case, is '*sports*'.

```
select keyword from keywords
  where parent_id = (
  select id from keywords
    where keyword=''sports '')
```

**Retrieving Motion Information of an Event:** This example shows how to retrieve motion IDs of all motion records for '*dancing*' event. These IDs can be used in later steps to retrieve actual motion data.

```
select m.id from motion m,
  annotations a, amc_file af,
  keywords k
  where m.id=a.motion_id
  and a.keyword_id=k.id
  and af.motion_id=m.id
  and k.keyword=''dancing ''
```

**Retrieving Synchronized C3D Data:** In this example, we show how to retrieve synchronized data based on *syn_frame* value. The synchronization frame, *syn_frame*, is used to overcome synchronization problem amongst different sensor systems, which occurs when a single motion is simultaneously recorded by multiple motion capture devices. The data-type of the *syn_frame* attribute is *real* and stores synchronization time in seconds. In the presence of this time, retrieving synchronized data is very easy and straight forward as shown in the following query.

```
select * from c3d_data
  where c3d_file_id=1 and frame >
    (select sync_frame*frame_rate
     from c3d_file where id=1)
```

**Retrieving AMC Data:** In this query we extract all AMC data for *'throwing'* event where actor is *'mm'* and source of mocap data is *'HDM05'*. This is a complex query as it involves multiple joins among various entities such as *actors, motion, amc_file* etc.

```
select * from amc_data
 where amc_file_id in
   (select af.id from amc_file af,
   motion mo where
   mo.source =''HDM05''
   and mo.id=af.motion_id
   and mo.id
     in (select m.id from motion m,
       annotations a,
       actors_motion am
       keywords k, actors ac
       where m.id=a.motion_id
         and a.keyword_id=k.id
         and ac.id=am.actors_id
         and am.motion_id=m.id
         and ac.name=''mm''
         and k.keyword=''throwing''))
```

# 5   Database Performance Evaluation

## 5.1   Performance Optimization

Before we evaluate the performance of the presented database scheme, we give some insights of the steps taken for optimization of the database structure.

The size of the database on hard disk is approximately 61 GB after parsing and inserting data from all ASF/AMC and C3D files for both *HDM05* and *CMU*. Entities *amc_data* and *c3d_data* are the largest entities having approximately *90 million* and *130 million* records respectively. Hence, an optimization policy is required in order to minimize database search and retrieval time and maximize the system's response time. Indexing is one of the widely used optimization techniques in relational database management systems. PostgreSQL provides several built-in indexing algorithms such as

**Table 4.** A comparison of performance optimization with and without indexing. The data search and retrieve time has substantially decreased by introducing binary tree based indexes.

| Retrieving | Execution Time (ms) | |
|---|---|---|
| | Not Indexed | Indexed |
| All Actors | 51 | 11 |
| All Keywords | 13 | 10 |
| Annotations of an Event Group | 12 | 11 |
| Motion Information of an Event | 111 | 30 |
| Synchronized C3D Data | 244,770 | 18,029 |
| AMC Data | 217,795 | 8,132 |

B-tree, Hash, GiST and GIN [26]. PostgreSQL uses B-tree as default indexing algorithm [26]. We have created indexes using B-trees on primary keys of both tables. We have also created indexes using B-trees on foreign keys to minimize search and retrieval time.

The trade-off of using indexes is slow data insertion as indexes are updated upon each insertion. However, mocap data sets are not frequently updated so one can compromise on slow data insertion over fast find and fetch. Alternatively, indexes can be dropped during insertion to speed up the insertion process and can be regenerated afterward. We have executed all queries listed in the section *Retrieving Collections* 4.2 with and without indexing and the results are presented in Table 4. The comparison clearly indicates substantial decrease in data search and retrieve time after introducing B-tree based indexes.

**Table 5.** Database performance in terms of execution time

| Entity (Size) | Total Records | Fetched Records | Trials Count | Exec Time (ms) | Fetch Criteria (Event, Actor, Source) |
|---|---|---|---|---|---|
| amc_data (28 GB) | 164x10⁶ | 2,581 | 1 | 202 | T-pose, bd, HDM05 |
| | | 237,771 | 2 | 8,134 | Throwing, mm, HDM05 |
| | | 751,042 | 20 | 24,576 | Walking, bd, HDM05 |
| | | 1,505,390 | 13 | 51,182 | Dancing, All Actors, HDM05 |
| | | 126,701 | 12 | 4,293 | Walk, 07, CMU |
| | | 522,058 | 19 | 17,656 | Modern Dance, 05, CMU |
| | | 744,662 | 7 | 26,081 | Swimming, 125, CMU |
| c3d_data (32 GB) | 230x10⁶ | 360,756 | 2 | 15,978 | Throwing, mm, HDM05 |
| | | 1,139,512 | 20 | 48,121 | Walking, bd, HDM05 |
| | | 2,196,786 | 13 | 98,768 | Dancing, All actors, HDM05 |
| | | 179,129 | 12 | 8,395 | Walk, 07, CMU |
| | | 738,082 | 19 | 34,928 | Modern Dance, 05, CMU |
| | | 1,052,798 | 7 | 47,209 | Swimming, 125 , CMU |

The performance of a database can be analyzed based on how much time it takes to search and retrieve records against simple and complex queries. As said earlier, we have a particularly large database with a disk size of approximately 61 GB. The two largest entities in our database are *'amc_data'* and *'c3d_data'* having a disk size of *28*

*GB* and *32 GB* respectively. In section 5.1, we have outlined our strategy to optimize performance of the two entities by means of indexing. To test the performance of the database, we have executed several queries on these two large entities. In general, we have found minimum search and retrieval time when the retrieved collections are small in count and maximum search and retrieval time when the retrieved collections are large in count.

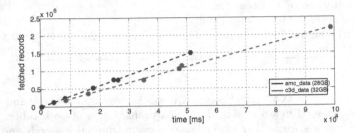

**Fig. 2.** Scatter plot of the timings, when querying the AMC and C3D datasets. We observed a linear relation between the number of retrieved records and the execution time.

In order to test the performance of the database, several fetch criteria are used to fetch data and the results are presented in Table 5. All tests are performed locally on the database server machine. The database took only *202* ms to fetch *2581* records of '*T-pose*' data of the actor '*bd*' for *HDM05*. The database took *4293* ms to fetch *126701* records of the '*Walk*' event of the actor '*07*' (12 trials in total) from *CMU*. One the other hand, it took *51182* ms to fetch *1505390* records of the '*dancing*' event for all actors (13 trials in total) from *HDM05*. During experimentation, we have observed that the execution time increases as the size of the retrieved records increases and a linear tendency is seen as shown in Figure 2. From this, we conclude that the performance of the database is optimal for small record sets. Most applications work in cycles of retrieving small chunks of data from the database and processing these records instead of retrieving the whole data at once. With small execution time (such as *202 ms*), it is possible to achieve interactive processing by fetching and processing data frame by frame.

**Complexity Analysis of SQL Queries:** The complexity of an SQL statement depends upon a number of factors such as number of tables involved, number of joins, unions, intersections, where/having clauses, sub-queries and so on. The complexity of an SQL statement directly effects its execution cost. The execution plan of any SQL statement can be analyzed in PostgreSQL using the '*explain*' command. "The execution plan shows how the table(s) referenced by the statement will be scanned by plain sequential scan, index scan, etc. and if multiple tables are referenced, what join algorithms will be used to bring together the required rows from each input table" [27].

*Retrieving Motion Information of an Event.* The SQL query of *retrieving motion information of an event* is given in number 4 of the section *Retrieving Collections* 4.2. A visual complexity analysis of this query is presented in Figure 3. This query retrieves motion IDs of all motion records for the '*dancing*' event. It consists of four inner-joins

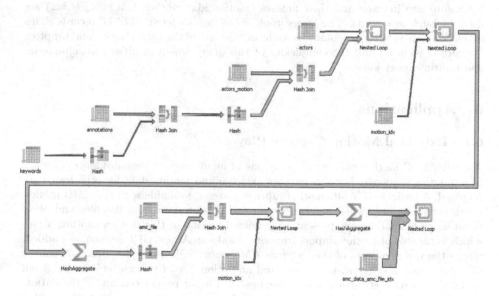

**Fig. 3.** Complexity analysis of the query 'retrieving motion information of an event' analyzed by PgAdmin Query Tool. This query involves inner-joins between four entities: 'keywords', 'annotations', 'amc_file', and 'motion'. It took 12 ms to fetch 13 records.

between entities: '*motion*', '*annotations*', '*amc_file*', and '*keywords*'. In order to relate entities, PostgreSQL uses indexes for those entities which are indexed and hash joins are used for non-indexed entities. In this example only '*motion*' entity is indexed so its index (*motion_idx*) is used to retrieve records. This query took *12 ms* to fetch *13* records.

**Fig. 4.** Complexity analysis of the query 'retrieving AMC data' analyzed by PgAdmin Query Tool. This query consists of two sub-queries and seven inner-joins between entities: 'keywords', 'annotations', 'actors_motion', 'actors', 'motion', and 'amc_file'. It took 8,066 ms to fetch 237,771 records.

*Retrieving AMC Data.* The SQL query of *retrieving AMC data* is outlined in number 6 of the section *Retrieving Collections* 4.2. This is one of the most complex queries in our schema. A visual complexity analysis of this query is presented in Figure 4. This query retrieves AMC data records of all '*throwing*' events performed by the actor '*mm*'. This query consists of two sub-queries and seven inner-joins between entities: '*keywords*',

**Fig. 5.** Extended version of ASF/AMC motion capture player originally developed by [28]. This extended version can be used to fetch data from the database and play it. In the left side figure, the rectangle (right side bottom) highlights the extended part. The user provides as input an 'actor name', an 'event', and an optional 'motion number'. The example shows three 'throwing' actions (left), two 'rope skipping' actions (center), and three 'skiing' actions (right). All motions are performed by the actor 'bd'.

'annotations', 'actors_motion', 'actors', 'motion', and 'amc_file'. Entities 'motion' and 'amc_data' are indexed and their indexes (motion_idx, motion_data_amc_file_idx) are used to retrieve records. The query took *8,066 ms* to fetch *237,771* records. This execution time is fairly acceptable considering the size of the entity 'amc_data' (approx. *90 million* records) and the complexity of this query which involves two sub-queries and multiple inner-joins.

# 6    Applications

## 6.1    Extended Motion Capture Player

An ASF/AMC motion capture player is one of many ways to visualize motion frames as animation sequence. For this purpose skeleton and motion data for each frame are required. A basic ASF/AMC motion capture player is available with the CMU motion capture dataset [28], which reads skeleton and motion data from flat files and plays them as an animation. We present an extended version of this motion capture player which is capable of motion import from our database. A new GUI element was added, where the variable parts of the 'retrieve AMC data' SQL query (see Sec. 4.2) can be filled. As input an 'actor', an 'event', and an optional 'motion number' can be given. A database search is carried out for the specified input parameters and if the data is found, it is loaded into the player. A user can then use various control buttons provided in the player to play the animation. Multiple motions can be loaded and played at the same time. With this type of interface it is simple to search for individual motion sequences without having knowledge, or even touching the actual motion capture files. Figure 5 shows three different types of motions loaded in the extended motion capture player. In the left side figure, the rectangle (right side bottom) highlights the extended part. The example shows three 'throwing' actions (left side figure), two 'rope skipping' actions (center figure), and three 'skiing' actions (right side figure). All motions are performed by the actor 'bd' in these examples. This simple example already shows how simple selections can be made on the basis of SQL queries. More sophisticated data visualization techniques [22,29] could use such selections to allow rapid drill down to important parts of motion data sets for further exploration and analysis.

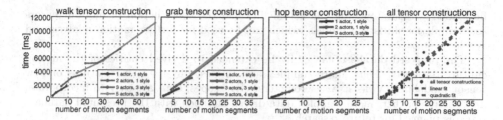

**Fig. 6.** Scatter plot of the timings, when querying the amc dataset to construct motion tensors. We observed a linear relation between the number of fetched motion segments, but no dependency in the number of actors, motion classes or styles.

## 6.2   Automatic Generation of Statistical Models

Another set of techniques that can benefit from the connection to a relational mocap database is the automatic construction of statistical models. Such models are used for analysis tasks such as motion classification and segmentation or motion synthesis tasks where motion sequences should be developed that fulfill certain user specified constraints. To show the effectiveness of our approach, we show the automatic construction of motion tensors, that have been shown to be useful for motion synthesis [7,8]. To this end, we fetched data from the database for various actors in the same number of repetitions for multiple motion classes that belong to various styles of a motion. Krüger et al. [7] introduced so called *natural modes*, that belong to different actors, repetitions or motion styles. In the original works these example motions were selected by hand carefully. By using a data retrieval function, which is written in procedural language for the PostgreSQL (PL/pgSQL), we fetch the individual motions for construction of the multi-modal model. The function takes as input *actor name*, *motion class*, and *number of repetitions* and retrieves related data from the database. Using this approach the construction of each tensor model, as described by Krüger et al. [7], needed less than ten seconds. Larger sets of motions, including up to 5 actors and 4 motion classes could be retrieved in about 12 seconds. The actual motions for tensor construction where taken from the HDM05 motion capture database. For the *walk*-tensor examples motions from the motion classes: *walkRightCircle4StepsRstart*, *walk4StepsRstart*, *walkLeftCircle4StepsRstart* were used. For the *grab*-tensor the classes *grabHighR*, *grabMiddleR*, *grabLowR*, *grabFloorR* were retrieved. And for the *hop*-tensor the classes *hopRLeg2hops*, *hopBothLegs2hops*, *hopLLeg2hops* where used. The annotations from the classes where taken from the so called *cut-files* subset which is described in the documentation [2] of the HDM05 data set. Overall we observed that retrieval times depend linear on the number of fetched motion segments instead of the number of actors or motion classes (See Fig. 6). Thus, large data sets can be the basis for an efficient construction of statistical models and therefore for a bunch of new applications in motion analysis and synthesis.

## 7   Conclusion and Future Work

In this paper, we presented a relational database scheme for mocap data sets. According to this scheme, a database has been created using the open source PostgreSQL RDBMS, and motion capture data from HDM05 and CMU datasets. The functionality has been shown in two applications: A simple player where specific motions can be loaded without touching the actual files and the construction of motion tensors as example for statistical models, which can be used for further motion analysis and synthesis steps.

The proposed database can easily be extended to derived entities of the motion capture data sets. Thus, more complex annotations, physics based features or new sensor modalities (Videos, Accelerometers, Gyroscopes, etc.) are easy to incorporate. To combine further, more complex data-driven methods with the presented database setup is planned for future research.

# References

1. CMU: CMU Motion Capture Database (2003). http://mocap.cs.cmu.edu/
2. Müller, M., Röder, T., Clausen, M., Eberhardt, B., Krüger, B., Weber, A.: Documentation Mocap Database HDM05. Technical Report CG-2007-2, Universität Bonn, June 2007
3. Sigal, L., Balan, A., Black, M.: Humaneva: Synchronized video and motion capture dataset and baseline algorithm for evaluation of articulated human motion. International Journal of Computer Vision 87(1–2), 4–27 (2010)
4. Guerra-Filho, G., Biswas, A.: The human motion database: a cognitive and parametric sampling of human motion. In: 2011 IEEE International Conference on Automatic Face Gesture Recognition and Workshops (FG 2011), pp. 103–110, March 2011
5. Zhou, F., la Torre, F.D., Hodgins, J.K.: Hierarchical aligned cluster analysis for temporal clustering of human motion. IEEE Trans. on Pattern Analysis and Machine Intelligence 35(3), 582–596 (2013)
6. Vögele, A., Krüger, B., Klein, R.: Efficient unsupervised temporal segmentation of human motion. In: 2014 ACM SIGGRAPH/Eurographics Symposium on Computer Animation, July 2014
7. Krüger, B., Tautges, J., Müller, M., Weber, A.: Multi-mode tensor representation of motion data. Journal of Virtual Reality and Broadcasting 5(5), July 2008
8. Min, J., Liu, H., Chai, J.: Synthesis and editing of personalized stylistic human motion. In: Proceedings of the 2010 ACM SIGGRAPH Symposium on Interactive 3D Graphics and Games, I3D 2010, New York, NY, USA, pp. 39–46. ACM (2010)
9. Min, J., Chai, J.: Motion graphs++: A compact generative model for semantic motion analysis and synthesis. ACM Trans. Graph. 31(6), 153:1–153:12 (2012)
10. Baumann, J., Wessel, R., Krüger, B., Weber, A.: Action graph: a versatile data structure for action recognition. In: GRAPP 2014 - International Conference on Computer Graphics Theory and Applications, SCITEPRESS, January 2014
11. Kovar, L., Gleicher, M.: Automated extraction and parameterization of motions in large data sets. ACM Transactions on Graphics 23(3), 559–568 (2004). SIGGRAPH 2004
12. Lin, Y.: Efficient human motion retrieval in large databases. In: Proceedings of the 4th International Conference on Computer Graphics and Interactive Techniques in Australasia and Southeast Asia, pp. 31–37. ACM (2006)
13. Basu, S., Shanbhag, S., Chandran, S.: Search and transitioning for motion captured sequences. In: Proceedings of the ACM Symposium on Virtual Reality Software and Technology, pp. 220–223. ACM (2005)
14. Liu, G., Zhang, J., Wang, W., McMillan, L.: A system for analyzing and indexing human-motion databases. In: Proceedings of the 2005 ACM SIGMOD International Conference on Management of Data, pp. 924–926. ACM (2005)

15. Müller, M., Röder, T., Clausen, M.: Efficient content-based retrieval of motion capture data. In: ACM Transactions on Graphics (TOG), vol. 24, pp. 677–685. ACM (2005)
16. Forbes, K., Fiume, E.: An efficient search algorithm for motion data using weighted pca. In: Proceedings of the 2005 ACM SIGGRAPH/Eurographics Symposium on Computer Animation, pp. 67–76. ACM (2005)
17. Arikan, O., Forsyth, D.A., O'Brien, J.F.: Motion synthesis from annotations. ACM Transactions on Graphics (TOG) **22**(3), 402–408 (2003)
18. Krüger, B., Tautges, J., Weber, A., Zinke, A.: Fast local and global similarity searches in large motion capture databases. In: 2010 ACM SIGGRAPH/Eurographics Symposium on Computer Animation, SCA 2010, pp. 1–10. Eurographics Association, Aire-la-Ville, July 2010
19. Wang, P., Lau, R.W., Zhang, M., Wang, J., Song, H., Pan, Z.: A real-time database architecture for motion capture data. In: Proceedings of the 19th ACM International Conference on Multimedia, pp. 1337–1340. ACM (2011)
20. Awad, C., Courty, N., Gibet, S.: A database architecture for real-time motion retrieval. In: Seventh International Workshop on Content-Based Multimedia Indexing, CBMI 2009, pp. 225–230. IEEE (2009)
21. Ramanan, D., Forsyth, D.A.: Automatic annotation of everyday movements. In: Neural Information Processing Systems (2003)
22. Bernard, J., Wilhelm, N., Krüger, B., May, T., Schreck, T., Kohlhammer, J.: Motionexplorer: Exploratory search in human motion capture data based on hierarchical aggregation. IEEE Trans. on Visualization and Computer Graphics (Proc. VAST) (2013)
23. Lin, E.C.-H.: A research on 3d motion database management and query system based on kinect. In: Park, J.J.J.H., Pan, Y., Kim, C., Yang, Y. (eds.) Future Information Technology - II. LNEE, vol. 329, pp. 29–35. Springer, Heidelberg (2015)
24. Xsens: Products - Xsens, July 2013. http://www.xsens.com/en/general/mti
25. Robbins, K.L., Wu, Q.: Development of a computer tool for anthropometric analyses. In: Proceedings of the International Conference on Mathematics and Engineering Techniques in Medicine and Biological Sciences (METMBS 2003), pp. 347–353. CSREA Press, Las Vegas (2003)
26. PostgreSQL: Documentation: 9.0: Index Types, July 2013. http://www.postgresql.org/docs/9.0/static/indexes-types.html
27. PostgreSQL: Documentation: 9.0: EXPLAIN, July 2013. http://www.postgresql.org/docs/9.0/static/sql-explain.html
28. Barbič, J., Zhao, Y.: ASF/AMC Motion Capture Player, July 2013. http://mocap.cs.cmu.edu/tools.php
29. Wilhelm, N., Vögele, A., Zsoldos, R., Licka, T., Krüger, B., Bernard, J.: Furyexplorer: visual-interactive exploration of horse motion capture data. In: Visualization and Data Analysis (VDA 2015), February 2015

# Immersive and Interactive Simulator to Support Educational Teaching

Marcelo de Paiva Guimarães[1,2] (✉), Diego Colombo Dias[3], Valéria Farinazzo Martins[4], José Remo Brega[5], and Luís Carlos Trevelin[3]

[1] Universidade Aberta do Brasil, UNIFESP, São Paulo, SP, Brasil
marcelodepaiva@gmail.com
[2] Programa de Mestrado em Ciência da Computação, FACCAMP,
Campo Limpo Paulista, SP, Brasil
[3] Universidade Federal de São Carlos, UFSCAR, São Carlos, SP, Brasil
diegocolombo.dias@gmail.com, trevelin@dc.ufscar.br
[4] Faculdade de Computação e Informática,
Universidade Presbiteriana Mackenzie, São Paulo, Brasil
valfarinazzo@gmail.com
[5] Universidade Estadual Paulista "Júlio de Mesquita Filho", Bauru, SP, Brasil
remobrega@gmail.com

**Abstract.** Visualization and manipulation of a discipline's content integrated with educational planning can enhance teaching and facilitate the learning process. This paper presents a set of tools developed to explore educational content using a multi projection system with high immersion and interaction support for group learning and a support run on Internet browsers for individual learning. The objects visualized are enriched with multimedia content such as video, audio, and text and can be used in different educational proposals. In this work, these tools were populated with content for teaching computer architecture for computer undergraduate students.

**Keywords:** Virtual reality · Immersive visualization · Teaching · Learning · Multimedia

## 1 Introduction

It is becoming increasingly common for students in educational environments to use smartphones, laptops, and tablets with high processing capacity and visualization. This has motivated studies on integrating these devices with educational practices in classrooms, which might allow more effective support for the new digital culture age. Teachers need to understand the modern technological society and how this current digital culture can be favorable to the learning process, bringing interactivity and autonomy to the process of knowledge acquisition.

At present, teachers can take advantage of digital technologies in educational practices, promoting real knowledge gain according to the level of development, skills, and abilities of students in training [1]. Therefore, mere technology adoption is

© Springer International Publishing Switzerland 2015
O. Gervasi et al. (Eds.): ICCSA 2015, Part V, LNCS 9159, pp. 250–260, 2015.
DOI: 10.1007/978-3-319-21413-9_18

not enough; this requires planning and appropriate tools. The main features of the digital learning culture point to new needs in student training [2, 3, 4, 5, 6], for example, how to organize and assign meaning and sense to information researched by learners, in order to build knowledge and develop the capacity of learning management, knowledge, and training.

Simulation is an example of a technological tool that can be used by teachers to get closer to the actual digital culture age. By means of this, it is possible to represent situations that would be impossible without this computational solution, even in real laboratories [7, 8]; one example is an expedition inside an erupting volcano [9]. Using simulators, the users enter a virtual place and visualize, manipulate, and exploit some data in real time [10]. For that, they use their senses, particularly the natural three-dimensional movements of the body. A simulation activity can strengthen, facilitate, and consolidate the understanding of involved concepts during a study. Moreover, such activities must be supported by a well-defined pedagogical strategy to be able to achieve good results in the educational process. Simulation can also be a precious tool in group work, mainly in situations that involve decisions during an activity such as, for example, choosing the route in a driving simulation. Different students can test and evaluate various hypotheses and thus understand their decisions and those decisions' consequences. Moreover, adoption of simulators in a classroom tends to encourage and promote the learning process, making it an exciting way of learning.

There are several kinds of simulators [11, 12], for instance, static simulators, which requires no whole-body movement or significant physical effort by the students; another example is dynamic simulators, which the users move through the immersive and interactive virtual environment, providing a greater degree of freedom. The immersive simulators discussed in this paper are dynamic, and they are based on virtual reality, a computer area able to create synthetic environments that enable the exploration of basic senses of the human body such as vision, hearing, touch, and smell. The main features of these tools are immersion, interactivity, and involvement. Immersion has the capacity to hold the user's attention; interactivity is related to the capacity of the system to respond to user actions; and the involvement is aimed at user engagement in the activity. User immersive interaction is achieved, for instance, by using 3D stereoscopic display devices, touch feeling devices (haptic devices), and devices that capture the user's movements. User engagement with simulations is achieved with planned actions during the simulation, for example, setting challenges to be achieved.

This paper presents a set of educational tools developed to support well-defined pedagogical strategies that promote learning situations that are not arbitrary. The first tool is an immersive and interactive simulator running on a multi projection system, a CAVE-like (Cave Automatic Virtual Environment [13]) system called miniCAVE. In this environment, teacher and students work in a group to simulate some educational content. The other tool shown is also a simulator, but it runs on web browsers (e.g., Firefox, Chrome, and Explorer), and it is aimed at promoting student self-study. It is also part of the set of two other tools that are used for the preparation of the simulation.

The paper is organized as follows: section 2 discusses the use of simulations in teaching; section 3 presents the research method; section 4 presents the developed computational tools; section 5 describes the process of creating educational content for these tools; and finally, section 6 covers some conclusions.

## 2    Using Simulations for Teaching

The use of simulators in the educational context is not new; it has already been applied in various situations such as the teaching of physics [14], chemistry [15], biology [16, 17], and medicine [18, 19]. Guimarães and Gnecco [20] showed virtual environments for astronomy education. Dias and colleagues [21] developed a system to support the teaching of dental structures using semantic descriptions in virtual environments. Souza and colleagues [22] have created a simulator for the operating systems discipline.

The tools presented in this paper are different from the others in several aspects, including: the high degree of immersion (3D) and interaction by means of input devices such as Kinect, the Wii remote, a keyboard, and a mouse; the possibility of developing activities in groups or individually; customization that addresses educational content, and integration with multimedia resources (video, audio, and text). The support that allows the semantic description (meaning) of the models is part of such tools. This feature makes the organization of the content flexible; for example, in a computer motherboard simulation, the teacher can simply select the processor and can easily assign it (through links) with some multimedia content available on the Internet. Thus, the developed tools described in this paper go beyond a simple viewer of 3D objects; they are a solution that allows a teacher to plan and execute lessons according to his or her educational goals.

Until recently, the effective use of virtual reality applications in the educational context had as barriers the high price of the equipment and the small amount of available software. However, technological advances have reversed this situation; currently, it is possible to build virtual reality laboratories at an affordable cost for educational institutions [23]. Equipment such as computers and projectors are capable of generating good quality images, and appropriate software can now be purchased in the marketplace [20].

The tools presented in this paper provide the use of high quality and easy to use simulations in the educational context. The goal is to provide solutions that allow teachers to focus their work on educational goals and not on technological aspects.

Figure 1 depicts the use of our tools. Initially, the content is addressed by the teacher; afterward, the teacher defines the educational strategy and content and then performs the group simulation (at an institution). After this, the students perform their self-study (at home). And finally, the learning is evaluated and is used as input into a new teaching-learning cycle. Thus, it is expected that students will play the role of coauthors of the learning situations by means of scenario simulations guided by immersion, interactivity, and involvement. Viewing scenarios should be able to stimulate interest and, consequently, enhance the learning activity [1].

The success of learning to use our tools depends on this cycle, which requires a clear definition of goals and steps taken during the use of the resources involved. For this, additional complementary resources such as formal presentations, readings, and discussions are required in the classroom. Only in this way can education become efficient and effective, enabling the student to apply such knowledge in the real world.

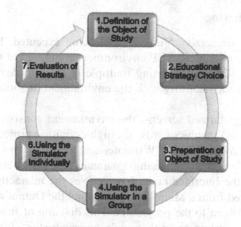

**Fig. 1.** Teaching-learning cycle

# 3    Research Method

Most software for immersive and interactive environments is designed to display 3D models and is generally tailored to specific problems, for instance, astronomy or physical education. So we projected the tools from the beginning to be adapted to various educational contexts. For this, we used Unified Modeling Language (UML). The agile method of extreme programming [24, 25] was used to develop the set of tools. The teamwork was done by the authors of this paper, who are from the computer and education areas. The tools were developed using the Java language, which allowed the development of web and local tools in a similar way, since Java applications are able to run on browsers (as applets). Finally, we used the content of the computer architecture discipline to perform the validation.

# 4    Using Simulations for Teaching-Learning Process

This section presents the developed tools, which are:

- Immersive viewing: this is a tool used by the teacher to provide 3D immersion and interaction with objects during a group activity. It is executed on multi projection systems, such as CAVE-like systems. This tool depends on specific equipment such as screens and projectors, so it is best executed in an educational institute that has a virtual reality laboratory. It runs on a commodity cluster that has 3D graphic cards.
- Simple viewing: this is a tool used by students to perform self-study. It runs on Internet browsers, so it can run on a personal computer.
- Multimedia editor: this is a tool used by the teacher that allows the association of multimedia content with 3D models.
- Semantic editor: this is a tool used by the teacher to create semantic descriptions about every 3D model visualized.

## 4.1    Immersive Viewing

Figure 2 depicts the Immersive Viewing tool being executed. In this example, the application is running in a miniCAVE environment composed of three side walls with passive stereoscopic 3D and supporting multiple interaction devices. To ensure that everyone can interact satisfactorily with the environment, it should have a maximum of 12 people using it.

In addition to the polarized screens, the environment consists of a six-node computer cluster that has 3D graphics cards, six high-definition projectors, and interaction devices such as Microsoft Kinect, Wiimote, and keyboard/mouse. There are three kinds of nodes: the Coordinator node that guarantees the distribution and synchronization of information; the Interface node that receives the interactions of users, for example, entries captured from a Microsoft Kinect; and the Output nodes, which process the images and send them to the projectors. This division of functionality allows the assigning of tasks according to each node's characteristics; for example, a mobile device, which has low processing capacity, can be responsible for the acquisition of user interaction data.

**Fig. 2.** Computer architecture class – video card

Besides offering immersion, this tool has multimodal support for interaction. Teachers and students can navigate freely in the environment, with the opportunity to choose what they want to view and interact with. The user interaction generates changes to the application state, generating feedback. For example, if the student tries to fit some component into a wrong slot, the component becomes red to alert them. The users can examine the structure and properties of the 3D objects. No special knowledge is required for users to manipulate the interface.

The mounting cost of a display laboratory as shown in Figure 2 is low when compared with the most used solutions in the past decade [20]. However, some studies argue that it is still impossible for some institutions to have these kinds of environments, since they still involve necessary equipment, trained teachers, applications, and physical space. Considering this context, it is possible to simplify it; for example, the visualization can be performed in a classroom using just one projector.

## 4.2    Simple Viewing

The Simple Viewer (Figure 3) allows students to run simulations on browsers on their own personal computers. It was designed to run in a Web environment, aiming to omit all installation steps by the user. Thus, the user needs only a browser to visualize and manipulate the 3D virtual models presented in the classroom.

It is possible to visualize the same content in the Immersive Viewer, but with limited interaction resources, immersion, and involvement. For example, this tool only supports anaglyph stereoscopy, and the execution of associated videos should be available in a web video server.

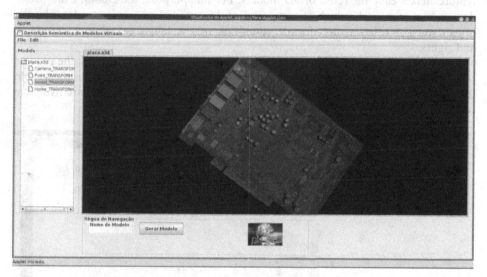

**Fig. 3.** Simple Viewer running on a browser

## 4.3    Multimedia Editor

The multimedia content editor is associated with each part of the displayed model, as is depicted in Figures 2 and 3. Using this, for example, the teacher can associate a video to a capacitor. When the capacitor is selected, the video is displayed. Thus, planning lessons should take into account the object and the multimedia content previously associated with it.

The model can be decomposed into parts, for example, a video card 3D model, for a given class. The teacher can determine that the class will only be addressed with a GPU (graphics processing unit), which allows the flexible development of the class. Thus, the same video card 3D model can be used with various approaches.

The 3D models must be created in a decomposed form or in shaped parts to allow the teacher to plan a class according to a specific subject. The semantic description tool takes advantage of this decomposition to describe the model and to separate the objects into parts. Thus, the teacher is free to tailor the displayed content to the learning content, including changing the multimedia content associated with the class focus. The system supports different content types such as presentation (ppt, pps, and pptx), text (txt, pdf, doc, docx, and html), image (jpg, bmp, gif, and png), and video (avi, mov, jpeg, and wmv). The objects are stored in X3D format (Extensible 3D Graphics).

### 4.4    Semantic Editor

The Semantic Editor promotes the use of description in 3D environments, using metadata to select, extract semantic objects, and generate queries. This creates a vocabulary of terms that facilitate discovery, retrieval, and integration among objects. This feature makes easy the reuse of 3D models. For this purpose, the Dublin Core Metadata model was used, which is composed of 15 optional properties to describe the object, such as title, creator, format, identifier, origin, language, relation, coverage, and rights.

Figure 4 depicts the Semantic Editor interface. With organized objects, the process of planning and constructing classes is facilitated. The semantic description can help the teachers to find other models (submodels) and types of content that are available on the Internet via a search engine.

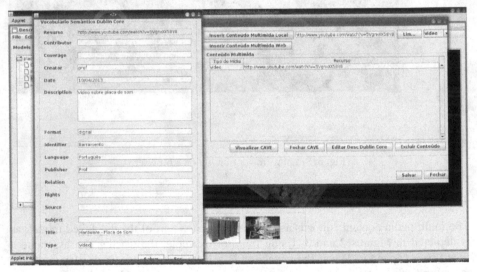

**Fig. 4.** Content description and aggregation of parts of a 3D model

## 5    Creation of Content

Before creating some 3D models, it is necessary to define the educational goals that will permeate the teacher education plan, following the educational strategies associated with the potential and limitations of the tools. Figure 5 depicts an example of a strategy that we developed for the computer architecture discipline for undergraduate students in courses related to computing. Initially, the teacher sets the objective content, which is the understanding of the functioning of the main parts of a video card, and sets the strategy to be used, consisting of the following: 1. Present to the classroom the main parts of a video card; 2. Use Immersive Viewing, which allows the mounting of a virtual video card by teacher and students; 3. Use Simple Viewing, where each student can conduct self-study about the video card and can access the associated multimedia content with it (video and text); 4. Execute an evaluation of the learning content.

**Fig. 5.** Case of study – computer architecture

We chose to use these tools to teach the computer architecture discipline because the viewers allow the simulation of concepts whose viewing is not possible in the real world, such as the energizing process of a computer motherboard, performed via an animation model. It should be emphasized that the tools are prepared to work with any 3D content, that is, they are not linked to a specific discipline.

With the teaching objectives and technical-pedagogical strategies defined, the creation of 3D models is the next phase (Figure 6). In phase 1, a technological evaluation of the feasibility of the simulation occurs, which is crucial for the creation of 3D models. For example, it takes into account the available computing resources (computers, projects, CAVEs, or others) and whether the model will be activated or not. In phase 2, template creation, two approaches are possible: the modeling of 3D objects or the use of models already created.

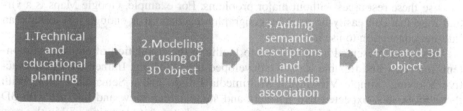

**Fig. 6.** Process of educational content generation

The advantage of modeling objects is that every member of the teaching staff involved can determine how the objects should be, but this kind of modeling requires more time to run (even months), and a technical team of modelers is required. On the other hand, when a ready 3D object is obtained, the educational content should be designed in accordance with its predetermined characteristics; however, the time at this stage will be spent in finding the right objects, which may or may not happen.

A crucial and difficult point to solve is when a teacher needs a crew to perform the 3D modeling of objects, since 3D objects, whether free or paid, may not be sufficient for the pedagogical approach adopted. This modeling takes time and makes the teacher dependent on such services. However, the trend is that in the future, there will be more 3D objects available on the Internet, since the number of 3D object repositories has been increasing. In phase 3, the teacher and/or technical staff add the semantic description to the objects, performing association (creating links) with the multimedia content. Finally, the 3D models are created in step 4.

The generation of an effective range of 3D content models linked to educational content is a challenge for everyone, because several hurdles must still be overcome, such as the identification of educational content that can take advantage of an environment such as that developed in this project. This challenge exists because teachers do not yet know enough about the potential of immersive and interactive simulation, and there are few user-friendly tools. Another barrier that limits the advance in the use of these simulations is the gap between the specifications of content and preparation. In addition, even if the cost of implementing these simulations is lower than solutions of the past decade, institutions still need the financial viability to incorporate this educational context.

## 6    Conclusions

There is interest in educational simulations that facilitate the teaching-learning process. However, an effective use of simulators based on virtual reality that is affordable and easy to use must be provided. This requires a solution where the teacher him-/herself can draw up all the simulations simply based on content. But the fact is that the everyday use of technology in the classroom is not so simple, because it may often fail, and it requires specific training and much time to prepare content for it. Ideally, the teacher's lesson preparation time should be spent primarily on educational planning and educational content and not on technical issues such as 3D object modeling. However, the scenario has changed in recent years, and more and more teachers can use these resources without major problems. For example, Google Maps is a virtual map that can easily be used by geography teachers; it is enough just to have an Internet connection to use it.

This paper presented a set of tools to facilitate the simulation of educational content in an easy and flexible way. We developed the following: Immersive and Interactive Viewing, Simple Viewing, a Multimedia Editor, and a Semantic Editor. With these tools, it is expected that teachers and students can view and interact with 3D models, enhance the understanding of content, and stimulate learning within the new digital culture learning context. It is important to highlight the ease of content creation, as it is a key factor in using this tool effectively in the classroom, since the teacher should spend his/her time on the pedagogical rather than on the technological issues. In addition, the flexibility of its use should also be stressed, since this tool can be run on a browser or even on CAVE-like systems.

In future research, we plan to improve the tool set, adding new features such as support for Augmented Reality content. We also plan to add new content and perform evaluations of the tool set's use in different courses.

# References

1. Vygotsky, L.S.: The problem of age (m. hall, trans.). In: Rieber, R.W., vol. 5, pp. 187–205 (1998)
2. Robelia, B., Greenhow, C., Hughes, J.E.: Learning, teaching, and scholarship in a digital age: Web 2.0 and classroom research - what path should we take "now"? Educational Researcher **38**(4), 246–259 (2009)
3. Lai, K.-W.: Digital technology and the culture of teaching and learning in higher education. Australasian Journal of Educational Technology **27**(Special issue, 8), 1263–1275 (2011)
4. Saljo, R.: Digital tools and challenges to institutional traditions of learning: technologies, social memory and the performative nature of learning. Journal of Computer Assisted Learning **26**(1), 53–64 (2010)
5. Thomas, D., Brown, J.S.: A New Culture of Learning: Cultivating the Imagination for a World of Constant Change. CreateSpace Independent Publishing Platform (2011)
6. Wang, S., Hsu, H., Reeves, T., Coster, D.: Professional development to enhance teachers' practices in using information and communication technologies (icts) as cognitive tools: Lessons learned from a design-based research study. Computer Education **79**, 101–115 (2014)
7. Amirian, M., Lindner, S., Trabulsi, E., Lallas, C.: Surgical suturing training with virtual reality simulation versus dry lab practice: an evaluation of performance improvement, content, and face validity. Journal of Robotic Surgery **8**(4), 329–335 (2014)
8. Smith, S., Ericson, E.: Using immersive game-based virtual reality to teach are-safety skills to children. Virtual Reality **13**(2), 87–99 (2009)
9. Boudreaux, H., Bible, P., Cruz-Neira, C., Parham, T., Cervato, C., Gallus, W., Stelling, P.: V-volcano: addressing students' misconceptions in earth sciences learning through virtual reality simulations. In: Bebis, G., et al. (eds.) ISVC 2009, Part I. LNCS, vol. 5875, pp. 1009–1018. Springer, Heidelberg (2009)
10. Kuhlen, T.W., Hentschel, B.: Quo Vadis CAVE: Does Immersive Visualization Still Matter? IEEE Computer Graphics and Applications **34**(5), 14–21 (2014)
11. Aykent, B., Merienne, F., Guillet, C., Paillot, D., Kemeny, A.: Motion sickness evaluation and comparison for a static driving simulator and a dynamic driving simulator. Proceedings of the Institution of Mechanical Engineers, Part D: Journal of Automobile Engineering, 1–2 (2014)
12. Jaeger, B.J., Mourant, R.R.: Comparison of Simulator Sickness Using Static and Dynamic Walking Simulators. Proceedings of the Human Factors and Ergonomics Society 45th Annual Meeting, 1896–1900, October 2001
13. Cruz-Neira, C., Sandin, D.J.: The CAVE: audio visual experience automatic virtual environment. Magazine Communications of the ACM **35**(6), 64–72 (1992)
14. Franco, L.R., Raimann, E., de Souza, R.R., de Souza Ribeiro, M.W.: Force and motion: virtual reality as a study instrument of alternative conceptions in dynamics. In: XIII Symposium on Virtual Reality (SVR), May 23–26, 2011, Uberlandia, Brazil, pp. 89–95 (2011)
15. Sampaio, P., Mendonça, R., Carreira, S.A.: Learning chemistry with VirtualLabs@Uma: a customizable 3D platform for new experimental protocols. Multimedia Tools and Applications **71**(3), 1129–1155 (2014)
16. Muhamad, M., Zaman, H.B., Ahmad, A.: Developing virtual laboratory for biology (VLab-Bio): a proposed research conceptual framework. In: 2011 International Conference on Electrical Engineering and Informatics (ICEEI), vol. 1, pp. 1–6, June 15–17, 2011

17. Lee, J., Quy, P., Kim, J., Kang, L., Seo, A., HyungSeok, K.: A collaborative virtual reality environment for molecular biology, ubiquitous virtual reality. In: International Symposium on Ubiquitous Virtual Reality, ISUVR 2009, pp. 68–71, July 8–11, 2009

18. Ahmed, K., Keeling, A.N., Fakhry, M., Ashrafian, H., Aggarwal, R., Naughton, P., Darzi, A., Cheshire, N., Athanasiou, T., Hamady, M.: Role of virtual reality simulation in teaching and assessing technical skills in endovascular intervention. J. Vasc. Interv. Radiol. **21**(1), 55–66 (2010)

19. Schreuder, H.W.R., Oei, G., Maas, M., Borleffs, J.C.C., Schijven, M.P.: Implementation of simulation for training minimally invasive surgery. Tijdschrift voor Medisch Onderwijs **30**(5), 206–220 (2011)

20. Guimarães, M.P., Gnecco, B.B.: Teaching astronomy and celestial mechanics through virtual reality. Computer Applications in Engineering Education **17**(2), 196–205 (2009)

21. Dias, D.R.C., Brega, J.R.F., de Paiva Guimarães, M., Modesto, F., Gnecco, B.B., Lauris, J.R.P.: 3D semantic models for dental education. In: Cruz-Cunha, M.M., Varajão, J., Powell, P., Martinho, R. (eds.) CENTERIS 2011, Part III. CCIS, vol. 221, pp. 89–96. Springer, Heidelberg (2011)

22. Souza, C.C.D., Medeiros, T.R., Gadelha, R., Sousa, T., Silva, E.L., Azevedo, R.R.D.: Um ambiente integrado de simulação para auxiliar o processo de ensino/aprendizagem da disciplina de sistemas operacionais. In: Proceedings of Simpósio Brasileiro de Informática na Educação, 2011, pp. 406–414 (2011)

23. Carozza, L., Bosche, F., Abdel-Wahab, M.: Robust 6-dof immersive navigation using commodity hardware. In: Proceedings of the 20th ACM Symposium on Virtual Reality Software and Technology, VRST 2014, pp. 19–22. ACM, New York (2014)

24. Beck, K.: Embracing change with extreme programming. Computer **32**(10), 70–77 (1999)

25. Beck, K., Andres, C.: Extreme Programming Explained: Embrace Change (2Nd Edition). Addison-Wesley Professional (2004)

# Unity Cluster Package – Dragging and Dropping Components for Multi-projection Virtual Reality Applications Based on PC Clusters

Mário Popolin Neto[1,4]($\boxtimes$), Diego Roberto Colombo Dias[2],
Luis Carlos Trevelin[2], Marcelo de Paiva Guimarães[3],
and José Remo Ferreira Brega[4]

[1] Federal Institute of São Paulo – IFSP Registro, São Paulo, SP, Brazil
mariopopolin@ifsp.edu.br
[2] Computer Science Department, Federal University of São Carlos – UFSCAR,
São Carlos, SP, Brazil
{diegocolombo.dias,trevelin}@dc.ufscar.br
[3] Computer Science Master Program – FACCAMP, Open University of Brazil –
Federal University of São Paulo – UNIFESP, São Paulo , SP, Brazil
marcelodepaiva@gmail.com
[4] Computer Science Department, São Paulo State University – UNESP,
Bauru, SP, Brazil
remo@fc.unesp.br

**Abstract.** Virtual Reality applications created using game engines allow developers to quickly come up with a prototype that runs on a wide variety of systems, achieve high quality graphics, and support multiple devices easily. This paper aims to present a component set (Unity Cluster Package) for the Unity game engine that facilitates the development of immersive and interactive Virtual Reality applications. This drag-and-drop component set allows Unity applications to run on a commodity PC cluster with passive support for stereoscopy, perspective correction according to the user's viewpoint and access to special servers to provide device-independent features. We present two examples of Unity multi-projection applications running in a mini CAVE (Cave Automatic Virtual Environment)-like (three-screens) system ported using this component set.

**Keywords:** Virtual reality · Unity game engine · Multi-projection · PC cluster

## 1 Introduction

Game technology can be used to create Virtual Reality (VR) applications, since both areas exploit technological advances in diverse fields, such as image synthesis and interaction devices [6]. Game engines have emerged providing greater interactivity and graphics performance [16], which has led researchers to explore these tools for game creation in applications for CAVE™ [8] (Cave Automatic

© Springer International Publishing Switzerland 2015
O. Gervasi et al. (Eds.): ICCSA 2015, Part V, LNCS 9159, pp. 261–272, 2015.
DOI: 10.1007/978-3-319-21413-9_19

Virtual Environment)-like systems [19]. These multi-projection applications created with game engines typically are found in arts [7,18], simulators [4,14] and visualization systems [17].

The exponential growth of processor capabilities and the emergence of dedicated graphics hardware have enabled the use of PC clusters to run immersive and interactive applications. There are some libraries that support the development of immersive and interactive applications to run on PC clusters (for example, libGlass [1] and FreeVR [21]). However, they require integration with external rendering solutions. Current game engines integrate a set of reusable modules, such as Graphics and Network, but do not have original support to create VR applications to run on PC clusters; they are adapted to meet this demand by means of multi-projection frameworks.

The main contribution of the present research project is to simplify the design of immersive and interactive VR applications based on PC clusters, freeing the developer from having to know all the details of the individual technologies. We present the Unity Cluster Package, a drag-and-drop component set for the Unity game engine that allows the development of immersive and interactive VR applications based on PC clusters in an easy way. We also show two Unity demos that were ported to run in a 3-screens multi-projection system. The package extends these applications to support multi-projection, passive stereoscopy, perspective correction according to the user's viewpoint and access to device servers. We developed this component set for Unity because it is one of the most popular game engines with a custom rendering engine, an extensive documentation and an easy-to-use editor, and it is multi-platform [11].

The remainder of the paper is organized as follows. Section 2 contains the background work related to this research. Section 3 describes the Unity Cluster Package and the main components to use the Unity game engine in the development of immersive and interactive VR applications. Section 4 explains the port of two Unity demos for a 3-screens multi-projection system. Section 5 discusses final considerations. Section 6 presents conclusions and future works.

## 2   Related Work

Although Paul Rajlich's CAVE Quake II was probably the first implementation with a game engine (Quake Engine) for an immersive system [19], the framework CaveUT[TM] [12] became more popular allowing the development of Unreal[TM] Engine applications [7,18] for CAVE[TM]-like systems. The use of game engines to develop multi-projection applications depends on the level of access to their internal modules, since these tools are not only a set of reusable modules, but also a layer that connects and manages them [3]. Multi-projection frameworks for game engines, such as CaveUT[TM] and its successor CaveUDK [19] (Unreal® Engine), and CryVE [13] (CryEngine® 2), use the internal modules to develop immersive and interactive virtual environments for multi-projection systems.

---

[1] http://sourceforge.net/projects/libglass/

As well as the use of game engines that have multi-projection frameworks, VR integration libraries can be combined with game engines with accessible modules, such as the use of FreeVR with the open source Delta3D, for the development of an immersive simulator [14]. Although open source game engines are an option, they do not exhibit some properties provided by commercial game engines, such as the possibility of development for multiple platforms and rendering quality. Hardware adapters that split and distribute the video signal also can be used, such as the Matrox DualHead2Go [2] used in the Virtual Reality Theatre [20]. This approach is based on hardware and it does not require any type of software, but increases the area occupied by a single pixel resulting in loss of image resolution.

PC clusters based on the Master-Slave [22] rendering model are the most frequently used when dealing with game engines for multi-projection applications development [4, 7, 14, 17, 18, 23]. The multi-projection frameworks for game engines, such as CaveUT[TM] [12], CaveUDK [19] and CryVE [13], provide support to PC clusters. These frameworks are similar regarding the usage of the game engine network module for communication and synchronization among cluster nodes applications. The utilization of the network internal module for these purposes has become a trend, and it is also being used for local solutions [17]. The Unity Cluster Package presented in this paper was developed based on the Unity network module.

CaveUDK [19] is a natural successor of CaveUT[TM] (Unreal® Engine 2), extending the Unreal® Development Kit (Unreal® Engine 3) for CAVE[TM]-like systems. CaveUDK supports stereo images by active stereoscopy and interaction device from VRPN (*Virtual Reality Peripheral Network* [25]) servers via an integrated module (C++ DLL). This framework is an open source initiative, but requires Unreal® Development Kit purchase. CryVE [13] aims for a low-cost implementation that sits on top of CryEngine® 2; however, despite offering an extensible solution (open source), it does not provide stereo images or VRPN access, and it uses an out-of-date game engine, since there is no record from code recompilation of this solution for the CryEngine® 3.

CaveUDK and CryVE are game engine extensions in open source compiled code, usually C++ DLLs, providing code reusability for developers. The game engine purchase is also a characteristic of both frameworks. The MiddleVR [3] plugin can be used with the popular Unity game engine to develop applications for different VR systems, such as HMD (Head-Mounted Display) [5] and CAVE[TM] [23]. This plugin supports PC clusters, active and passive stereoscopy, and VRPN servers, but it is a closed commercial solution and depends on Unity Pro to provide the PC cluster feature. The Unity Cluster Package is free of charge, and it can be used in both Unity game engine versions (Pro and Free). This package provides reusable drag-and-drop components to develop Unity applications for PC clusters, allowing exploitation of the Unity qualities in multi-projection systems.

---

[2] http://www.matrox.com/graphics/en/products/gxm/
[3] http://www.middlevr.com/middlevr-sdk/

Fig. 1. Unity Cluster Package

## 3   Unity Cluster Package

Unity Cluster Package is a Unity package (file extension .unitypackage) that was designed to overcome the difficulties faced by developers in creating immersive and interactive VR applications. Figure 1 shows the layers of components an immersive and interactive virtual reality PC cluster application created with Unity, consisting of: (1) Virtual Reality Applications, immersive and interactive VR applications; (2) Unity Cluster Package, drag-and-drop components to support VR applications development; and (3) Unity, the game engine Unity.

### 3.1   Multi-projection System

The Unity Cluster Package can be used in various multi-projection systems, and a MiniCAVE [9,10] is our target system here. Figure 2 presents the MiniCAVE, which consists of three screens (2.5 m x 1.5 m each) arranged at an angle of 30 degrees to each other. For each screen, there are two projectors (BenQ W1000 HD) connected to two dedicated computers (Intel Core i7 8GB RAM), with NVIDIA FX 1800 as the graphics card. Polarized lenses are used to provide passive stereoscopic projection. A total of six computers are used for the projections, and they are connected to a server computer by a gigabit switch. The Unity Cluster Package provides a coherent, seamless and contiguous view from the PC cluster (six rendering nodes and one server node) applications in the MiniCAVE.

### 3.2   Unity

Unity is a commercial game engine created and maintained by Unity Technologies that has a Pro and Free version [1]. This popular game engine is composed of a custom rendering engine (OpenGL or DirectX) and an editor with intuitive workflows. Unity belongs to the Modular Framework Engine group [3], which provides a good relation between customization and complexity. Developers can implement scripts in JavaScript, C# and Boo to build the game logic, and these

**Fig. 2.** MiniCAVE: PC cluster (seven computers and gigabit switch), three screens, projectors and polarized lenses

scripts can be attached to scene objects using prefabs. Prefab is a Unity mechanism that allows storage of objects with their properties and components, working as a template which can be shared and reused. Unity is also a multi-platform game engine, which enables the development of games for Windows, Linux, Mac OS, iOS, Android, web browsers, and consoles [1].

### 3.3 Unity Cluster Package: Drag-and-drop Components

The Unity Cluster Package is easily imported into the Unity Editor and it allows the developer to build immersive and interactive VR applications, dragging and dropping the available components as needed. This package is user-friendly, and not only suitable for new applications, but also for legacy code. This is an important feature, as most available solutions for these applications require a substantial amount of code; some of them require a complete change of the application design.

The Unity Cluster Package's main goal is to allow easy development of immersive and interactive VR applications based on PC clusters through well-designed and high configurable components, which are Unity prefabs that arrange scene objects and C# scripts according to their functionality. There are three main components: the *Multi Projection Camera*, the *Node Manager*, and the *Device Manager*.

**Multi Projection Camera.** The *Multi Projection Camera* is a customized virtual camera for the development of multi-projection applications. It arranges a PreRender script, and the scene objects *ProjectionPlane* (plane) and *UserHead*

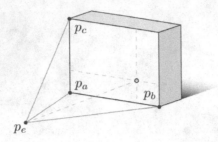

**Fig. 3.** Points used to calculate the projection matrix: Pa, Pb, Pc, and Pe - Adapted [15]

(sphere) to calculate the projection matrix [15], where the points Pa, Pb, and Pc (Figure 3) encode the size, aspect ratio, position and orientation of the screen. This projection matrix allows perspective correction according to point Pe.

The PreRender script calculates the projection matrix obtaining the points Pa, Pb, and Pc in the virtual environment through the *ProjectionPlane*. The *User-Head* position is taken as the point Pe, providing a motion parallax metaphor. By moving the object *UserHead* according to any head-tracker device, the motion parallax is achieved, offering a better perception of the virtual environment.

**Node Manager.** The *Node Manager* uses the Unity network module to implement the Master-Slave [22] rendering model, in which each cluster node runs the same application with different virtual camera configurations, and all user input interactions are treated by the master node. This component is a scene object with a special script attached to start the application according to the cluster node type, where each slave node (Unity Client) is connected to the master node (Unity Server), as shown in Figure 4.

The script also instantiates the *Multi Projection Camera* in the master node application using the Unity network module; in this way, all slave node applications have an instance of this custom virtual camera. To achieve image coherence, the *Multi Projection Camera*'s position and rotation must be synchronized. The Unity network module provides the NetworkView component to perform scene objects synchronization. The *Multi Projection Camera*'s position and rotation are changed on the master node and synchronized through the StateSynchronization functionality provided by the NetworkView.

Slave node applications are assigned to the screen setting up the *Multi Projection Camera* by means of the points Pa, Pb, and Pc in a coordinate system with the origin in the center of the front screen. The *Node Manager* obtains the cluster node specification through an XML configuration file that contains the points Pa, Pb, and Pc, the node type (master or slave), and the IP address and connection port to the Unity Server.

Figure 4 shows the points Pa, Pb, and Pc for the MiniCAVE, where the Slave #3 application is assigned to the front screen with Pa = (-8,-5,0), Pb = (8,-5,0), and Pc = (-8,5,0). The Slave #3 configuration file is shown in Figure 5. The

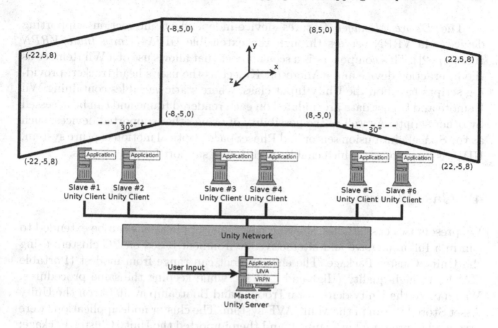

**Fig. 4.** Unity Cluster Package for the MiniCAVE

*screen* tag parameters *stereo* (true or false) and *eye* (right or left) specify for which eye the application addresses the projection. In order to achieve passive stereoscopic projection, the Slave #3 and Slave #4 address the projection for the same screen but for different eyes: Slave #3 for the left eye (eye = "left") and Slave #4 for the right eye (eye = "right").

```
<node type="slave">
    <server ip="192.168.1.13" port="25000"/>
    <screen stereo="true" eye="left">
        <pa x="-8" y="-5" z="0"/>
        <pb x="8"  y="-5" z="0"/>
        <pc x="-8" y="5"  z="0"/>
        <pe x="0"  y="0"  z="5"/>
    </screen>
</node>
```

**Fig. 5.** Slave #3 configuration file

**Device Manager.** The Unity game engine provides the Input class to access data from the supported interaction devices (mouse, keyboard, joystick, and mobile devices) through static variables and functions. All input variables are reset on each rendered frame.

The *Device Manager* features device-independent interaction supporting devices via VRPN servers through the extensible UIVA (*Unity Indie VRPN Adapter* [2]). This component is a scene object that allows use of a Wii Remote as the interaction device and a Microsoft Kinect as the user's head tracker, providing scripts based on the Unity Input class, where static variables containing Wii Remote and Kinect data are updated on each rendered frame and can be accessed by other scripts. As well as the possibility of using other supported devices, such as the SpacePoint Fusion sensor and PhaseSpace's optical motion capture system, UIVA is open source, which enables addition of support for new devices.

## 4   Case Studies

We present two case studies, showing how Unity applications can be extended to run in a full immersive and interactive environment based on PC clusters using the Unity Cluster Package. The environment can range from modest (Portable CAVE) to high-quality (High-end CAVE), while keeping the same procedures. We adapted the Unity demos StarTrooper and Bootcamp available on the Unity Asset Store [4] to run in the MiniCAVE system. The cluster node applications were arranged as presented in Figure 4, and then imported the Unity Cluster Package; its components are used by dragging and dropping into the scene creation window in the Unity Editor. The Unity demos adaptation involved three basic steps:

1. Add the *Node Manager*, which starts the cluster node application and instantiates the *Multi Projection Camera* according to the cluster node type.
2. Add the *Device Manager*, which enables access to the Wii Remote and Kinect data updated on each rendered frame.
3. Define the scene objects that need to be synchronized among the cluster node applications through the Unity network module, where only the master node application must perform state changes on these objects.

StarTrooper in the MiniCAVE (Figure 6–(a)) is a circular arena where the aircraft flight and missile launch systems are provided by a Wii Remote. Regarding the third step, the StarTrooper game objects 'aircraft', 'missile', and 'ring' are synchronized by the NetworkView component from the Unity network module. Bootcamp Explore (Figure 6-(b)) enables first person navigation in the Bootcamp scenario using a Wii Remote and Nunchuk in the MiniCAVE. The Unity standard scripts for first person controller were modified and attached to the *Multi Projection Camera*, enabling the Wii Remote and Nunchuk as controller devices, and using the NetworkView component for synchronization (third step). Motion parallax was implemented in the Bootcamp Explore, assigning the *User-Head* position from the *Multi Projection Camera* to the user's head position obtained from the Kinect.

The master node is responsible for running the UIVA, the VRPN server for the Wii Remote [2], and the VRPN server FAAST (*Flexible Action and*

---

[4] https://www.assetstore.unity3d.com/en/

**Fig. 6.** Unity demos adapted for the MiniCAVE

*Articulated Skeleton Toolkit* [24]) for the Kinect. Some changes were made in the UIVA in order to support the Wii Remote additional peripheral controller Nunchuk. The Unity demos adaptation can be viewed at this video [5].

The performance of these demos adapted for the MiniCAVE is presented in Table 1. It shows the frames per second (FPS) average in relation to the number of rendered triangles. The performance is satisfactory, considering that VR applications should have more than 30 FPS. Unity Cluster Package provides a good relationship between frame rate and rendered triangles. CaveUDK gives a frame rate of 50 FPS for 200 K to 600 K rendered triangles [19]. CryVE, with its medieval museum virtual tour, provides a frame rate less than 20 FPS [13].

**Table 1.** Unity demos adaptations – performance

| Unity Demo Adaptation | FPS Average | Rendered Triangles |
|---|---|---|
| StarTrooper | 566 | 12K |
| Bootcamp Explore | 58 | 237K |

## 5   Final Considerations

Some comparisons can be made between the Unity Cluster Package and other solutions (Table 2). The reusability of our solution resides in its components, which can be used by dragging and dropping into the scene creation window

---

[5] https://www.youtube.com/watch?v=nvVDAvdw_k8

**Table 2.** Solutions comparison

| Solution | Type | Game Engine | Stereoscopy | VRPN | Version |
|---|---|---|---|---|---|
| Our Package | Open Source | Unity | Passive | Yes | Free/Pro |
| MiddleVR | Commercial | Unity | Active/Passive | Yes | Pro |
| CaveUDK [19] | Open Source | Unreal® Engine | Active | Yes | Pro |
| CryVE [13] | Open Source | CryEngine® 2 | – | No | Pro |

in the Unity Editor. Multi-projection frameworks for game engines, such as CaveUDK and CryVE, are extensions as open source compiled code (DLLs), requiring programming knowledge from developers, and game engine purchase (Pro), similar to MiddleVR with the Unity game engine. The Unity package presented in this paper is easily imported into the Unity Editor, and can be used in both Unity versions (Free and Pro).

Through polarized lenses (passive stereoscopy), such as in the MiniCAVE, stereo images can be achieved with common graphic cards and projectors; on the other hand, active stereoscopy requires specific hardware, such as NVIDIA Quadro 5000 and the shutter glasses used in the CaveUDK. Access to interaction devices from VRPN servers can be considered as a standard, since this is found in both commercial (MiddleVR) and open source (CaveUDK and Unity Cluster Package) solutions.

## 6    Conclusion and Future Works

This paper presented the Unity Cluster Package, a flexible and extensible solution for creating immersive and interactive VR applications using the Unity game engine. This package allows extension of Unity applications to support multi-projection systems based on PC clusters, passive stereoscopic projection and perspective correction according to the user's viewpoint. The integration with the extensible UIVA enables the use of the wide range of interaction devices supported by VRPN servers. In the case studies we used a MiniCAVE to test our solution; however, this is not a restriction since the package is suitable for other multi-projection systems. Unity Cluster Package is open source and it is available via the Unity Cluster Package Sourceforge project page [6].

As future work, we will exploit the Unity game engine in the development of immersive and interactive VR applications, and also investigate the use of mobile devices (tablets and smartphones) as cluster nodes and interaction devices.

**Acknowledgements.** The authors would like to thank to CAPES Foundation, a body of the Brazilian Ministry of Education, for financial support. Mário Popolin Neto and Diego Roberto Colombo Dias were recipient of scholarships from CAPES.

The adaptations in this paper used the Unity demos StarTrooper and Bootcamp available on the Unity Asset Store. All rights to the graphical and audio assets belong to Unity Technologies and are not used here for commercial purposes.

---

[6] https://sourceforge.net/projects/unityclusterpackage/

# References

1. Create the games you love with Unity. http://unity3d.com/unity, (accessed March-2015)
2. Unity Indie VRPN Adapter - UIVA. http://web.cs.wpi.edu/~gogo/hive/UIVA/, (accessed March-2015)
3. Anderson, E.F., McLoughlin, L., Watson, J., Holmes, S., Jones, P., Pallett, H., Smith, B.: Choosing the infrastructure for entertainment and serious computer games - a whiteroom benchmark for game engine selection. In: 2013 5th International Conference on Games and Virtual Worlds for Serious Applications (VS-GAMES), pp. 1–8, September 2013
4. Backlund, P., Engstrom, H., Hammar, C., Johannesson, M., Lebram, M.: Sidh - a game based firefighter training simulation. In: 11th International Conference on Information Visualization, IV 2007, pp. 899–907 (2007)
5. Beimler, R., Bruder, G., Steinicke, F.: Smurvebox: a smart multi-user real-time virtual environment for generating character animations. In: Proceedings of the Virtual Reality International Conference: Laval Virtual, VRIC 2013, pp. 1:1–1:7. ACM, New York (2013). http://doi.acm.org/10.1145/2466816.2466818
6. Bouvier, P., De Sorbier, F., Chaudeyrac, P., Biri, V.: Cross benefits between virtual reality and games. In: Proceedings of the Computer Games and Allied Technology 2008, CGAT 2008 - Animation, Multimedia, IPTV and Edutainment, pp. 186–193 (2008). www.scopus.com
7. Cavazza, M., Lugrin, J.L., Pizzi, D., Charles, F.: Madame bovary on the holodeck: immersive interactive storytelling. In: Proceedings of the 15th International Conference on Multimedia, MULTIMEDIA 2007, pp. 651–660. ACM, New York (2007). http://doi.acm.org/10.1145/1291233.1291387
8. Cruz-Neira, C., Sandin, D.J., DeFanti, T.A.: Surround-screen projection-based virtual reality: the design and implementation of the cave. In: Proceedings of the 20th Annual Conference on Computer Graphics and Interactive Techniques, SIGGRAPH 1993, pp. 135–142. ACM, New York (1993). http://doi.acm.org/10.1145/166117.166134
9. Dias, D.R.C., Brega, J.R.F., Trevelin, L.C., Popolin Neto, M., Gnecco, B.B., de Paiva Guimaraes, M.: Design and evaluation of an advanced virtual reality system for visualization of dentistry structures. In: 2012 18th International Conference on Virtual Systems and Multimedia (VSMM), pp. 429–435, September 2012
10. Dias, D.R.C., Brega, J.R.F., Lamarca, A.F., Popolin Neto, M., Suguimoto, D.J., Agostinho, I., Gouveia, A.F.: Chemcave3d: Sistema de visualização imersivo e interativo de moléculas 3d. In: Workshop de Realidade Virtual e Aumentada. WRVA, Uberaba (2011)
11. Hu, W., Qu, Z., Zhang, X.: A new approach of mechanics simulation based on game engine. In: 2012 Fifth International Joint Conference on Computational Sciences and Optimization (CSO), pp. 619–622, June 2012
12. Jacobson, J., Hwang, Z.: Unreal tournament for immersive interactive theater. Commun. ACM 45(1), 39–42 (2002). http://doi.acm.org/10.1145/502269.502292
13. Juarez, A., Schonenberg, W., Bartneck, C.: Implementing a low-cost cave system using the cryengine2. Entertainment Computing 1(3–4), 157–164 (2010). http://www.sciencedirect.com/science/article/pii/S1875952110000108
14. Koepnick, S., Norpchen, D., Sherman, W., Coming, D.: Immersive training for two-person radiological surveys. In: Virtual Reality Conference, VR 2009, pp. 171–174. IEEE (2009)

15. Kooima, R.: Generalized Perspective Projection. http://aoeu.snth.net/static/gen-perspective.pdf, (accessed March-2015)
16. Lewis, M., Jacobson, J.: Game engines in scientific research. Commun. ACM **45**(1), 27–31 (2002). http://doi.acm.org/10.1145/502269.502288
17. Louloudi, A., Klugl, F.: Visualizing agent-based simulation dynamics in a cave - issues and architectures. In: 2011 Federated Conference on Computer Science and Information Systems (FedCSIS), pp. 651–658, September 2011
18. Lugrin, J.L., Cavazza, M., Palmer, M., Crooks, S.: Artificial intelligence-mediated interaction in virtual reality art. IEEE Intelligent Systems **21**(5), 54–62 (2006). http://dx.doi.org/10.1109/MIS.2006.87
19. Lugrin, J.L., Charles, F., Cavazza, M., Le Renard, M., Freeman, J., Lessiter, J.: Caveudk: a vr game engine middleware. In: Proceedings of the 18th ACM Symposium on Virtual Reality Software and Technology, VRST 2012, pp. 137–144. ACM, New York (2012). http://doi.acm.org/10.1145/2407336.2407363
20. Schou, T., Gardner, H.J.: A wii remote, a game engine, five sensor bars and a virtual reality theatre. In: Proceedings of the 19th Australasian Conference on Computer-Human Interaction: Entertaining User Interfaces, OZCHI 2007, pp. 231–234. ACM, New York (2007). http://doi.acm.org/10.1145/1324892.1324941
21. Sherman, W.R., Coming, D., Su, S.: Freevr: honoring the past, looking to the future, vol. 8649, pp. 864906–864906-15 (2013). http://dx.doi.org/10.1117/12.2008578
22. Staadt, O.G., Walker, J., Nuber, C., Hamann, B.: A survey and performance analysis of software platforms for interactive cluster-based multi-screen rendering. In: Proceedings of the Workshop on Virtual Environments 2003, EGVE 2003, pp. 261–270. ACM, New York (2003). http://doi.acm.org/10.1145/769953.769984
23. Steptoe, W., Steed, A., Slater, M.: Human tails: Ownership and control of extended humanoid avatars. IEEE Transactions on Visualization and Computer Graphics **19**(4), 583–590 (2013)
24. Suma, E., Lange, B., Rizzo, A., Krum, D., Bolas, M.: Faast: the flexible action and articulated skeleton toolkit. In: 2011 IEEE Virtual Reality Conference (VR), pp. 247–248, March 2011
25. Taylor, II, R.M., Hudson, T.C., Seeger, A., Weber, H., Juliano, J., Helser, A.T.: Vrpn: a device-independent, network-transparent vr peripheral system. In: Proceedings of the ACM Symposium on Virtual Reality Software and Technology, VRST 2001, pp. 55–61. ACM, New York (2001). http://doi.acm.org/10.1145/505008.505019

# Author Index

Printed in the United States
By Bookmasters